praise for

From Knowledge to

"Everything needed by those of us who are concerned about climate change is finally collected in one place. The problem, the consequences, and the solutions are all contained in this book. John Perona has done a great service to climate activism."

—Mark Reynolds,
Executive Director, Citizens' Climate Lobby

"Science, public policy, and politics all come together in this book and show us how to survive and eventually thrive while ending the climate crisis."

—Bill Bradbury,
Oregon Secretary of State

"Scientists and activists seeking the best ways to help reverse the threats of climate change can often become paralyzed with indecision in the face of a dizzying number of policy options, all seemingly inadequate. Environmental biochemist John Perona offers an ambitious, articulate, and surprisingly optimistic road map in his new book, From Knowledge to Power. At once a basic tutorial on climate science, a catalog of energy policies, a polemic against the oil industry, and a hopeful collection of short- and long-range solutions to the climate crisis, this unique book couldn't be more timely. It inspires readers to think boldly about how they can become more engaged, drawing on Perona's unusual personal journey as a biochemist, chemical engineer, legal scholar, and political activist."

—Dr. Jonathan Fink,
Professor of Geology & Director of the Digital City Testbed Center, Portland State University

"From Knowledge to Power presents explanations of Earth's climate and atmospheric chemistry and how they affect society and the natural systems on which society depends in a way that is admirably clear and accessible to non-technical audiences. Additionally, the book explains how contemporary climate policy, including policy that is related to equity, follows from strong scientific evidence and consensus. This compelling volume will serve as a comprehensive, multidisciplinary resource for audiences with diverse interests and backgrounds."

—Dr. Erica Fleishman,
Director of the Oregon Climate Change Research Institute & Professor, College of Earth, Ocean & Atmospheric Sciences, Oregon State University

From Knowledge to Power

The Comprehensive Handbook for Climate Science & Advocacy

John Perona

From Knowledge to Power: The Comprehensive Handbook for Climate Science & Advocacy
© 2021 John Perona

ISBN 13: 9781947845299

Ooligan Press
Portland State University
Post Office Box 751, Portland, Oregon 97207
503.725.9748
ooligan@ooliganpress.pdx.edu
www.ooliganpress.pdx.edu

Library of Congress Cataloging-in-Publication Data
Names: Perona, John, 1961- author.
Title: From knowledge to power: the comprehensive handbook for climate science & advocacy / Dr. John Perona.
Description: First edition. | Portland : Ooligan Press, 2021.
Identifiers: LCCN 2021009372 (print) | LCCN 2021009373 (ebook) |
ISBN 9781947845299 (trade paperback) | ISBN 9781947845282 (ebook)
Subjects: LCSH: Climatic changes. | Global warming--Prevention.
Classification: LCC QC902.8 .P47 2021 (print) | LCC QC902.8 (ebook) |
DDC 363.738/7470973--dc23
LC record available at https://lccn.loc.gov/2021009372
LC ebook record available at https://lccn.loc.gov/2021009373

Cover design by Morgan Ramsey
Interior design by Michael Shymanski
Graphics by Callie Brown, Morgan Ramsey, & Michael Shymanski

Printed in the United States of America

For the young people in the Perona, Bozzalla, Hood, Wohl,
Zanzucchi, Zanzucchi-Washington, Washington, and Roberts clans,
and for all the other young people in my life,
and for my students.

"When you realize the Earth is so much more than simply your environment, you'll be moved to protect her in the same way as you would yourself. This is the kind of awareness, the kind of awakening we need, and the future of the planet depends on whether we're able to cultivate this insight or not."

—Thích Nhất Hạnh,
Statement on Climate Change for the United Nations

"The climate is a common good, belonging to all and meant for all."

—Pope Francis, Laudato Si´

about the author

John Perona earned a BS in Chemical Engineering from Rutgers University (1983) and PhD in Molecular Biophysics and Biochemistry from Yale (1989). He has also earned a JD (Santa Barbara College of Law, 2008) and LLM in environmental and natural resources law (Lewis & Clark University, 2016). After earning his doctorate, Dr. Perona conducted postdoctoral studies in enzyme chemistry at the University of California, San Francisco, prior to joining the faculty of the Chemistry Department at UC Santa Barbara in 1994. In 2011, he moved to his current position on the faculties of Portland State and Oregon Health & Science Universities.

Dr. Perona's accomplishments include publication of over 100 peer-reviewed articles and reviews in Biochemistry and related fields, and critically published analyses of the law and policy of groundwater management, genetic engineering of agricultural crops, and bio-diesel development. His teaching experience includes classes in environmental chemistry and the application of synthetic biology to solve environmentally challenging problems.

Since 2013, Dr. Perona has been active in a number of climate advocacy groups, especially the Citizens' Climate Lobby, which petitions Congress to enact an aggressive economy-wide price on carbon, and the Metro Climate Action Team in Portland, Oregon, whose work focuses on ensuring compliance with Governor Kate Brown's 2020 executive order directing state agencies to reduce and regulate greenhouse gas emissions. He has worked with several nonprofit environmental law firms and legal clinics, advocating against expansion of fossil fuel infrastructure.

acknowledgments

This book was catalyzed by the many dedicated citizen lobbyists it has been my privilege to work with and learn from. I am especially grateful to the staff and volunteers of the Citizens' Climate Lobby, who meet the realities of Washington politics with civility and grace, and have done so much to promote Congressional engagement with climate change—especially through the lean years before it finally emerged as the defining issue that it is. In particular, I cannot adequately convey my thanks to Tamara Staton and Daniela Brod, leaders of CCL's efforts in Oregon, for their personal support and guidance in propelling my modest contributions to this work. For their valuable input on the carbon pricing chapter, logistical support for my community seminar series, which helped inspire the book, and ever-present good humor and camaraderie in Portland and Washington, D.C., I also especially thank Francine Chinitz, Margaret Eickmann, Chelsea Maricle, Kirstin Meneghello, KB Mercer, Brian Ettling, Walt Mintkeski, Barry Daigle, and Eric Means.

I also drew inspiration from many other citizen volunteers, especially the remarkable Portland Metro Climate Action Transportation (MCAT) team led by Jane Stackhouse and Rich Peppers. When Oregon governor Kate Brown issued her recent comprehensive executive order accelerating agency efforts on climate change, MCAT and many other advocacy teams jumped into action across the state. The work of engaging in rulemaking processes and testifying on key legislation requires in depth engagement and preparation, and the successful efforts of these supercommitted individuals show how much can be done when the willpower to foster change is collectively brought forth. If this book inspires even a small number of new volunteers to join in and act likewise, the effort of writing it will have been justified.

Many other generous folks helped a great deal along the way. Steve Ghan of the Pacific Northwest National Laboratories critically reviewed almost every chapter, bringing a lifetime of expertise in climate science to help hone the content and organization of the material. Kyle Meyer, an expert on geochemistry and my close colleague at Portland State, tolerated many questions on the fine points of paleoclimatology and crunched the data for figures. Ethan Seltzer offered invaluable advice on the publishing process, clued me in to the fact that two chapters is actually plenty to engage potential publishers, and introduced me to Ooligan Press. Mark McLeod, a fixture in the local advocacy community, connected me with local climate groups and reminded me early on of the importance of incorporating social justice into climate policy. For many other conversations, inspirations, figures, crucial corrections, and reviews of chapters, I am grateful to Mary and Ralph Perona, Will Musser, Doug Nichols, David Lea, Catherine Gautier-Downes, Richard Turnock, Sabrina Fu, Lea Mbengi, Frank Granshaw, Pat DeLaquil, and Rachel Slocum.

My grounding for this book began to form some fifteen years ago, when I was privileged to be able to work with Linda Krop and her colleagues at the Environmental Defense Center in Santa Barbara. There I gained my first glimpses into environmental law while also taking night classes at the local law college. Later, my capacity to think critically about the intersection of climate science, law, and policy was greatly boosted by the Environmental and Natural Resources Law faculty at Lewis & Clark University, who (however inexplicably) admitted a middle-aged science professor to their prestigious Master of Laws program, allowing me access to some of the finest minds in the field. I am especially grateful to Melissa Powers for teaching me climate and energy law, an academically rigorous and invaluable experience that gave me the necessary foundation to write several of the later chapters.

This book would have certainly never reached anything resembling its final form without the sustained enthusiasm and hard work of the student publishing team at Ooligan Press, especially Callie Brown, Julie Collins, Morgan Ramsey, Emma Wolf, and Michael Shymanski. Many of the figures are products of their savvy in computer-aided design, and their brilliant and comprehensive developmental edit of my muddled first draft gave me key insights for properly organizing a good chunk of the material in the middle chapters. The team also made extraordinary efforts in factchecking, tracking down citations, detailed formatting of notes, creating figure legends and tables, and building chapter layouts. Any remaining errors, of course, are entirely my responsibility.

To Rusty, Brian, Brian, Sean, John Henry, Daniel, Kai, and Ben: thank you, men—you have been there for me through thick and thin. To my mother Maria: Mom, it was only with the benefit of your superhuman efforts that I've managed to attain the success that I have. And to Jennifer, the love of my life, thank you for bringing your wisdom and unfailing support to this project, and for always believing in me, no matter what.

contents

Glossary can be found at
www.fromknowledgetopower.com/glossary

Preface

In 2019, the term "climate emergency" multiplied by at least a hundred-fold on websites in English-speaking nations around the world, leading the Oxford English Dictionary to designate it Word of the Year.[1] Such a dramatic increase in just one year heralds a welcome shift in consciousness about global warming—a sharp elevation in the sense of immediacy and urgency that many now feel about the issue.

Use of the word "net-zero" has also been surging. This is another hopeful sign because the term refers to an emerging consensus as to what our climate change policy should strive for: reducing human greenhouse gas emissions by 90 percent by 2050 and compensating for remaining emissions by removing excess carbon dioxide from the atmosphere.[2]

Some of this urgency derives from the accelerating human costs of recent extreme weather events, especially heat waves, hurricanes, and wildfires. These events are among the first clear responses of the Earth system to the temperature increase caused by human activities. Accelerating extreme weather is like the proverbial canary in the coal mine—an unambiguous sign that something is fundamentally awry in the natural environment that we all depend on.

Although the critical need for action has only recently been recognized by the general public, it has been central to climate scientists' thinking for decades. For example, renowned chemist and inventor James Lovelock has long used the Gaia metaphor to illustrate how the mineral parts of the Earth system are integrated with the biosphere to generate a climate stable enough to withstand external shocks, such as changes in solar radiation or volcanic eruptions.[3] Gaia, the Earth goddess from Greek mythology, is the nourishing mother who created that system. But in Lovelock's potent version of the myth, Gaia also has a vengeful side, and she will wreak havoc on those who disrupt her creation.

In scientific language, this means that if our greenhouse gas emissions exceed a critical level, the added heat may overwhelm natural balances, flipping the climate into a "hothouse" world with much higher average temperatures and sea levels – a world far less favorable to human flourishing. Earth's rocks and fossils contain ample evidence of hothouses caused by massive natural carbon releases, but humans are now releasing carbon at a rate ten times faster than at any time in the past 66 million years.[4] This carbon release is the ultimate legacy of the fossil fuel era, and its geologic signature will remain long after the heating eventually stops.

In a sense, we are fortunate that extreme weather is so evident, since it signals the need for urgent action in a way that the many less visible effects on Earth's natural balances cannot. By moving decisively to a net-zero carbon economy by mid-century, it is still possible to limit damages and maintain Gaia's creation in a recognizable state. This is the great project of our time, and our obligation to future generations. And since we all now take part in the fossil fuel economy, all of us can play a role in bringing it to a definitive end.

This book offers a comprehensive guide to climate science, policy, and politics in the United States, written specifically for citizen advocates. It comes at a propitious moment: a new president who is squarely confronting the full magnitude of the global warming crisis, and a rare consensus among many Democrats about the kinds of policies that need to be put in place. This moment is without precedent in modern politics. President Biden's early efforts are a sea change from the disaster of the Trump era, but they also signal a shift from the Obama administration, which supported the natural gas industry and relied heavily on top-down regulations to cut carbon dioxide emissions. The Biden team does not compartmentalize climate change but treats it as a condition that must be integrated into all policies. Coordinated from a central White House office, federal efforts to combat global warming are now embedded in Coronavirus relief spending, infrastructure rebuilding plans, land and coastal ocean conservation, and a myriad of other programs.[5]

All of this is promising, yet the Biden administration cannot tackle climate change alone. Certainly, the damage done during the Trump years will be reversed through executive orders—which can be implemented rapidly—as well as by rewriting agency regulations on issues such as automobile mileage standards, which is a lengthier process that will probably have to surmount legal challenges. But executive actions and regulations are not enough to solve the climate crisis because they can be reversed by subsequent administrations that are hostile to their intent. Lasting change requires legislation, and that means negotiation between Democrats and Republicans. In some cases, Democrats may be able to enact legislation alone, but their narrow majorities in the 2021-2022 House and Senate leave almost no margin for dissent, and those hailing from conservative states often do not view climate change with the same urgency as their progressive colleagues from deep blue parts of the country.[6]

The structure of the federal system in the US, which divides power between the national government and state and local authorities, is another reason why we should not expect the Biden team to solve the crisis on its own. Congress and the president can adopt economic and environmental policies that set the stage for deep emissions reductions, but a great deal of the follow-through must occur in the states, which can either enthusiastically adopt or resist the changes. Fossil fuel development, in particular, largely takes place on state and privately held lands, and the enduring reliance of some states on oil and gas revenues will require a great deal of effort to dislodge, regardless of what the federal government does.

This is where citizen advocacy comes in. We should all petition our federal representatives to act, yet most opportunities for direct involvement are closer to home. This is a good thing. It means that relationships can be more easily forged among advocates united in a common cause, and that influential local businesses, community organizations, faith groups, and other stakeholders can lend their institutional clout. The power of an engaged citizenry is not often recognized, but it has played a central role in several successful campaigns for change. Such victories for the Left include same-sex marriage and the right of Guantanamo detainees to legal representation, and on the Right, an expanded right to private gun ownership.[7] In each case, grassroots movements at the local level grew and eventually gained enough influence among decision-makers to drive political change.

A prominent example of citizen advocacy for a healthy climate is the widespread opposition to the Keystone XL oil pipeline. If built, this pipeline would transport heavy oil from dirty "tar sands" deposits in Alberta, Canada, to the US Gulf Coast.[8] Since it was first proposed in 2008—and through a seemingly endless series of environmental reviews—the pipeline has ping-ponged from eventual opposition by the Obama administration, to support from President Trump and, most recently, revocation of a key permit by President Biden. Citizen advocates from all walks of life have called attention to the consequences of the pipeline, which include ecological devastation as well as a spike in greenhouse gas emissions enabled by the use of the oil. Their actions, together with a barrage of lawsuits, built strong Democratic opposition and substantially slowed progress during the Trump administration. There is now good reason to think that the pipeline will never be built.

As this example and others we will explore demonstrate, citizen advocacy is instrumental to creating the political will to follow through on President Biden's plans at all levels of government. To date, only a small fraction of Americans are participating in this effort. However, an enormous reservoir of potential new advocates exists in the estimated 53 million Americans who are alarmed about climate change. These alarmed citizens—and an even larger group who identify as concerned—are the primary audience for this book.[9]

Citizen advocacy for a healthy climate is much more likely to be successful if advocacy groups organize and work together toward common goals. Historically, this has been a challenge for the climate movement, most prominently displayed in clashes between progressive and center-left advocates. Progressives, led by Green New Deal champion Alexandria Ocasio-Cortez and advocacy groups such as the youth-led Sunrise Movement, argue for a transition that emphasizes justice, which would restore and strengthen historically marginalized groups that have borne a disproportionate share of climate change impacts. The center-left, in contrast, has been more focused on impacts to the Earth system—treating climate change mainly as a scientific problem. They favor technical or market-based solutions without paying specific attention to disparities in impact on particular populations, or on ensuring that those groups have a seat at the policymaking table.

In the months before the 2020 election and empowered by a unified awareness that the climate stakes could not be higher, a number of coalitions formed to put forth policy proposals that seem to have yielded a substantial consensus.[10] Net-zero emissions by 2050 has emerged as the dominant organizing principle, with a strong emphasis both on setting tough emissions reductions standards tailored to particular economic sectors, and on very large public investments in infrastructure, land management, and other areas that impact the climate. In a clear victory for progressives, environmental justice is also prominent in these plans. There are, however, still significant areas of disagreement among advocates, especially with respect to policies that many economists and climate scientists see as necessary and effective but have historically been championed more by political centrists or conservatives. These policies include carbon pricing, nuclear power and carbon capture and storage approaches for large-scale drawdown of atmospheric carbon dioxide.

The early executive orders by President Biden clearly embrace the main themes of this consensus. Moving forward, unity among Democrats is crucial because of strong Republican resistance to much of the agenda—a good deal of which is driven by a well-organized network of donors with roots in the fossil fuel industries.[11] This coalition, unfortunately, can be counted on to oppose virtually any substantive climate policy proposal. As is so often the case on many issues, federal legislation addressing climate change may be shaped by the small number of Representatives and Senators that remain at the political center. If these efforts turn out to require significant compromise from the net-zero policy blueprint, climate advocacy in the states will assume even more importance.

This is the outlook in the early months of the Biden administration. It follows an effective eight-year hiatus, from 2010 to 2018, during which Congress paid little to no serious attention to climate change. This period opened with the failure of the Senate to pass a comprehensive federal carbon pricing bill,[12] and ended with the rise of the Green New Deal and the recapture of the House of Representatives by Democrats in 2018. The hopeful shift in climate change politics is still very recent, and the landscape may certainly change again. But what will not change are the dynamics of the Earth system, the nature of the effective strategies that drive the carbon-free energy transition, and the key role of engaged citizen advocates to bring about change. These three topics define the scope of this book.

The book is divided into ten chapters and includes an Interlude that describes the contours of the necessary carbon-free energy transition. I drew on many years of my laboratory's research on the chemistry of the biosphere, legal training and research into climate change policy, and experiences in the climate advocacy community, primarily through the Citizens' Climate Lobby—a nonpartisan grassroots group that advocates for federal carbon pricing legislation. This book offers climate education in the service of advocacy, and it fulfills the need for a comprehensive guide that provides an entry point for engaged citizens who wish to help build the political will to solve the problem of climate change.

The first four chapters describe the past, present, and potential future of Earth's climate system. The narrative assumes that the reader has no background in science. These chapters start with first principles and offer thorough explanations of all the key concepts as they are understood by climate scientists today. My approach to these chapters emerged from a series of monthly evening lectures I delivered to the environmental advocacy community in Portland, Oregon, in 2017-2018. From this experience, I was exposed to the keen desire of citizen advocates to fully understand the scientific data that underlie today's policy debates. The seminars were my education in appreciating the deep level of understanding that many citizen advocates are hungry for.

Interwoven in the scientific narrative are descriptions of some of the most egregious ways that climate change deniers have tried to distort the meaning of the data. Many of these individuals and groups are now avoiding outright denial in favor of more subtle arguments that undermine the need for urgency or exaggerate the costs of taking action.[13] But a grasp of the more direct denialist tactics is important because it shows how easy one can be misled on a subject as complex as climate change. This can make a difference when engaging with well-meaning people who are simply uninformed or confused about what the science says.

The Interlude sets the stage for the second half of the book by offering a summary of many of the policy roadmaps for achieving net-zero emissions by 2050 in the US. These are aggressive plans that do nothing less than remake the entire energy economy of the country, with consequences that propagate to every level of society. The Interlude is followed by a chapter that surveys the landscape of US climate advocacy and offers information about groups and resources that provide support for lay citizen advocates. Some helpful research that informs advocacy strategies is also summarized here. Opportunities for advocacy that are linked to specific policies are then interwoven throughout the remainder of the book.

Chapters 6 through 10 explore the policy landscape for the carbon-free energy transition in detail. Chapter 6 stands somewhat apart because it is devoted to an exploration of the fossil fuel industry, especially strategies for accelerating its contraction and ultimate demise. Although the recent rapid growth and plummeting costs for solar and wind power provide the essential foundation for the energy transition, attention to fossil fuels is crucial because these sources must be replaced by renewables. While coal-fired power plants in the US have been closing at record rates, recent growth in the solar and wind industry has been matched by increased oil and gas production; the fraction of total US energy provided by fossil fuels has thus remained nearly the same over the past decade. Curtailing fossil fuel production through advocacy for divestment and against the construction of new infrastructure is an area of engagement and subsequent victory that the climate advocacy community can be especially proud of.

The approach to the final four chapters might be summarized as, "All of the above—except fossil fuels." The dedicated chapter on carbon pricing reflects its unique role as a foundation for other efforts. Even a relatively modest, economywide carbon price would selectively disadvantage coal, oil, and natural gas in proportion to how much carbon they emit, making all other policies that promote carbon-free energy more effective. The consideration of all carbon-free energy sources includes nuclear power, because maintaining its present 20 percent share of the electricity grid balances the intermittency of solar and wind power while cutting down the amount of additional expensive energy storage that is needed in the short term. Just as important is the emphasis on carbon capture and storage, which includes both natural land management and large-scale industrial approaches that draw down atmospheric carbon dioxide. This is necessary because many decades were lost to climate change inaction while greenhouse gas levels continued to increase, making these approaches now essential to stabilizing temperature at a safe level.

A good understanding of climate science yields this inspiring message for advocates: We have the power to stabilize the climate by limiting warming to under 2°C (3.6°F) compared to preindustrial temperatures. In contrast, skeptical accounts that misrepresent the scientific literature and underestimate climate change damages lead to complacency instead of action. Similarly, books promoting the false doom and gloom perspective that climate chaos is already inevitable deliver the message that advocacy efforts will be fruitless. Think of this book as the middle ground—the antidote to both those extremes.[14]

Many people wonder about the relationship between individual actions that we may take to reduce our personal climate footprints and engaged group advocacy, which this book describes. Some have suggested that too much emphasis on personal actions is a mistake because it lets the fossil fuel companies off the hook and may even undermine support for effective national policy. Yet, in cutting the link between individual and collective action we risk missing a key opportunity. *Sierra* magazine editor Jason Mark puts it this way: "When we take personal responsibility for our actions, we deepen our commitment to environmental sustainability. Living in accord with one's political vision is a way of laying the foundations for the world we want to see—to engage in a kind of 'prefigurative politics' that makes the future into now."[15]

So, by all means, let's dust off the bicycle, eat a bit less red meat, electrify our homes and cars at the next opportunity, sell off those fossil fuel stocks, and reduce air travel. And if you are inspired to also help shift our national priorities, this book will show you that there is a great deal to do and will help prepare you to take action.

Let's get started.

Chapter 1
Earth's Climate System

From the indigenous North American people, who first learned how to live in harmony with our land, to naturalists, poets, and politicians—like John Muir, Walt Whitman and Al Gore—Americans have been blessed to have many eloquent voices speaking on the importance of environmental stewardship.[1] Most scientists came later to this calling, but moved by the climate crisis, are overcoming natural reticence and recognizing that their advocacy adds a crucial dimension to the conversation.[2] We should listen closely to all these voices with both our heads and our hearts. In today's America, though, it is the hard-headed, no-nonsense voice of science, now under sustained attack by those who find its truths *inconvenient*, that most needs to be amplified.[3]

For advocacy, a good understanding of climate science is enormously beneficial because it empowers us to speak with confidence about the urgency of the problem. It can be tempting to skip the science, which sometimes appears difficult, and to jump right to solutions. But while the need to act rapidly is clear to climate advocates, many influential people seemingly lack this urgent perspective. Lawmakers, business executives, and community leaders—the individuals we must engage—are immersed in the health and economic crises brought on by the Coronavirus pandemic, and in the long-standing challenge of racial and social justice that is intertwined with both of these. In this context, the urgency of global warming, emerging from the basic science, is easy to overlook. But while slow-moving, climate change is inexorable. We are at a critical juncture, and all of our voices are vital.

Given the importance of understanding the science of climate change, this chapter begins with a straightforward description of the climate system. The second section explains how greenhouse gases naturally trap the heat emitted from the Earth's surface to create a warmer, livable environment. The last section details why carbon is the most important element to the Earth's climate, describing its various forms and how it naturally cycles through the atmosphere, land, and oceans. A key takeaway from this section

Climate Components

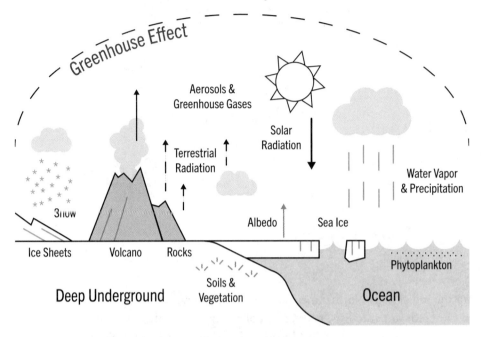

Figure 1.1: Components of Earth's climate system—land, oceans, biosphere, ice, and atmosphere. Dotted lines depict outgoing radiation (Earthlight), which emanates from both the surface and the atmosphere. The sun and deep underground are outside the climate system.

is the persistence of heat-trapping atmospheric carbon dioxide over many centuries and millenniums, demonstrating the potentially devastating impact on future generations if we fail to act.

1. Describing the Earth

How can we describe Earth's climate in a way that takes advantage of our common experience and what we already know? Let's begin where the pioneering 19th century Earth and environmental scientists also started—we'll take a look at the Earth system, describe what we see, and then think a little about how this system might respond to external influences. Our observations can be categorized in many different ways, but the consensus today is to divide the climate system into five parts: the atmosphere, oceans, ice, land surfaces, and the biosphere (Figure 1.1).[4] Among these, the biosphere has the unique property of *being alive*—the climate system includes essential functions for the biology of Earth's organisms.[5]

The Earth's atmosphere is a thin layer of gases that extends from the surface to the boundary with outer space.[6] Most of the climate effects we are concerned with happen in the lowest atmosphere layer, the *troposphere*, which extends up from the surface for 5–10 miles (for reference, Mt. Everest is 5.5 miles high). Above this is the *stratosphere*, which extends 30 miles from the surface. The stratosphere is important because it contains the protective *ozone layer* as well as particles from volcanic eruptions that remain suspended for long periods of time.

There is a greater concentration, or pressure, of gases close to the surface, and most of us have experienced how the atmosphere thins out with increased altitude (in mile-high Denver, Colorado, the pressure is 82 percent of what it is at sea level—already enough to send some of us puffing). Free circulation of the air means that the atmosphere is well-mixed, so the emissions of gases from the surface become distributed everywhere within a few weeks to a few months. Although the effects of air pollution can be local—for example, the daily smog in the Los Angeles basin mostly stays there—in regards to long-term climate, the whole atmosphere is freely inter-connected. The atmosphere is a *global commons*, a resource shared by everyone, and essential to making Earth a livable planet.[7]

Earth's atmosphere consists mainly of two gases: nitrogen and oxygen, which together make up about 99 percent of the entire mixture.[8] Most of the remaining 1 percent is an inert gas called argon. Oxygen plays a central role in the chemistry of the atmosphere and is essential for breathing, while nitrogen gas is important for its role in the overall *nitrogen cycle*, by which this element moves around among the atmosphere, land, vegetation, and oceans. It's important to note that none of these gases play a role in the *greenhouse effect*, which allows heat to be trapped near Earth's surface and thus leads to global warming.[9] Instead, most of the naturally occurring greenhouse gases—*carbon dioxide*, *methane*, *nitrous oxide* and *ozone*—are present only in very low amounts. There are also a few greenhouse gases that are entirely of human creation, like the *hydrofluorocarbons* (HFCs) used today as refrigerants (Box 1.1).

Surprisingly, the most important gas for greenhouse warming is water vapor, which gives the atmosphere its humidity. While water vapor is not well-mixed it does vary throughout the Earth's surface (think of the difference between deserts and jungles), at a level of 1–3 percent. The amount of water held in the atmosphere increases as the Earth warms, but only in response to increases in the other greenhouse gases.[10] So, you don't need to call the global warming police when your neighbor waters her lawn, because that water will just evaporate and rain out again as part of the natural cycle.

It is easy to conceptualize this effect for yourself. If you have ever camped in the desert, you probably noticed that it got quite cold at night, even in the summer. It doesn't cool off nearly as much at night in Florida. The low humidity of the desert atmosphere means that when heat escapes the surface after the Sun goes down, there is much less water vapor to absorb it and radiate it back down to you.

Let's move on to the other parts of the Earth system. Saltwater oceans occupy 70 percent of the Earth's surface and contain the vast majority (97 percent) of our water; the remaining water can be found in ice, and in both surface and underground lakes and rivers. The most important role the oceans play in the climate system is to absorb heat from the Sun (which occurs primarily in the tropics) and to distribute that heat to the North and South poles.[11] The Atlantic Gulf Stream, which starts in the Gulf of Mexico and keeps Northern Europe warm despite its high latitude, is the most well-known part of this circulation. Another crucial role that oceans play in the climate system is to take up and dissolve gases from the air—especially carbon dioxide. After carbon dioxide dissolves in the oceans, it no longer contributes to the greenhouse effect.[12]

The *ice* component of Earth's climate system, also called the *cryosphere*, consists of three main parts. Sea ice floats over the liquid ocean surface and is found only near the North and South poles. The other two parts are the large continent-sized ice sheets found on land in Greenland and Antarctica, and the mountain-glacier ice that occurs at high altitudes in many parts of the world. All exposed ice has a bright white color, which allows it to reflect a large percentage of the incoming sunlight that strikes it, keeping Earth's surface cooler than it otherwise would be.[13]

Finally, the *land* and *biosphere* form the last two crucial parts of the climate system. A great deal of solid carbon is stored in both the living biosphere and the nonliving parts of land–in soils and limestone rocks, for example. Similar to the oceans, these parts of Earth's surface also take up carbon dioxide from the atmosphere. Some of this uptake occurs very slowly, over what we can think of as geological time. Vegetation, however, takes up carbon dioxide quickly through photosynthesis, which occurs on an annual summer-winter cycle.

This function of vegetation is crucial and is the reason why climate scientists place so much emphasis on maintaining the size and health of forests. A good deal of photosynthesis also occurs in the oceans, for example through microscopic algae called *phytoplankton* (Figure 1.1).

Knowing the five parts that make up the Earth's climate system gives us the grounding we need to start exploring climate change. What's important to recognize is that we have so far described *the system*. The next step is to look at what is outside the system that influences how the five parts interact with one another. There are just two natural factors that are of importance—the Sun and the *deep underground*, the great part of the subsurface that humans do not access (Figure 1.1). The deep underground exerts its effects via volcanic eruptions and, in the very long term, through geologic forces. Appreciating the role of these factors will put us in a much better position to understand how humans have changed the climate, what consequences our actions have already wrought, and what it all means for the future.

2. The Sun and the Greenhouse Effect

Earth's position in the solar system is sometimes described by the Goldilocks principle. Earth is not too hot (like Venus), nor is it too cold (like Mars). Instead, Earth is *just right*—it's the perfect distance from the Sun to allow a temperature suitable for life as we know it to flourish. This seems obvious enough—the principle really just says that planets are hotter if they are closer to the Sun. However, this leads us to an important question: is the distance from the Sun the only thing that determines a planet's temperature, or are there other influential factors?

Since the 19th century, physicists have been developing and improving measurements of how much radiation from the Sun actually strikes the Earth. Given the temperature of the Sun's surface, the distance between the Sun and Earth, and the tiny fraction of the Sun's total output that hits the Earth, it is possible to apply basic physics to calculate what Earth's average surface temperature should be. If these were the only crucial factors, that calculation ought to reasonably match the observed average temperature of 59°F, or 15°C (Box 1.2). But it does not. Instead, the calculation says that the average surface temperature of Earth should be a frigid 5°F, or -15°C. Something is missing.

That missing something is the atmosphere. The -15°C number comes from what climate scientists call the "bare rock" model of Earth.[14] Although incomplete, it is still very useful to our thinking because it shows that additional factors besides the Earth-Sun distance, the surface temperature of the Sun, and the Earth's diameter help determine temperature. The next step, then, is to understand something about how the atmosphere is able to trap some solar heat at the surface, thereby raising the temperature by 30°C. This observed feature of our planet—that it is much warmer than the bare rock model predicts—is called the *natural greenhouse effect*. This principle is the key to unlocking the entire problem of global warming that concerns us.

When sunlight hits Earth's atmosphere and surface, some of it is reflected straight back into space. For example, most of the Sun's radiation that happens to strike any part of

Earth's white ice is reflected right back out. This also happens when sunlight strikes some types of clouds: in this case, part of the incoming radiation never even makes it to the Earth's surface. None of this reflected sunlight warms the Earth at all. On the other hand, when the solar radiation strikes the dark ocean, most of it is not reflected but absorbed. The ocean accepts the radiation and takes it in, which causes it to warm up. Taking into account the many different variations in the Earth's atmosphere and surface—water, ice, fields, forests, etc.—it becomes clear that every aspect of Earth's climate system reflects or absorbs sunlight to its own particular extent.

Climate scientists have calculated that, of the total incoming sunlight that hits all of Earth, 30 percent is reflected and 70 percent is absorbed. This 30 percent is called the reflectivity, or *albedo*, of the Earth. This is a phenomenon we all likely have personal experience with: if you work outside on a hot sunny day, and you wear a dark T-shirt, you will be warmer than if you were wearing a white T-shirt. The dark shirt absorbs more heat than the white one does.

We have just learned that the 70 percent of the Earth system that absorbs sunlight warms up as a result. That extra heat is transferred throughout the surface and atmosphere (for example, by winds and ocean currents), so the parts of Earth that reflected sunlight warm as well. But notice next that the sunlight, of course, keeps on coming in. Something else must be at work, or else Earth would continue to warm and warm, and indeed would have disintegrated a long time ago. To prevent this, much of the absorbed sunlight is re-emitted back out into space. We can clearly see this phenomenon, using a specialized camera on an orbiting satellite. The Earth is actually glowing: it is emitting radiation (Earthlight) just like the Sun (Figure 1.2).[15] The big difference here is that the Earth, being much cooler than the Sun, emits much lower energy radiation, called *infrared radiation*, which is too weak to see with the unaided eye. You can, however, feel this exact type of radiation by standing under a heat lamp.

Not all of the absorbed sunlight is re-emitted back to space. Here is where the atmosphere comes in. After Earthlight is emitted from the planet's surface, some of it is intercepted by greenhouse gases, which in turn emit even weaker radiation in all directions. Some of this weaker radiation does make it out to space, but a lot stays inside the Earth's climate system, making it permanently warmer than it would be if there were no atmosphere at all (Figure 1.2). How much warmer? We discussed this above—30°C of additional warming as compared to that frigid bare rock.

A good analogy for the natural greenhouse effect happens any night when we go to sleep in a cool room under a blanket. Let's say the room temperature is 60°F and you are most comfortable at 70°F. You choose a certain blanket that absorbs some of the heat that your body radiates out. The blanket re-radiates that heat back down, so the temperature underneath is 70°F, just as you like it. This system—you, the room, and the blanket—is in an *energy balance*. The temperature under the blanket is warmer than the room, and it is not changing. Given the thickness of the blanket and the temperature outside it, a certain constant fraction of your body heat is radiated back to you, and a certain amount is lost to the room.

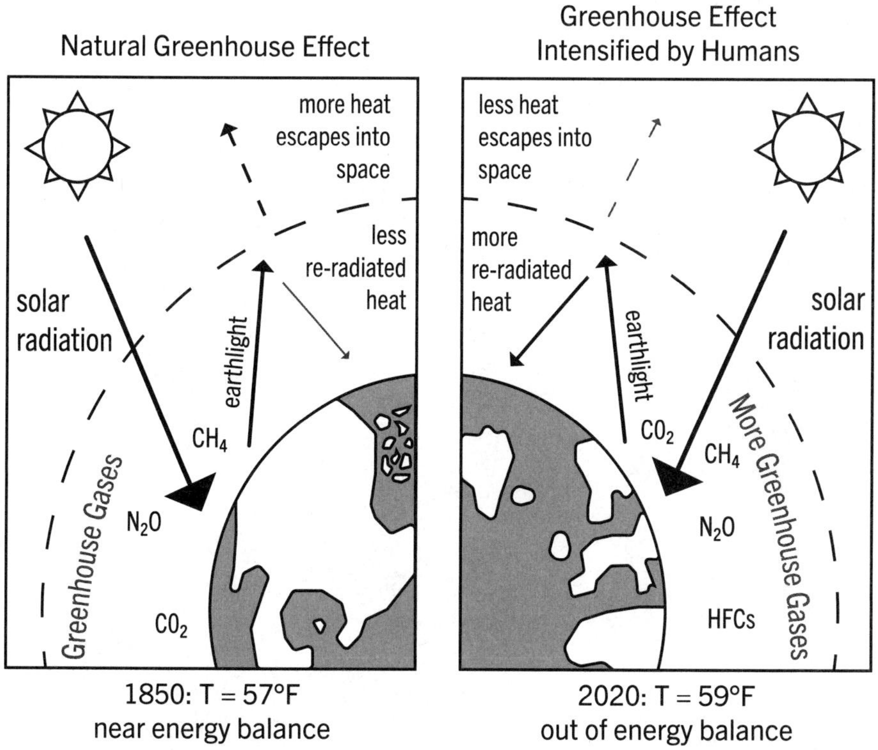

Figure 1.2: Natural (left) and human-enhanced (right) greenhouse effect. Compared to 1850, the amount of absorbed and reradiated Earthlight has increased, while the amount of heat escaping to space has decreased. Based on Figure 33.1 of the US National Climate Assessment, see https://nca2014.globalchange.gov/report/appendices/climate-science-supplement.

With the blanket you chose, the temperature underneath remains at 70°F. Similarly, if the "blanket" of greenhouse gases in the atmosphere were to remain constant, then, in the absence of any other human or natural perturbations, the average Earth surface temperature would also not change—it would stay constant at the present 59°F. The input energy from sunlight would be balanced by the output energy in the form of Earthlight, and the temperature underneath the atmosphere is then determined by how thick the greenhouse blanket is.

Now suppose you add another blanket to the bed. Your local system will adjust—more of your escaping body heat will be radiated back from the thicker blankets, and soon a new energy balance will be reached. Now, however, the temperature under the blanket will be warmer—let's say it is 80°F. It is starting to get uncomfortably warm with the extra blanket, and so we might want to take it back off. In our bedroom, this is no problem. But when we continue to add to Earth's greenhouse gas blanket, needless to say, the situation is more problematic.

Since 1850, when the average surface temperature of Earth was only 57°F, the temperature under Earth's atmosphere has increased by 2°F (about 1.2°C), mostly from our additions of carbon dioxide and other greenhouse gases to the blanket (Figure 1.2). Greenhouse gas emissions and deforestation continue, and the Earth, already far out of energy balance, becomes more so with each passing day. The *anthropogenic greenhouse effect*—the global warming that humans are responsible for—continues to increase.[16] This is our situation today.

3. The Carbon Cycle

The *carbon cycle* is a complex subject, no question about it.[17] But it is crucial in understanding climate change because the two most important molecules that trap heat— carbon dioxide and methane—are both built around a carbon atom (Figure 1.3). In this

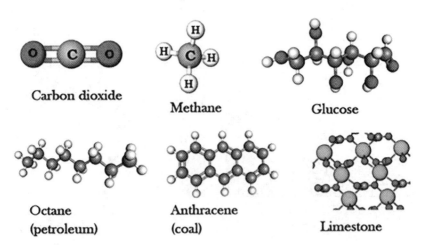

Carbon dioxide

Methane

Glucose

Octane
(petroleum)

Anthracene
(coal)

Limestone

Figure 1.3: Forms of carbon. Carbon atoms are depicted in light gray, oxygens in darker gray, and hydrogens in white. Glucose ($C_6H_{12}O_6$) is an abundant carbohydrate (sugar) in cells. In calcite (limestone), the atoms form a repeating crystal. The large spheres in limestone depict calcium atoms. Octane (C_8H_{18}) is a common component in crude oil and refined gasoline. Anthracene ($C_{14}H_{10}$) is abundant in coal.

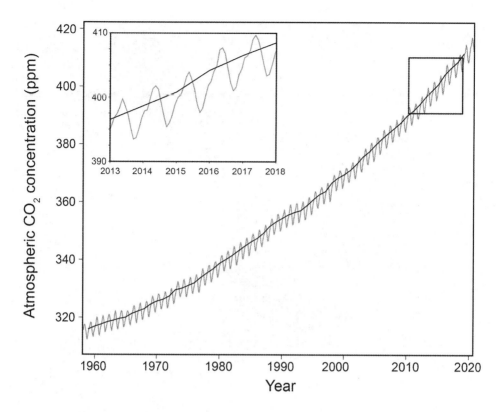

Figure 1.4: Keeling record of atmospheric carbon dioxide concentration from 1958–2020. Concentration units are given as parts per million (ppm). Plotted from data available at: https://scripps.ucsd.edu/programs/keelingcurve/.

section, we will explore how carbon, in its many different forms, moves around among the parts of Earth's climate system.

Some of the carbon dioxide from fossil fuel burning persists in the atmosphere for a very long time before circulating back to the land and oceans. This basic fact may be the single most difficult aspect of climate change to grapple with. It means that our actions today will have consequences far into the future, impacting the lives of our descendants who will have had nothing to do with creating the problem. Understanding and effectively communicating this can be a powerful way to build the argument for urgency in dealing with the climate threat.

To fully understand how our actions today will affect the carbon cycle far into the future, it is first crucial to gain an appreciation for the diversity of molecules that contain carbon. The different forms of carbon depend on which atoms it combines with to form these larger molecules—especially hydrogen, oxygen and other carbons. Some carbon-containing molecules (carbon *compounds*) have just one carbon atom in them and are so light that they are found as gases in the atmosphere. These are the heat-trapping molecules mentioned before: carbon dioxide (one carbon atom joined to two oxygen atoms) and methane (one carbon atom joined to four hydrogen atoms). In contrast, chains of carbon atoms joined to hydrogen, oxygen, and other elements form molecules that are not gases, but are included as part of the solid land and vegetation components of the Earth's climate system. This form of carbon is also found in underground oil and coal deposits. Carbon's remarkable ability to join with itself and other atoms, in literally millions of combinations, is the chemical basis for all life on Earth (Figure 1.3).[18]

That's enough chemistry for now. To dive into the carbon cycle, we need to discuss what might be the single most famous data set in all of climate science—the plot of atmospheric carbon dioxide concentration over time, beginning in 1958 and continuing to the present day. These measurements are made at the top of the Mauna Loa volcano in Hawaii. They were initiated by Charles David Keeling and are continued now at the Scripps Institute of Oceanography. This plot of how atmosphere carbon dioxide concentration changes over time is referred to, deservedly, as the Keeling Curve (Figure 1.4).[19] The Mauna Loa measurements complement many other datasets collected at research facilities all over the world, and our understanding of the carbon cycle is based on the sum of this knowledge.

The vertical axis of the curve is labeled as carbon dioxide (CO_2) concentration (ppm), where "ppm" means *parts per million* (Box 1.3). As fossil-fuel based technologies advanced over the past century and a half, the atmosphere's carbon dioxide concentration steadily increased. Consider that in 1850, at the beginning of the Industrial Revolution, the

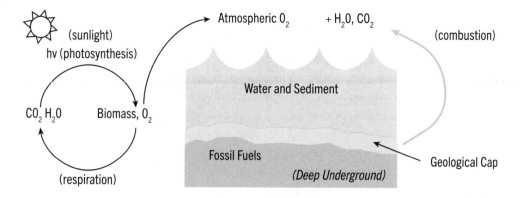

Figure 1.5: Impact of fossil fuel burning on Earth's climate system. The photosynthesis and respiration cycle (left) accounts for the yearly cycles in the Keeling curve (Figure 1.4). The gray arrow at right depicts fossil fuel excavation and burning, which delivers carbon dioxide and water vapor to the atmosphere.

atmospheric level registered about 280 ppm, and by 1958 the concentration had increased to 315 ppm. By late spring of 2021, however, there were 419 ppm of CO_2 in the atmosphere, which means that for every one million molecules of the atmosphere gas mixture, 419 were CO_2 (for comparison, about 780,000 are nitrogen). It is easy to see why carbon dioxide is talked about as a "trace" gas.

The increase from 280 ppm in 1850 to 419 ppm in 2021 occurred because carbon dioxide is the byproduct formed when carbon-containing fossil fuels, such as coal, oil, and natural gas, are burned. The term *anthropogenic greenhouse effect* refers to the additional warming caused by this fossil fuel burning and by changes to the land—like industrial agriculture—brought about by human activities. It is important to distinguish this from the *natural greenhouse effect* described above (compare the left and right panels in Figure 1.2).

We can see that the atmospheric concentration of carbon dioxide is steadily increasing—in fact, it is sobering to note that the slope of the Keeling curve is getting steeper in recent years. However, there are two other aspects of the curve that provide insight into the natural carbon cycle. The first is that the curve goes through yearly up-and-down cycles, with a sharp decrease in the Northern hemisphere summer (May to October) followed by a somewhat slower increase from October through May of the next year.

The second feature to note on the Keeling curve is that the rate of increase in atmospheric carbon dioxide is not smooth, but fluctuates significantly, from as little as a 1 ppm increase in some years, to over 3 ppm in others (see the inset in Figure 1.4). These fluctuations occur despite the fact that fossil fuel carbon dioxide emissions have been rising at a fairly constant rate during the whole timespan of the Keeling curve. Why, then, is there such a strong variation from year to year in the amount of carbon dioxide that accumulates in the atmosphere?

Let's start with the yearly up-and-down cycle, which actually reflects nothing less than the *breathing of the Earth*. To understand this idea, first consider that most of the land on Earth is in the Northern hemisphere, and almost all the vegetation on Earth is on land. The Northern hemisphere summer is therefore the time of year when the vegetation on Earth is in highest abundance, and it is also when the concentration of atmospheric carbon dioxide is sharply decreasing.

The connection between Keeling curve fluctuations and Northern hemisphere vegetation can be explained by the biochemical process of *photosynthesis*,[20] which starts when leaf pigment molecules, called *chlorophylls*, absorb the energy from sunlight and use it to extract carbon dioxide from the atmosphere. Carbon atoms from the extracted carbon dioxide are then joined together, making larger *carbohydrates*, or sugars (Figure 1.3). The metabolism of the cell then distributes this carbon where it is needed, helping to build the other important cell constituents: proteins, fats, and the molecules of heredity, DNA and RNA. Any plant cell has a large variety of carbon-containing compounds in it which form part of the substance of the plant, like the tough cell walls, or that are dissolved in its juices. Scientists collectively refer to all of this carbon in the cells' large molecules as *fixed carbon*.

This explains how the abundance of vegetation in the Northern hemisphere summer decreases the concentration of carbon dioxide in the atmosphere from June to October. Looking again at the Keeling curve, we see that the difference in the carbon dioxide level between the peak and trough of the curve is typically about 6–8 ppm (Figure 1.4, inset).

More photosynthesis in the Northern hemisphere, caused by the increased vegetation in this area during the summer months, explains the downward curve of the yearly carbon dioxide cycle. The upward curve involves the production and release of carbon dioxide originating from the fixed carbon in plants, which occurs in animals (and other life forms) that get their metabolic energy from eating plants. This is the mirror image process to photosynthesis, known as *respiration*. A byproduct of photosynthesis is oxygen gas, which has accumulated in the atmosphere to its present level of about 21 percent. Animals, including humans, breathe (respire) the oxygen, combining it with fixed carbon from eating the plants. This process releases carbon dioxide while providing energy to the animal cells.

Respiration dominates over photosynthesis in the period from November through May, when there is less vegetation in the Northern hemisphere. Unlike photosynthesis, the amount of respiration doesn't vary much throughout the year, so the up-and-down feature of the curve is controlled by how much plant photosynthesis is taking place (Figure 1.5, left side). The ocean phytoplankton photosynthesize at the same rate year-round, and don't contribute to the seasonal variation.[21]

It is worth pausing a moment to note that respiration is the same chemical process that occurs when gasoline is burned in an internal combustion engine. Gasoline contains larger carbon compounds, just like the fixed carbon in the plant cell. In the engine these carbon compounds are combined with the oxygen in air to liberate energy that powers the car, while producing carbon dioxide as a byproduct.[22]

Here is the crux of the anthropogenic global warming problem. We are extracting fossil fuels from the deep underground, not from any part of Earth's surface climate system. When we burn those fuels, we are adding new carbon into the climate system in the form of carbon dioxide (Figure 1.5, right side). That carbon dioxide then enters the cycle of photosynthesis and respiration, but the key point is that the total amount of carbon in all five parts of Earth's climate system has now increased. This explains why the Keeling curve trends upward from 1958 until today. It is beyond all doubt that the increasing level of atmospheric carbon dioxide since the beginning of the Industrial Revolution is of human origin.[23]

Next, we'll look at why the increase in the rate of carbon dioxide buildup in the atmosphere varies so much from year to year. A close study of the Keeling curve shows that in some years, the level of carbon dioxide rises much more than in others. The easiest way to see this is to compare carbon dioxide ppm levels at their peaks in May of each year (Figure 1.4). For example, the monthly average of carbon dioxide levels at Mauna Loa increased by 3.6 ppm between May of 2015 and May of 2016. By comparison, this monthly average increased by only 1.4 ppm between May of 2017 and May of 2018. This may seem like a trivial detail, yet it amounts to a huge variation of 250 percent. And with further investigation, we will gain a much deeper understanding of how the carbon cycle works.

Imagine the entire atmosphere as a reservoir—like a bathtub—but one in which water is continually coming in and going out, as if the tap and drain were always open (Figure 1.6).[24] The level of water in a tub is determined by how much enters from the tap, compared to how much goes out through the drain. This remains true if our imaginary tub has multiple taps and drains. If the combined amount coming in through all the taps equals the amount going out through all drains, the level of water in the tub stays the same. In climate science, the taps are called *sources*, and the drains are known as *sinks*.

The same principle applies to carbon dioxide in the atmosphere. If the amount of carbon dioxide goes up from one year to the next, it is because inputs exceeded outputs in that year. And if it goes up a lot in year 1 and less in year 2, as the Keeling curve shows, it must be that in year 1 there was either more input or less output (or some combination of the two). We know that the amount of new atmosphere input of carbon dioxide from fossil fuel burning has been increasing by a fairly constant amount, and that the uptake and release of carbon dioxide from the oceans does not exhibit large, yearly variations.[25] The cryosphere is not a significant source or sink of carbon dioxide. Thus, significant

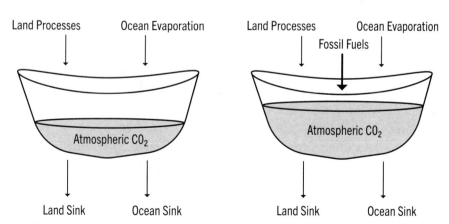

Figure 1.6: Sources and sinks for atmospheric carbon dioxide. The left side depicts the preindustrial Earth, with 280 ppm of carbon dioxide in the atmosphere and no fossil fuel carbon input. The right side depicts the 2020 Earth, with elevated atmospheric carbon dioxide and input of carbon dioxide from fossil fuel combustion.

year-to-year differences in how much the atmosphere ppm level increases must come from changes in the amount of the carbon held on land.

The land surface is extremely complex, and we do not have a thorough understanding of how it absorbs and releases carbon dioxide. One complication is that after carbon dioxide is fixed by plants during photosynthesis, only some of it returns to the atmosphere by respiration. A great deal of it instead sticks around on the land for much longer. For example, a lot of this carbon is stored in the soils for varying amounts of time, but the precise amounts depend on local conditions such as the soils' moisture content. Some of this carbon gets buried on the way to becoming peat or coal, some runs off into the oceans, and some is metabolized in complex ways by single-celled bacteria that are found almost everywhere. In Figure 1.5, the two arrows coming from Biomass show that the newly fixed carbon is either respired or has other fates that keep it sequestered on the land or in the ocean.

Another factor in year-to-year variation in the land sink is that different plants carry out photosynthesis at different rates. Further, the increasing carbon dioxide in the atmosphere prompts some plants to grow faster (this is called *carbon dioxide fertilization*), but others to grow slower—depending on the soils' moisture content and other factors. At the same time, humans are influencing the land carbon sink dramatically. An example is planting or cutting down forests or using agricultural practices that retain varying amounts of carbon in soil.

The bottom line is that while we do know that the amounts of atmospheric carbon dioxide going to the land and ocean sinks are roughly equivalent, we know this mainly because we can measure the amounts in the atmosphere and ocean pretty accurately, so the difference must be going to the land (this is referred to as the *missing land sink* in the climate research literature). More research is needed to gain the detailed knowledge necessary to come up with more effective land management practices that retain as much carbon on the land as possible.[26]

We have been discussing carbon transfers from an atmosphere-centric point of view, but not all carbon transfers necessarily have to go through the atmosphere. For example, carbon on land is transferred directly to the oceans when it runs off in rivers. Over very long time periods, some carbon in land and oceans is transferred to the deep underground to either form coal deposits (from land) or oil and gas deposits (from marine environments). Thinking back to our bathtub analogy, the same principles can be applied to other parts of Earth's climate system, and also to any component that moves between these parts. Examples of this are the well-described nitrogen and sulfur cycles by which these elements move around Earth's climate system.

The amount of carbon involved in all these movements is enormous. The entire atmosphere contains about 800 billion tons of carbon, which is about three trillion tons of carbon dioxide (see Box 1.3). A little over 10 percent of this amount dissolves in and evaporates from the oceans every year (a little bit more dissolves than evaporates, to account for how the amount in the ocean is increasing). Even greater quantities of carbon in other forms are present in the deep underground and in ocean sediments (Color Plate 1).[27]

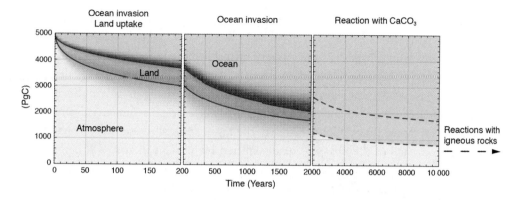

Figure 1.7: Fate of atmospheric CO_2 from human activity. The curves show the decay of an initial pulse of CO_2 at time zero. Very long term reactions with igneous rocks (right) remove the remaining amount over hundreds of thousands of years. From Ciais, "Carbon and Other Biogeochemical Cycles", FAQ6.2, Figure 2, p. 545. Used by permission.

Short and Long Timescales of the Carbon Cycle

There is only one fundamental concept left to learn about the carbon cycle, and that is the idea of *residence time*, which is how long the carbon dioxide sticks around in the atmosphere before it is taken up by the land and ocean sinks. The best way to comprehend residence time is to think about the amount of carbon dioxide that humans emit over the course of a year, mostly from burning fossil fuels. In 2019, that was 37 billion metric tons, or 37 gigatons.[28] What is going to happen to all this carbon, and how long will those processes take?

It is reasonable to think that since all carbon dioxide molecules are alike, they will all go into land and ocean sinks at the same rate, with just one residence time. However, the problem with this notion is that each sink takes up the carbon dioxide with its own set of distinct processes. Also, both land and ocean have multiple distinct sinks, and the rate at which the carbon dioxide leaves the atmosphere is controlled by the details of what happens in each case. This means every sink has its own separate residence time.

Let's first take a look at how sinks and residence times work in a simpler case: the greenhouse gas methane. In sharp contrast to carbon dioxide, methane has just one major sink that removes about 90 percent of it from the atmosphere. This sink consists of a well-defined set of chemical reactions that occur under the influence of sunlight, which ultimately converts the methane to carbon dioxide. Because this is the predominant way that methane leaves the atmosphere, we can approximate the full picture by defining just one residence time, which turns out to be nine years.[29] Defined in this way, the residence time is the average amount of time a methane molecule persists, so some of the methane lasts for shorter or longer than that.

What about carbon dioxide? We have described the annual cycle of photosynthesis and respiration that shows up in the up-and-down fluctuations of the Keeling curve, and photosynthesis is certainly one important carbon dioxide sink. As the Keeling curve implies, some of the carbon dioxide going into this particular sink has a residence time of about one year, though on average, carbon dioxide molecules in the photosynthesis and respiration cycle remain on land for several years. This is the shortest residence time of all the carbon dioxide sinks.

Carbon sinks with longer residence times on land include vegetation that decays over a period of years without being eaten by animals. Another example is passage of carbon through the soils, which also takes years to decades. The dissolving of carbon dioxide into the surface layer of the ocean and some uptake by shallow ocean sediments also have relatively short residence times (Color Plate 1). The result from all these sinks is that about one-third to one-half of the carbon dioxide emitted by humans in 2018 will be removed from the atmosphere within a few decades.[30] These relatively fast exchanges of carbon between the land, ocean, and atmosphere comprise what climate scientists call the *fast carbon cycle* (Figure 1.7).[31]

If these fast carbon dioxide sinks were the whole story, then the effects of global warming, while still severe, would not persist very far into the future once we make the transition away from fossil fuels to a fully renewable energy economy. Unfortunately, Earth's climate system just doesn't work that way. The problem is that the capacity of the fast land and ocean sinks to absorb carbon dioxide is simply not great enough to take up all the anthropogenic emissions. The rest of our emissions are instead taken up by additional sinks which have much longer carbon dioxide residence times; together, these sinks make up what is called the *slow carbon cycle*.

One part of the slow carbon cycle comprises a residence time of centuries to tens of millenniums, by which the carbon dioxide that dissolved at the ocean surface reacts with another carbon compound: the calcium carbonate that forms the shells of many marine organisms. Another way that oceans sequester carbon dioxide is through the action of photosynthetic, carbon-fixing phytoplankton, which eventually sink to the deep sea, taking their newly fixed carbon with them. This makes the phytoplankton what marine ecologists call a *biological pump*—an effective means of sequestering carbon away from the surface Earth system for time periods of about 200–1500 years.[32] However, after these sinks have collectively operated on our 2019 emissions, 10–25 percent of the carbon dioxide will still remain in the atmosphere.[33]

At this point it may be hard to care much about how the Earth system manages to remove that last 10-25 percent, what climate scientists call the *long tail* of carbon dioxide residency in the atmosphere.[34] We are so far into the future already that the point about the longevity of carbon dioxide has surely been made well enough. However, this is also important because some scientists have proposed that we might be able to speed up the natural process enough to remove some of the excess carbon dioxide in a much shorter period of time.[35] The longest-term sinks for carbon dioxide in the slow carbon cycle are the *weathering reactions*. This includes the physical breakdown of rocks to increase their surface area and the uptake of carbon dioxide into the rocks, which incorporates that carbon into minerals.[36] This weathering is mediated through rain or surface water and will remove almost all the rest of humanity's 2019 carbon dioxide emissions within a couple of hundred thousand years or so.

All the carbon on the Earth's surface and in fossil fuels makes up less than 1 percent of the total pool (Color Plate 1). Most carbon instead resides in deep rocks and sediments, and over extremely long periods of time, geologic forces allow it to exchange with the surface. This happens both when Earth's tectonic plates collide, and when these plates spread apart on the ocean floor. Plate collisions take carbon out of the surface system and return it to the deep underground, while plate spreading creates new ocean floor, with previously buried carbon moving into the surface cycle. This very deep carbon cycle operates on a timescale of tens of millions of years and does not influence carbon dioxide residence time in the atmosphere. However, these deep Earth processes can produce volcanic eruptions, which are relevant to the climate because they can eject large amounts of dust into the atmosphere. Volcanoes and hot springs also eject some carbon dioxide into the atmosphere, but only in small amounts.[37]

The longevity of carbon dioxide in the atmosphere has significant implications for our efforts in climate advocacy. It means that even very sharp emissions reductions will not have detectable effects on slowing the temperature rise for at least a few decades.[38] This delayed response is an intrinsic feature of the Earth climate system known as *climate inertia*. In this respect the climate system is like a large ocean liner that has built up a head of steam to go in its present direction. One may certainly hit the brakes, but the forward motion will still take a long time to stop.[39]

Climate inertia poses a challenge for climate advocates because it is natural for many people to expect fast results once actions are taken. It might then be necessary to explain that the basic workings of the climate system mean that patience is required. In fact, even sharp cuts in climate pollutants that don't have long atmospheric residence times will not yield immediately obvious benefits: the large amount of natural variability in the system can mask the effects of emissions reductions in the shorter term. The large annual differences in how much carbon dioxide is taken up by the land sink are an example of this natural variation. The danger, of course, is the potential for political pressure to repeal hard-won policy victories when immediate results do not appear. This is where science-savvy advocacy can be applied to preserve the gains.

Chapter 2

Earth Out of Balance

The title of this chapter evokes Vice President Al Gore's classic book, *Earth in the Balance.*[1] The idea of restoring the Earth's balances can have powerful resonance in advocacy since it connects to themes of sustainability and living in harmony with the natural world. Climate scientists have a specific way of thinking about balances in terms of the energy streams that the Earth receives and emits back into space. As we saw in the first chapter, humans have reduced the outgoing energy stream by adding greenhouse gases and other climate pollutants to the atmosphere, as well as by transforming the land surface to convert fixed, sequestered carbon into carbon dioxide gas. Our focus in this chapter is to gain a more detailed appreciation of what we have done. This is an important inquiry because many policy and technology solutions are targeted at specific sources of the imbalance.

We will begin by looking further at the natural forces that, over very long geological time periods, have altered the energy streams and thus contributed to changes to the climate. By reading the Earth's rocks and fossils, biologists and geologists have been able to piece together records of temperature and atmospheric gases extending millions of years into the past. Those records overlap with modern measurements using sophisticated instruments that show in detail how air, land, and ocean temperatures have changed in response to recent human activities. Unfortunately, both the contemporary and geological temperature records have been disputed —sometimes by well-meaning skeptics who are genuinely confused, but also by denialists who harbor a political agenda. This opposition is also part of our story, which, as climate advocates, we may find necessary to engage with.

1. Natural Influences on Climate

Earth is in an energy balance with its surroundings when the amount of sunlight energy coming in equals the Earthlight energy radiating back out. In this state, the average temperature stays constant and is determined by the natural greenhouse effect. To understand the role of human activities in disrupting this balance, it is helpful to look first at the influences, or *climate change drivers*, that have been occurring naturally since the planet was formed. As we've seen, the two natural drivers outside the climate system are the Sun and the deep underground.[2]

Climate change drivers can be compared by their relative contributions to upsetting the energy balance, the issue we care most about. Color Plate 2 shows the most common diagram that climate scientists use to depict this.[3] It is taken from the fifth report of the *Intergovernmental Panel on Climate Change (IPCC)*, a group founded by the United Nations to provide regular reviews of climate science.[4] The figure contains an apparently formidable amount of detail, but we will unpack it slowly over this chapter, and all of the new ideas will follow from what we've already learned.

Climate scientists look at drivers by way of a concept called *radiative forcing*, which measures the capacity to drive, or *force*, a change in the input or output of radiation to the Earth. The length of the bars in Color Plate 2 shows the cumulative forcing from each driver from 1750 through 2011. The term W m-2 (sometimes written W/m²) indicates the units of this radiation transfer, which are watts per square meter of Earth's surface; one square meter is a bit less than 11 square feet (Box 2.1). Most of the drivers, including all the greenhouse gases, decrease how fast the Earthlight energy radiates out. However, a few drivers instead change the Earth's albedo, while variations in sunlight, of course, alter the rate at which energy enters our climate system.

Box 2.1: Energy and power

Energy is often defined as "the capacity to do work." Something that is full of energy can accomplish more work. For example, low-pressure steam might spin a rotor slowly, while high-pressure steam—with more energy—spins it faster. Energy is also readily converted from one form to another. When sunlight reaches Earth, the energy in the light is converted into heat, which increases temperature and drives the winds and ocean currents.

An example of an energy unit is the *calorie*. Counting calories is a good way to keep track of how much food energy we ingest. Milkshakes have a fixed, high amount of energy, and carrot sticks have a fixed low amount. These energy quantities are static, and we only have to add up calories to know if we are sticking to our diet or not.

In contrast, *power* means energy input or output over time. Since sunlight and Earthlight are continuously entering and leaving the Earth climate system, it makes sense to express the balance not in terms of plain energy but rather in terms of the rates at which energy comes in and goes out. Power is measured in units of *watts*. A 100-watt light bulb is more powerful and thus brighter than a 50-watt bulb because it outputs more energy each minute it is on.

A good practical way to see the distinction between energy and power is to look at an electric bill, where the amount owed depends on the number of *kilowatt-hours* (kWh) consumed. One *kilowatt* is 1000 watts, so *kilowatt* is a power unit, meaning energy per time. The kilowatt-hour number on the bill is thus the rate of energy consumption multiplied by how many hours consumption goes on. Energy per time, multiplied by time, yields plain energy. The time unit has been canceled out, and *kilowatt-hour* is just a measure of plain energy.

The deep underground affects climate in two ways. Most important are volcanic eruptions, which deliver large amounts of ash, dust, and sulfur into the atmosphere.[5] This contributes to the "mineral dust" and "Sulphate" parts of "Aerosols and precursors" (Color Plate 2). Note that the blue and yellow bars depicting these effects extend leftward from zero on the horizontal axis. This means that the aerosols have a cooling effect because they reflect solar radiation. A modern-day example was the June 1991 eruption of Mt. Pinatubo in the Philippines, which caused global temperatures to drop by about 0.6°C. There have been many such eruptions over the last millennium, though the cooling effects don't last long since the ash settles out after a few years.[6]

Volcanoes and hot springs also deliver a small amount of carbon dioxide from the deep underground into the atmosphere. The amounts are so small, however, that they are not included in Color Plate 2. If volcanoes did contribute significantly to atmospheric carbon dioxide, we would see regular increases in the ppm level after eruptions. No such spikes are observed in the Keeling curve or other data sets—they show only steady increases, with year-to-year variations from the photosynthesis and respiration cycle and other aspects of the land carbon sink.[7]

Changes in sunlight intensity can also alter the climate. For example, sunspots wax and wane over an 11–year cycle, and the presence of fewer sunspots is connected to a decrease in the amount of sunlight that reaches the Earth. Despite this regular variation, there has been little to no trend towards increased sunlight since measurements began in the late nineteenth century (Color Plate 3).[8] The bottom line with respect to natural influences is clear: the deep underground contributes to cooling the Earth, while changes in sunlight have made no contribution to warming. Natural forcings are not responsible for the global warming we have experienced.

2. Carbon Dioxide and Temperature

The most fundamental consequence of Earth's energy imbalance, which in turn drives all further impacts, is an increase in temperature throughout the Earth's climate system (Figure 2.1). By early 2021, the average surface temperature had increased by about 1.2°C, or 2°F, compared to the average temperatures in the time period from 1850–1900. 2020 and 2016 are tied for the warmest years on record, and 2016–2020 was the warmest five-year stretch since the beginning of modern-day temperature measurements, and—as we'll see soon—for a long time before that.[9]

Figure 2.1 shows that the overall, average warming measured across the Earth's surface is increasing in jumps and starts, similar to how atmospheric carbon dioxide goes up by different amounts each year. However, even though carbon dioxide is the major driver of global warming, the overall climate system is complex enough that there is no direct connection between year-to-year carbon dioxide buildup and temperature increase. So, if the atmospheric ppm number for carbon dioxide doesn't increase too much in a particular year, it doesn't mean that the temperature increase will also be small. You can see this by comparing the Keeling curve (Figure 1.4) with this temperature data (Figure 2.1). For example, increases

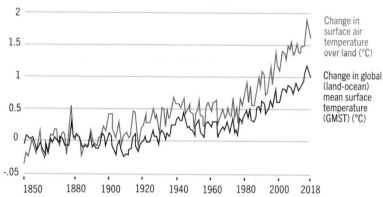

CHANGE IN TEMPERATURE rel. to 1850-1900 (°C)

Change in surface air temperature over land (°C)

Change in global (land-ocean) mean surface temperature (GMST) (°C)

Figure 2.1: Changes in land and ocean surface temperatures over time. From IPCC, Summary for Policymakers, Climate Change and Land. Used with permission.

in temperature were relatively small from about 1998 to 2014, even though atmospheric carbon dioxide levels continued to rapidly climb during that period.

Another key fact is that warming is accelerating: since the 1980s, the average amount of warming per decade is 0.18°C, more than twice as much as the overall average of 0.07°C per decade spanning the period from 1880 until today.[10] The oceans, the land, and the atmosphere are all warming, but it is the oceans that actually capture most of the heat trapped by greenhouse gases.[11] This is because oceans have a very high *heat capacity*, a measure of how much energy has to be added to warm something by a given amount.[12] We know that the temperature increase in the oceans nearly matches that on land and in the atmosphere near the surface. The high heat capacity of the oceans means that they have taken up more than 90 percent of the excess trapped heat, even though they have warmed somewhat less than the land (Figure 2.1). In contrast, the low heat capacity of the atmosphere means that it warms more easily, needing less input of heat energy.

The trapping of so much heat in the oceans turns out to be important to climate policy objectives, which are often stated in terms of staying below a certain maximum temperature increase, like 2°C. Putting a stop to fossil fuel burning will reduce the oceans' uptake of heat, further warming the atmosphere. Fortunately, this will be roughly balanced by greater uptake of carbon dioxide by the land and oceans. After we reach net-zero emissions, the average surface temperature will stabilize and then slowly decrease.[13]

Now, let's look at how temperature increases around the world connect to geography. We have already seen that the surface air temperature over land is increasing faster than over the oceans (Figure 2.1). Another is that the northern latitudes are warming much faster than the rest of the world. Color Plate 4 shows a NASA analysis of warming across the Earth for the period 2015–2019, as compared to the average baseline temperature during 1951–1980. In Alaska and other parts of the Arctic, temperatures are already more than 2°C warmer than in preindustrial times, an increase that is twice the world average.[14]

The Arctic gets much less direct solar radiation compared to the tropics, so this may seem counterintuitive. Here is the explanation: as humans add to the atmospheric blanket of greenhouse gases, the extra trapped heat starts to melt the ice, changing the color of some of the surface from white ice to dark blue ocean (Figure 2.2). These new parts of dark ocean have a very low albedo—they absorb much more solar heat than they did when covered with bright, reflective ice.[15] So now the overall Arctic, with a bit more dark blue and a bit less white, absorbs more heat than it did before. The extra heat then melts a bit more white ice, which leads to a bit more absorption of heat by dark ocean, and so on. As time passes, more and more white surface is changing to blue, and the effect is amplifying; this is what climate scientists refer to as a *positive feedback loop*.

In general, a feedback loop occurs when one or more climate drivers (in this case, greenhouse gas emissions) generate a primary effect, which in turn influences how those drivers continue to work.[16] In the Arctic, the feedback is positive, which means that the ice melts faster and faster as it is surrounded by more and more dark blue ocean that absorbs heat. This is called the *ice albedo feedback*. If humans do not stop driving energy imbalance in the Earth system, the melting will simply continue to accelerate until all the ice is gone. The depiction in Color Plate 4 is a sobering display of the capacity of human activities to bring about a dramatic shift in the Earth's climate system.[17]

Denial of the Data

The recognition that global warming is due to the burning of fossil fuels set off an extreme reaction among some with vested financial interests in the industry, who have created a well-organized network of climate change denialism that has been operating in the US for

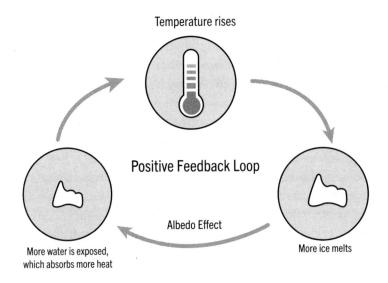

Figure 2.2: Depiction of the ice albedo positive feedback loop. The melting of some reflective ice in response to a rise in temperature exposes more dark-colored ocean, which in turn absorbs more heat.

more than three decades.[18] Certainly, climate change science can be complex and counterintuitive, sometimes defying common sense. For this reason, it can make sense to engage skeptical but well-meaning individuals who do not have an axe to grind. But there is little basis for genuine confusion in the modern temperature record that we have just looked at. And while many denialists have moved on to more subtle arguments, these data really form a key foundation for the entire field. This may be why they remain—for some—an irresistible target.

Denialists have challenged two aspects of the temperature records shown in Color Plate 3 and Figure 2.1. First, they claim that global warming is due to changes in sunlight. How can this claim be sustained in light of the data? Part of the answer is that arguments are made without recourse to the data at all, but rather by invoking the idea that since the Sun is responsible for creating a livable climate to begin with, it must also be the cause of any change in that climate. That is simply an appeal to common sense. We have to be wary about common sense when it comes to science, though, because there are many scientific ideas that appear counterintuitive but are nonetheless true. It is easy to be fooled, and the denialist with an anti-science agenda will exploit the unwillingness of some folks to engage critically on the subject.

Another denialist strategy is to focus only on the early part of the record, until about the 1960s. During that time, a gradual increase in sunlight intensity can be seen on top of the 11–year sunspot cycle, which appears to parallel the slope of the temperature curve (Color Plate 3). This is cherry-picking—focusing only on the data that is favorable to one's argument while ignoring the rest. In fact, careful analysis of sunlight data shows that, if anything, the Sun has cooled in the past few decades, at the same time that temperatures have been setting records.[19] That can be seen qualitatively in Color Plate 3, as well.

The other part of the modern-era temperature record that has been misinterpreted is the apparent pause in warming that can be seen from about 1998 through 2014 (this can be seen in Figure 2.1 for both the land and land-ocean records). Indeed, if we were to plot the data for just this period, it would look like a significant pause. From our vantage point today, any invocation of this argument would be more cherry-picking since the temperatures have since shot up strongly. But at the time, there was a great deal of reporting in the popular press about the so-called "hiatus in warming," with some editorial pages loudly proclaiming that the climate scientists were wrong, and we can burn all the coal we like.

Unfortunately for the coal companies, basic physics does not lie. When carbon dioxide accumulates in the atmosphere, it certainly traps heat, and that must be reflected in a temperature increase somewhere. So, if the average surface temperature stays flat or decreases, where does the extra heat go? Part of the answer—still under active investigation—is that it gets transferred deep into the oceans. Also, fewer measurements were made in the rapidly warming Arctic than elsewhere, which skewed the statistics downward. And we must also remember that there is always a great deal of natural variability in the Earth system. Subtle changes in ocean circulation, volcanic aerosol emissions, and many other factors also influence the temperature record.[20] The reality is that there was no hiatus, just the appearance of one caused by looking at only a subset of the data.

Table 2.1: IPCC statements about anthropogenic global warming

Report year	Statement
1990	Human signal unclear; comparable to natural variation
1995	Balance of evidence suggests discernible human influence
2001	Observed warming in last 50 years likely due to added greenhouse gas
2007	Very likely that human greenhouse gas increase caused most warming
2013	Extremely likely that human influence has been the dominant cause

To the climate scientist, an apparent slowdown in surface temperature increase is exciting and intriguing because it opens the possibility of new investigations that could result in better understanding of the whole climate system. But to the climate change denialist, the very same data show up instead as an opportunity to further push the agenda of the fossil fuel industry. There may be no better illumination of this fundamental split in motivations between scientists and denialists than what is revealed by their disparate responses to new data like this.

Before moving on from the subject of denialist attacks on the temperature record, we have to mention the famous "hockey stick" graph of temperature data extending over the millennium from 1000–2000 AD (Color Plate 5).[21] This is an icon of climate change because it shows the impact of human activities so clearly in the distinction between the "blade"—the instrumental record of the past few decades—and the "shaft"—all the prior data. For early centuries and through 1980, the temperature data were acquired by way of *climate proxies*, natural repositories of information. Commonly used proxies are tree rings, corals, and gas bubbles trapped in ice (see the next section). For example, careful analysis of tree rings gives information about climate because narrower rings are correlated with colder temperatures.[22] Analysis of rare atom *isotopes* in climate proxies greatly enhances the amount of detail and reliability of the information.

To create a full record, proxy temperature data is combined with the modern instrument record, which includes data taken from many sources all over the world. Since methods for making measurements vary, a careful process of data correction, termed reconstruction, is required.[23] This is the process that climate change denialists attack—claiming that the methods used are biased—and if only the data were analyzed differently, they would show no warming at all. But it is the denialist's approaches, rife with cherry-picking, misleading sound bites, and unfounded allegations of impropriety, that cannot be sustained.[24] Since the hockey stick data were published, many more reconstructions supporting its basic conclusion have appeared.[25]

The story of the increasingly robust temperature record is also the story of how the scientific community has become increasingly more confident in attributing global warming to human causes. This is documented over five successive IPCC reports spanning 1990–2013. The language in these reports is all the more convincing because it has been approved by the governments of almost all countries (Table 2.1).

3. Paleoclimatology

Paleoclimatology investigates the climates of the past Earth and what they tell us about climate change today. Scientists working in this area have learned how to read Earth's history in its rocks, sediments, and ice, and the emerging narrative is one of continuous change over billions of years.[26] Six or seven hundred million years ago, for example, geologic evidence shows that ice sheets extended nearly to the tropics, creating what is called the *snowball Earth*, when the whole planet may have come close to freezing. At that time, the great amount of surface water that was locked up in ice caused sea levels to be very much lower than they are today.[27]

In contrast, in epochs when there was a great deal of warming, much of the ice melted and sea levels were a lot higher. Good evidence for higher global sea levels in the deep past comes from marine sediments deposited simultaneously near the coasts and in the interiors of continents, at positions well above present sea level. The sediments are found in all parts of the world, demonstrating that sea level changes were global. From these data, we know there was a *hothouse Earth* in our past as well, corresponding to the Cretaceous era about 100 million years ago, when dinosaurs roamed. At that time, the sea level was about 300 feet higher than it is today—enough to submerge almost all of Florida. Atmospheric carbon dioxide levels then were four to ten times higher than the preindustrial concentration of 280 ppm.[28]

These insights are important because they tell us about the capacity of the Earth to harbor extreme climates, so distinct from today as to constitute what climate scientist Jim Hansen calls "a different planet."[29] The danger is that fossil fuel burning, if not curtailed, will tip us irreversibly in the direction of the hothouse.

Data concerning more recent times are also important for understanding what is going on with the climate today. Specifically, information that paleoclimatologists have acquired about the Ice Ages helps explain how natural drivers alone were able to cause large changes in past Earth climates. This addresses a common objection from both denialists and well-meaning skeptics: humans could not have caused global warming because *the climate has changed before*.[30]

The Ice Ages began a little less than 3 million years ago, the culmination of a prolonged cooling period in Earth's history that began 40 to 50 million years ago.[31] Rock debris and sediment deposits from the last glaciation that ended about 11,700 years ago are found throughout North America, reaching as far south as the Ohio Valley. This gives us clear visual evidence, allowing construction of maps that show how far the glaciers advanced.[32] At the last glacial maximum, about 18,000 years ago, ice covered 25 percent of Earth's land surface, compared to only 10 percent today.

A great deal of insight about past climates comes from the analysis of ice cores. In some remarkable technical achievements involving enormous efforts in very inhospitable conditions, climate scientists have drilled out thin cylinders of ice from Greenland, Antarctica and mountain glaciers that extend down through several kilometers (Color Plate 6). Annual layers of ice compacting over time provide a record of past climate conditions, with the most ancient time periods at the bottom of the cores.[33] Trapped air bubbles in the ice give information about atmospheric carbon dioxide and methane levels through time—the air can be extracted and analyzed in the same way as our present atmosphere.[34] Measurements of rare oxygen *isotopes* in the retrieved ice also gives information about atmosphere and ocean temperatures over time.[35] Like the tree ring data mentioned above, oxygen isotopes are also a form of proxy data that, when properly calibrated, yield reliable information about past Earth climates.

There were six or seven glaciations in the past 800,000 years, when ice sheets advanced, retreated, and advanced again, over and over. And in each case, drops in temperature corresponding to ice sheet advances occur together with decreases in the atmospheric levels of carbon dioxide and methane gases. Ice sheet retreat, by contrast, is always accompanied by increases in temperature, atmospheric methane, and atmospheric carbon dioxide. The ice core record can be read up to the present time, and the carbon dioxide measurements from the trapped ice since 1958 compared to the Keeling curve (Figure 1.4), where the data comes from very accurate instruments that sample the atmosphere directly. There is a close match, giving us confidence that the ice core experiments are reliable.

The repeating glaciation cycles are caused by astronomical influences called *Milankovitch cycles*. We all know about the day–night cycle and the seasons, which come from the spinning of Earth around its own axis (once per day), and the rotation of Earth around the Sun (once per year), respectively. But these familiar features of the Earth's motions are just the tip of the iceberg. Our planet also undergoes many other orbital variations that periodically repeat not over a day or a year, but over tens to hundreds of thousands of years.[36] These astronomical cycles occur because of complex and periodically changing gravitational attractions among the Earth, Moon, Sun, other planets, and moons of those planets. The details are complex, but the key takeaway is that the orbital variations result in long-term cycles of how much sunlight strikes the poles and the Equator during the different seasons of the year. And these cyclic variations in sunlight can be directly connected to the cyclic advances and retreats of the ice sheets, which also occur over the same long periods of time (Color Plate 6).

We have now learned enough to respond to those who think that past climate changes are reason enough to believe that humans are not responsible for today's climate change. Cyclical variations in Earth's orbit around the Sun changed the amount of input solar radiation over timescales of tens of thousands of years, corresponding to the advance and retreat of the ice sheets. And the ice core data show that when input solar radiation was high, both temperature and atmospheric carbon dioxide were high as well. Similarly, low sunlight levels in the past led to decreased temperature and low carbon dioxide levels.

The bottom line is that temperature and carbon dioxide were always connected—like they are today—but the key difference is that today, carbon dioxide accumulation is the driver, and the temperature response follows. In the past, temperature change from sunlight variation was the driver, and the atmospheric carbon dioxide level followed suit. In those times, before humans appeared, carbon dioxide evaporated from the oceans in times of high sunlight and dissolved back into the oceans when sunlight decreased again. This is a key way in which carbon dioxide levels could increase naturally, without excavation and burning of fossil fuels.

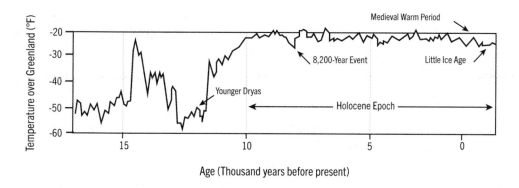

Figure 2.3: Temperature record from the final milleniums of the last Ice Age, through the development of human civilization. Credit: NOAA.

In Earth's climate system, then, initial changes in either sunlight or levels of heat trapping gases can be the driver for large scale alterations to climate. There is nothing saying that one of these drivers always has to precede the other. The fact that the climate has changed before says nothing about why it is happening now. The driver of the Ice Ages was sunlight changes from orbital variations, and the drivers today are anthropogenic fossil fuel burning and land use change.

We can draw at least two other important lessons from the paleoclimate data. First, the record of the past 10,000 years shows that human civilization developed during a time of exceptional temperature stability (Figure 2.3). This highlights the disruptive potential of warming, and emphasizes that, even with aggressive efforts to end fossil fuel use and promote renewable energies, it will be necessary to invest in adapting to those changes that we can no longer prevent.

The second lesson comes from noting that over the past 800,000 years, atmospheric carbon dioxide levels never exceeded 300 ppm even in the warm periods between Ice Ages (Color Plate 6). In fact, it has been millions of years since carbon dioxide concentrations were last at present levels. And during the last interglacial period, 125,000 years ago, Earth's average temperature was only 1–2°C warmer than today, but sea levels were 6–9 meters higher.[37] This is more than enough to submerge a great many US coastal communities.[38] We are fortunate that massive ice sheets take a long time to respond to warming, but these data suggest that large-scale melting may occur eventually if we do not halt fossil fuel burning and perhaps even take active steps to remove carbon from the atmosphere.[39]

4. Human Influences on Climate: Greenhouse Gases

All greenhouse gases work by the same mechanism: absorbing Earthlight and reradiating it in all directions. While some of this reradiated light escapes another portion stays within the Earth climate system and warms it up. Color Plate 2 shows that greenhouse gas emissions are by far the main component of the energy imbalance created by human activities, and that about 75 percent of the cumulative forcing since 1750 is from carbon dioxide.

Color Plate 2 offers helpful, overall context before we look more in-depth at each gas. The drivers listed in the top section of the figure are the well-mixed greenhouse gases—their concentrations at a given altitude are the same no matter where on Earth they are measured. Next are the short-lived greenhouse gases which, as the name implies, do not remain in the atmosphere for very long. Sometimes these gases react with each other and with other components of the atmosphere under the influence of sunlight. This is why a distinction is made between the "emitted compound" and the "resulting atmospheric drivers." Most aerosols cause cooling, and so they are depicted as extending to the left of zero. Finally, one part of the energy imbalance from changes in land use—its effects on albedo—is also included. Other effects of land use on climate change are significant but accounted for separately.

The IPCC report author teams make strong efforts to report the uncertainties in their analysis. You can see this in Color Plate 2 in two ways: as the error limits drawn on each of the data bars, and in a more subjective assessment provided in the rightmost column, where confidence levels vary from low (L) to very high (VH).[40] High and very high levels of confidence mean that there is both robust evidence and strong agreement of experts on the team preparing the report. While greenhouse gases are well understood, the effect of clouds on greenhouse warming is highly complex and remains among the most uncertain parts of climate science today.

We should take note of one key number that is not indicated in Color Plate 2—the rate of input radiation from the Sun, expressed in the same units of watts per square meter of Earth's surface. That number (240 W/m^2) has changed very little since humans have been disrupting the natural system. Without anthropogenic climate change, Earth would be close to an energy balance (only small natural factors would disrupt it), so about 240 W/m^2 of Earthlight radiation would be heading back out.

The meaning of the numbers along the horizontal axis of Color Plate 2 then gains important context: we see that the total anthropogenic radiative forcing (RF) from 1750 to 2011 is estimated at 2.2 W/m^2, a little less than 1 percent of the solar input. This means that by 2011, 2.2 W/m^2 less radiation was being emitted out to space compared to what would have been emitted had human industry not emerged. We can roughly estimate that the effect of humans on the Earth climate system is to upset the natural energy balance to the tune of 1 percent. The respective contributions of all the different drivers to this 1 percent are then easily estimated just by comparing the sizes of the bars.[41]

This may not seem like much of a perturbation, but it has been enough to cause a hefty global temperature increase of over 1°C. In order to feel the significance of the 1 percent, let's calculate the amount of power that an imbalance of 2.2 watts per square meter comes out to. The number is staggeringly high because, to get the total power, you have to multiply 2.2 W/m^2 by the total surface area of the Earth—over 500 trillion square meters. For the year 2017, the energy imbalance amounts to about 60 times the total energy output of the entire human race.[42]

Carbon dioxide emissions from fossil fuel burning dominates a good deal of our discussion in later chapters, so we will limit our attention here to indicating the main sources. In addition to fossil fuel burning, carbon dioxide also emanates from the use of Earth materials in industries—notably cement production. The limestone used to make cement consists of

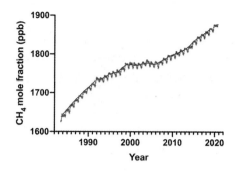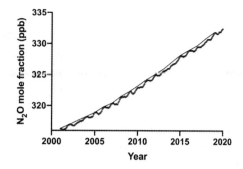

Figure 2.4: Recent anthropogenic increases in the concentration of both atmospheric methane (CH₄), shown left, and nitrous oxide (N₂O), shown right. In both panels, the concentration is given in parts per billion (ppb). Credit: NOAA.

calcium carbonate,[43] and the carbon dioxide portion of this mineral is emitted into the atmosphere during high-temperature heating in the manufacturing process. Today, cement manufacturing contributes about 8 percent of the total human input of carbon dioxide to the atmosphere.[44] The only other major source of carbon dioxide is land use change, especially deforestation and industrial agriculture. These activities break down fixed carbon, which can otherwise be stable for centuries, or even millenniums, on the land.[45]

Methane and Nitrous Oxide

There are three important things to know about any greenhouse gas. First, we need to establish the major human and natural sources of each gas, and how much of the increase in concentration is from human activity. Next, we must recognize how long they stick around in the atmosphere and where they go when they leave it. Thinking back to the discussion of the carbon cycle in the last chapter, this means understanding their respective residence times and sinks. The last important fact is the potency of each gas compared to carbon dioxide. This is different from comparing the sizes of the bars in Color Plate 2, which reports the cumulative effect of each driver. Instead, the issue here is how much more energy imbalance is created now by emitting, say, one new ton of methane, as compared to one new ton of carbon dioxide.

Let's start with methane, which accounted for 10 percent of the greenhouse warming from US emissions in 2018, and over 20 percent of the warming worldwide since the mid-eighteenth century.[46] Among the greenhouse gases, methane takes second place in its influence on climate change. The preindustrial concentration of methane was about 700 parts per billion (ppb), or 0.7 ppm—400 times less than the preindustrial carbon dioxide level of 280 ppm. However, today the level is nearing 1900 ppb (1.9 ppm), an increase of 260 percent, while the carbon dioxide level has increased by less than 50 percent over the same period from 1850 to today. After reaching a plateau that lasted about ten years, atmospheric levels of methane began increasing again in 2006 (Figure 2.4).[47]

Where has all this additional methane come from? Since methane is a fossil fuel (comprising about 80–90 percent of the *natural gas* we extract), it is reasonable to think that this is a major source. But when we burn natural gas to make electricity or light our stovetops, the greenhouse gas emitted is carbon dioxide, not methane. Nonetheless, fossil fuel operations do contribute to atmospheric methane levels due to leakage during oil and gas drilling processes, pipeline transport, and venting from coal mines. The Global Carbon Project estimates that 18 percent of atmospheric methane comes from fossil fuel operations (Figure 2.5), but recent ice core data that can better distinguish natural from anthropogenic contributions shows that the fossil fuel operations instead contribute 30 percent.[48] This is actually good news because it suggests that, with sound policy and better technologies for leak detection, methane levels can be more easily stabilized.[49] Cutting methane emissions from more diffuse sources, like agriculture and waste management, is often more challenging.

Most of the increase in atmospheric methane comes from wetlands, agriculture, and emissions from waste (Figure 2.5). Among the remaining sources, 5 percent comes from burning biomass—a human contribution—and about 6–7 percent is from other natural sources: permafrost, the deep underground, lakes, oceans, and termites.[50] Except for the pipeline leaks, all emitted methane is a byproduct from the metabolism of an ancient class of microorganisms called *methanogens*, which proliferate in environments lacking oxygen. The largest human contribution to higher atmospheric methane levels comes from greatly increasing the *anaerobic* (oxygen-lacking) habitat for methanogens, especially by raising cattle, growing rice, and building landfills (Figure 2.5).

Working to limit methane emissions from all human sources is a great way to get involved in climate advocacy. It is true that methane has a short residence time of just nine years, but this does not diminish its importance as an advocacy target. On a molecule-per-molecule basis, methane is far more potent than carbon dioxide at trapping heat. This explains how it could contribute over 20 percent to total global warming, despite its low atmospheric concentration. In fact, over a 20–year time frame, each new ton of

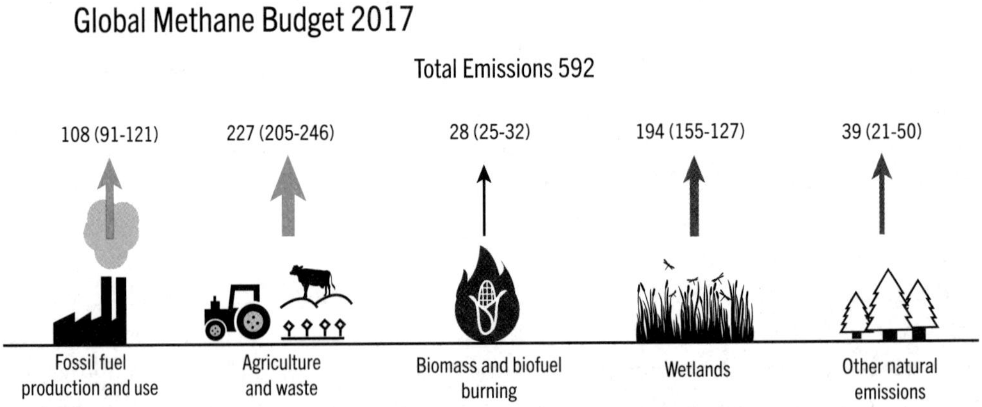

Global Methane Budget 2017

Total Emissions 592

108 (91-121)	227 (205-246)	28 (25-32)	194 (155-127)	39 (21-50)

Fossil fuel production and use	Agriculture and waste	Biomass and biofuel burning	Wetlands	Other natural emissions

Figure 2.5: Major sources of methane. Numbers are given in units of millions of metric tons emitted in the year 2017. Atmospheric methane concentrations are growing at a rate of about 17 million tons per year. Agriculture and waste, fossil fuel, and biomass burning are anthropogenic, while wetlands and other natural emissions are natural. Data from the Global Carbon Project, https://www.globalcarbonproject.org/methanebudget/20/files/GCP_MethaneBudget_2020_v2020-07-15.pdf.

Table 2.2: Properties of some greenhouse gases

Greenhouse gas	Residence time	Concentration	GWP[a]
Carbon dioxide	Undefined	415 ppm	-
Methane (100 yr)	9 years	1.9 ppm	28
Methane (20 yr)	9 years	1.9 ppm	86
Nitrous oxide	114 years	0.3 ppm	265
CFC-12	100 years	0.5 ppb	10,200
HFC 23	260 years	24 ppt	12,400

[a]Except for methane, as indicated, global warming potentials (GWP) are over 100 year timeframes. ppb means parts per billion and ppt indicates parts per trillion.

methane we emit is 86 times more potent at trapping heat compared to the corresponding amount of carbon dioxide (Table 2.2). Climate scientists say that the *global warming potential* of methane is 86. This is also described as the *carbon dioxide equivalent* of methane, written $CO_2(e)$ (Box 2.2).

The third most important greenhouse gas driver, after carbon dioxide and methane, is nitrous oxide (N_2O), commonly known as laughing gas. The concentration of this gas increased 22 percent since 1850 to reach a value of 0.33 ppm at the end of 2017 (Figure 2.4).[51] Both the atmospheric level and anthropogenic increase of nitrous oxide are smaller than carbon dioxide and methane. But the residence time and global warming potential of the gas are each high, so it still makes a significant contribution to warming (Table 2.2). In 2018, nitrous oxide contributed 7 percent of the warming from US greenhouse gas emissions.[52]

About 60 percent of nitrous oxide emissions are from natural sources, while 40 percent are anthropogenic. By far the most important anthropogenic source is agriculture, and a major contributor to that is the ammonia in fertilizer, which is converted to nitrous oxide by microorganisms as part of the global nitrogen cycle. Once nitrous oxide gas is formed in the soils it is then readily released into the atmosphere. Nitrous oxide is also generated as a byproduct of some industrial processes, but the amounts are relatively small. Like methane, control of atmospheric nitrous oxide levels will mainly require us to modify our industrial food system.[53]

Ozone, Halocarbons, and Short-Lived Pollutants

The role of ozone (O_3) in global warming is a complicated business. There is "good" ozone up in the stratosphere, where it forms the *ozone layer* and absorbs solar ultraviolet light, preventing too much from getting to the surface and causing skin cancer. But then there is also "bad" ozone in the troposphere near the Earth's surface, generated as a pollutant in urban smog. Ozone is not included in Table 2.2 because it is not well-mixed in the atmosphere, and so it can't be easily compared with the gases listed in the table. Instead, it is

Box 2.2: Greenhouse gas potencies

Some interesting science supports the concept of carbon dioxide equivalents. Just as incoming sunlight has components corresponding to all the colors of the spectrum, emitted Earthlight also features a spectrum of colors, but all invisible to the human eye. Within that Earthlight spectrum, each greenhouse gas only absorbs the outgoing light of specific colors, determined by how its atoms are bonded together to make the molecule. Methane and carbon dioxide are quite different gases, and so the Earthlight colors they absorb are distinct. Because there is much less methane than carbon dioxide in the atmosphere, a much smaller proportion of Earthlight radiation in methane's colors is absorbed. More of that Earthlight continues out into space.

Here's the tricky part. The reason this makes methane a more potent greenhouse gas than carbon dioxide is that the amount of Earthlight of a particular color that can be absorbed by a greenhouse gas depends on how much of that gas is already there. The first ton of greenhouse gas emitted absorbs a lot of Earthlight, but each successive ton absorbs less and less, until just about all the Earthlight of that color has been absorbed. At that point the atmosphere would be fully saturated with that gas, and further emissions of it could not cause more warming. This can't solve climate change, though, because we are nowhere near saturation yet for any greenhouse gas.

Here is an analogy that may help. Imagine adding a drop of bright blue dye to a glass of water and letting it fully blend in. The color of the water changes from clear to pale blue (see the Figure). As you add additional drops, the blue shade gets deeper. However, the biggest distinction in color came with the first drop, since the water was originally clear. The second drop makes the pale blue darker, but you get less bang for your buck in the visual effect. And as you add more drops, the amount of additional blueness keeps getting harder to see. The extra intensity of color depends on how much color was already there before you added each successive drop. Emitting more methane to the atmosphere today is like adding drops to the glass when the color is still pale blue, while emitting more carbon dioxide is like adding drops when the color is already much darker. The warming impact from one more ton of carbon dioxide is much less than from an additional ton of methane, because carbon dioxide is a lot closer to saturating its preferred Earthlight colors.

Since the residence time of methane is only nine years, it gets removed from the atmosphere quickly compared to most of the carbon dioxide. If we were to emit one ton each of methane and carbon dioxide today, and then wait 100 years to see the relative effects on trapping heat, we would find that the methane is 28 times more potent. But since the average residence time of methane is so short, as we get into the 100 years its contribution starts to really fall off. On a shorter time frame, which is much more relevant to policymaking, the $CO_2(e)$ for methane is higher—over 20 years, the $CO_2(e)$ is 86. We don't have 100 years to solve climate change; 20 years is much more like it. The fact that methane is 86 times more potent than carbon dioxide in this time frame elevates the importance of controlling its emissions as soon as possible.[78]

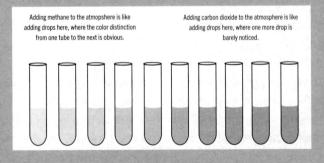

Adding methane to the atmosphere is like adding drops here, where the color distinction from one tube to the next is obvious.

Adding carbon dioxide to the atmosphere is like adding drops here, where one more drop is barely noticed.

concentrated in local areas, and destroyed by chemical reactions with other atmosphere gases within a few days or weeks. Nonetheless, ozone is still a greenhouse gas that makes a significant contribution to climate change, so we should know a little more about it.[54]

A look at Color Plate 2 shows that ozone is not listed among the emitted compounds in the leftmost column, but it does show up no less than five times in the second column, as a "resulting atmospheric driver." Four of the five ozone bars extend to the right, showing a worsening of the energy imbalance. The primary emitted compounds are methane and the three short-lived gases: *carbon monoxide* (CO), *non-methane volatile organic compounds* (NMVOC) and *nitrogen oxides* (NOx). The methane breakdown pathway (its main sink) generates ozone on the way to ultimately forming carbon dioxide,[55] which is why ozone shows up in the methane (CH_4) row of the table. And all three short-lived gases (which are actually collections of compounds for NMVOC and NOx) are automobile exhaust pollutants that react under the influence of sunlight and generate ozone as a hazardous component of urban smog.[56]

Low tropospheric ozone, while short-lived and produced by secondary reactions, nonetheless causes warming in the same way as the other greenhouse gases. The high stratospheric ozone layer is a bit more complicated. That ozone does act like a greenhouse gas, just like the low-level ozone does. But a good portion of it, especially over the South Pole, was lost through chemical reactions with a class of manmade compounds used as refrigerants, called *chlorofluorocarbons* (CFCs).[57] There are many different CFCs, but they all have similar chemical structures and climate effects. CFCs are examples of *halocarbons*: molecules with carbon joined to fluorine, chlorine, or bromine, and they are potent global warming agents in their own right (Table 2.2).

This ozone loss created the famous *ozone hole*, which caused a great deal of panic when it came to public attention in the 1980s. The effects were potentially catastrophic, and in response CFC's were banned all over the world by an international treaty called the Montreal Protocol.[58] The treaty was successful and the ozone hole is healing, although it will still take a long time because CFCs have long atmospheric lifetimes (Table 2.2). The treaty is a shining star, demonstrating that the international community can indeed come together to solve a critical environmental problem if the desire is there.

While CFCs were banned, their long lifetimes mean that they are still contributing to global warming today (Table 2.2; Color Plate 2). Their reaction with ozone is why the little green ozone bar on the CFC line in Color Plate 2 is pointing to the left: ozone has been depleted compared to preindustrial times. Today, CFC levels are declining and the ozone hole is recovering, so the left-pointing ozone bar is getting smaller every year. Solving the ozone hole problem helped global warming because it got rid of CFCs. Yet solving the ozone hole problem also worsened global warming because the ozone layer is recovering, and ozone is a greenhouse gas.

If all this were not complicated enough, the plot thickens again, and the last chapter of the story has only just been written. When CFCs were banned, the main replacement refrigerants were *hydrofluorocarbons* (HFCs), a related class of halocarbons that don't deplete ozone. So, everything is great, except that HFCs are also massively potent greenhouse gases, with global warming potentials that can exceed 10,000 (Table 2.2). Because the demand for refrigerants is projected to grow enormously in a warming world, it is imperative to replace HFCs with chemicals that neither destroy ozone nor function as greenhouse gases. Fortunately, such replacements are readily available, and after some delay, the US finally enacted legislation to phase out HFCs at the end of 2020.[59]

5. Aerosols and Clouds

The next set of drivers that are important to know about are the *aerosols*, which are small particles suspended in the atmosphere. We can often see them as a haze, especially over cities. The bars extending both left and right in Color Plate 2 show that these drivers have both cooling and warming effects. The only important aerosol that causes warming is called *black carbon*—more commonly known as soot—a byproduct of burning wood and fossil fuels that is not a gas but instead a very small black particle made of almost pure carbon.[60] Forest fires are also a major source of this pollutant.

Color Plate 2 shows that the effect of black carbon is quite large—possibly only second to carbon dioxide in trapping heat—but a key difference is that black carbon particles have very short atmospheric residence times, settling out within a few weeks. A major way that black carbon traps heat is by falling out of the atmosphere to land on bright surfaces, especially snow and ice. This decreases the albedo of the frozen surface, so it absorbs more sunlight than it otherwise would, speeding melting. Decreasing black carbon emissions, especially in diesel engines (a major culprit), would also benefit human health by reducing air pollution.[61]

The effects of aerosols show us that, while ending fossil fuel use creates a cleaner, more healthful environment, there are also collateral consequences to deal with. Nitrate and sulphate aerosols—byproducts of fossil fuel burning—have large cooling effects because they reflect solar radiation when suspended in the atmosphere (Color Plate 2). So as we eliminate fossil fuels, this cooling effect will go away, and there will be further warming.[62] Before pollution control measures were put in place in the 1970s, aerosol emissions from fossil fuel burning likely limited temperature increases from greenhouse gases, and this may partly account for why global temperatures did not rise much through the 1950s and 1960s (Figure 2.1 and Color Plate 3). Increased reflection of sunlight from atmospheric aerosols has been dubbed *global dimming* in the popular press.[63] The cooling effect of sulfur aerosols has also led some to propose their deliberate introduction into the atmosphere as a way to lessen global warming.[64]

Sulfur aerosols and other reflective particles are also important because of the way they interact with clouds. The formation of clouds is influenced by both natural aerosols, like blown dust from deserts, sea spray, or volcanic emissions, and anthropogenic aerosols, like the sulfur aerosols from fossil fuel burning. In Color Plate 2, the line labeled "Cloud adjustments due to aerosols" extends to the left, because the anthropogenic aerosols increase the proportion of smaller-sized particles in the clouds, and these small particles do the best job of reflecting sunlight, leading to cooling.

Clouds have complex effects on climate because they are able to cool the Earth by reflecting sunlight, but also able to warm it by absorbing re-emitted Earthlight. However, unlike all the greenhouse gases we discussed above, clouds can also absorb some incoming sunlight. We must keep in mind that clouds are not gases, but rather collections of small droplets of water and ice that continually form and disperse in the atmosphere, and they influence Earth's energy balance through mechanisms that are quite different from how the greenhouse gases work.

Whether the cooling or warming effect of a cloud wins out depends on the type of cloud and how high it is in the atmosphere. Overall, when averaged over the three major types (high cirrus clouds, low cumulus clouds that form localized towers, and low spread-out stratus clouds), the net influence of clouds remains uncertain. Complicating the picture further is that some types of clouds can amplify the warming effects of carbon dioxide, methane, and other greenhouse gases when their concentrations increase in the atmosphere.

Given all this complexity, it is not surprising that clouds are probably the least understood factor influencing Earth's climate, more responsible than perhaps anything else for the uncertainties in future climate predictions based on computer models. The study of cloud feedbacks is a crucial research area today.[65]

6. Land Use Change

The climate change drivers we have discussed so far all directly upset the energy balance. Greenhouse gases and aerosols are emitted and then warming or cooling effects follow from the increased or decreased absorption of Earthlight, or from changes to albedo. Land use change has also altered Earth's albedo, mainly through deforestation. This has had a cooling effect because the dark forest cover is replaced by lighter, more reflective grasslands and croplands (Color Plate 2). But the main way that land use change affects climate is by interfering with natural carbon and nitrogen cycles. Through deforestation, disruptive agriculture, and other means, some fixed carbon and nitrogen that would otherwise have remained on land for centuries or millennia ends up in the atmosphere as carbon dioxide, methane, or nitrous oxide, causing warming.

The climate consequences of changes to the carbon and nitrogen cycles are not included in Color Plate 2 because they can't be calculated in terms of alterations to incoming and outgoing radiation. Other approaches must be used to first quantify how human activities have affected the movements of carbon and nitrogen among the land, ocean, and atmosphere reservoirs, and then to assess the climate impacts of these movements. This is a challenging undertaking, and significant uncertainties remain in our estimates of how much land use change contributes to anthropogenic warming.

Human land use changes may have already begun to influence Earth's climate well before industrialization. The ice core data previously discussed shows that, beginning about 6000 years ago, the atmospheric concentrations of carbon dioxide slowly rose from 260 ppm to 280 ppm—the pre-industrial level.[66] There is also archaeological data demonstrating that large-scale transformation of Earth's land surface, especially deforestation, was already occurring 3000 years ago. Methane levels also rose, consistent with the start of rice cultivation and biomass burning.[67]

Land use change dominated anthropogenic climate change through the early decades of the twentieth century, but it was diminished in relative importance as fossil fuel burning accelerated. However, between 2009 and 2018, land use changes still accounted for 14 percent of the increase in atmospheric carbon dioxide.[68] The IPCC also estimates that 44 percent of methane and 82 percent of nitrous oxide emissions between 2007 and 2016 came from agriculture, forestry, and other land use changes.[69]

The largest losses of carbon from land use change have come from deforestation. This contributes to the global warming problem in two ways: much of the previously stored carbon ends up in the atmosphere as carbon dioxide, while at the same time, the ability of the land to act as a sink for fossil fuel emissions is diminished. Much of the carbon losses have come from the deliberate clearing of forest land for a variety of uses, including agriculture, logging, cattle husbandry, mining operations, and urbanization. More carbon is lost from uncontrolled wildfires and unsustainable forestry practices that promote fragmentation and encourage harvesting of the most carbon-dense regions.[70] In spite of all this, forests still cover about 30 percent of Earth's land area, and store twice as much fixed carbon as the atmosphere holds in carbon dioxide and methane.[71]

Agriculture also drives losses of carbon and nitrogen from the land. A major culprit is the industrial food system, which generates large amounts of carbon dioxide, methane, and nitrous oxide.[72] Production of fertilizers, which are an energy-rich form of fixed nitrogen, is particularly damaging. Industrial agriculture also demands large amounts of herbicides, insecticides, and fungicides—all of which also have to be manufactured at large scale using, at present, fossil fuel-intensive processes. And over time, the soil used in intensive farming is continually degraded and carbon loss accelerates. Much of the released carbon ends up in the atmosphere as carbon dioxide, although a portion of it may also be buried and removed from circulation for some time.[73]

Forestry and agriculture land use changes get most of the attention, which is appropriate because they offer the greatest opportunities for new practices that will allow more carbon and nitrogen to be sequestered on the land. However, we cannot overlook coastal wetlands, which store what has come to be called *blue carbon*. While relatively small in extent, these wetlands sequester very large amounts of carbon per acre—even more than a typical acre of terrestrial forest.[74] Urbanization, oil and gas mining, and other land exploitation caused the loss of over half the world's wetlands in the twentieth century alone. Peatlands are another large carbon reservoir, consisting of a marsh-like surface habitat that covers deep layers of compacted soil and plant debris.[75] In the US, they are most abundant near the East and Gulf coasts and in Alaska.

Finally, urbanization is another land use change that affects localized climates by creating urban heat islands.[76] This can produce long-term average temperature increases of 3–5°F or more than in surrounding rural areas. Urban heat islands result from dense construction with building materials that absorb heat, such as pavement and cement. The relative absence of soil and trees in cities also means that much less water is retained locally. This increases temperatures because liquid water takes up heat when it evaporates, cooling the surrounding air. Dense populations also generate significant amounts of waste heat from residential and commercial activities. Finally, the urban heat islands also worsen global warming in their own right by increasing demand for air conditioning and refrigeration. Improved urban design can go a long way toward mitigating these problems.[77]

Our inquiry into land use change is simple and powerful—by recognizing how present-day land use practices create warming, we open the possibility of reversing the effects by choosing to live differently. Similarly, the highly specific origins of many of the other climate pollutants also offer well-focused targets for advocacy. Whether we take on diesel engines generating black carbon, industrial agriculture spewing methane and nitrous oxide, or manufactured refrigerants spilling HFCs, there is little question that a great deal of important work is waiting to be done.

Chapter 3
Climate Models & Carbon Budgets

In the first two chapters, we looked at the many components of the climate system and how human activity has driven the Earth out of energy balance. Our purpose now is to tie this information together by understanding how this remarkable system functions as a larger whole. To do this, we must learn something about climate models—computer simulations that illuminate how processes in the atmosphere, oceans, cryosphere, land, and biosphere are connected to each other. We know that climate models are reliable because they are able to replicate both the climates of Earth's deep past and the contemporary changes driven by human activity.

A central purpose of climate models is to project what is likely to happen to the Earth system due to all the anthropogenic forcings. They allow us to see how our choices in the next few decades will determine the climate of the rest of the twenty-first century and beyond. Given the dominant role of fossil fuel burning, a key way to look at human activities in the upcoming decades is in terms of *carbon budgets*. These are estimates of how much coal, oil, and gas we can still afford to burn while staying within habitable temperature limits. The budgets unambiguously tell us that rapid transition to a carbon-free economy over the next three decades is necessary to limit warming to a relatively manageable 1.5–2.0°C.

1. A Cornucopia of Fossil Fuels

Earth possesses a large abundance of fossil fuels, all of which are nothing more than the remains of living organisms from bygone eras, compressed and transformed.[1] We will gain important insights about the carbon-free energy transition from a brief tour of these resources, looking at how they originated, where they are found, how much of each type we've already excavated and burned, and how much we think remains (Table 3.1). Fossil fuels are still by far the major energy source driving the global economy, making changes to the system unattractive to those who benefit financially from their use. Dislodging our hold on fossil fuels and replacing them with carbon-free energy sources is the most important technical and political challenge of the twenty-first century.

Coal, oil (petroleum), and natural gas differ most obviously in their material phases: solid, liquid, and gas, respectively. Although they are all *hydrocarbons* and burn with oxygen according to the same basic reaction, they differ greatly with respect to how much energy they yield.[2] Burning one ton of coal produces only about 75 percent of the energy obtainable from the same amount of natural gas, with oil intermediate between the two. The coal hydrocarbon also has a much higher ratio of carbon to hydrogen atoms compared to natural gas. Because of this, coal burning emits much more carbon dioxide per unit of energy production compared with natural gas, with oil again intermediate (Table 3.1).[3]

Coal deposits formed on land throughout Earth's past, but most of this resource was created from abundant, thick plants present during the hot and moist Carboniferous period, between 359 million and 299 million years ago. The plants broke down to form a complex polymer, *lignin*, which persisted in the soils because it is resistant to bacterial attack. As the lignin became buried and compressed, it was slowly converted to peat and then ultimately to coal.[4] Fossil fuels have been colorfully described as *buried sunlight*, since the Sun enabled life to develop in the first place.

Table 3.1: Fossil fuel reserves

Fossil fuel	C/H ratio[a]	CO_2 per unit of energy[b]	Burned so far[c]	Proven reserves[d]	R/P (world)	R/P (US)
Coal	1.4	1.67	171394 Mtoe[e]	738348 Mtoe	132 yrs	390 yrs
Oil	0.44	1.33	206928 Mtoe	242195 Mtoe	50 yrs	11 yrs
Natural gas	0.25	1	103007 Mtoe	171394 Mtoe	50 yrs	14 yrs

[a]Ratios for typical constituents: anthracene (coal), octane (oil) and methane (natural gas). Structures are depicted in Figure 1.3.

[b]Natural gas is arbitrarily assigned a value of 1, for comparison with oil and coal.

[c]Estimates for the years 1900-2018. Data compiled from various sources.

[d]Data for proven reserves and R/P from the BP Statistical Review of World Energy, 2020.

[e]Mtoe means million tons of oil equivalent, the amount of energy derived from burning one million tons of oil. It can be easily converted to more familiar energy units, like calories or kilowatt-hours (Box 2.1). Mtoe values can also be converted to tons of coal, barrels of oil, and cubic feet of gas; www.unitjuggler.com

Contrary to modern myth, oil did not form from the remains of dinosaurs. Both oil and gas originated in the oceans and in shallow inland seas that today form dry land. Hydrocarbons produced via photosynthesis by marine microorganisms called *phytoplankton* settle to the sediment at the bottom, where they, too, resist bacterial attack. As these compounds become more deeply buried below the seafloor they encounter higher temperatures and pressures, and undergo reactions that produce both natural gas and oil. The gas accumulates in pockets underneath impermeable rock, while oil is trapped in nearby porous rock layers.[5]

The origins of coal, oil, and gas in living organisms provides the proof that burning them has caused the steep increase in atmospheric carbon dioxide levels that we observe today. By comparing the precise subatomic makeup of the carbon atoms found in fossil fuels, present-day living things, oceans, and rocks, geochemists have been able to unambiguously demonstrate the human origins of global warming (Box 3.1).

Reserves and Resources

As we know, fossil fuels are not renewable, so they will eventually be depleted if we continue to mine them. For some time, the idea that the planet was running low on these fuels—especially oil—was of major concern, and many books on "peak oil" were written as a result.[6] Some even dared to think that while an imminent scarcity of oil resources and consequent price increases would produce a global economic shock, it could also force a rapid transition to renewable energy in the first half of the twenty-first century.

It did not turn out that way. The US fracking boom demonstrates that new technology can allow mining of resources that were previously inaccessible or too costly to develop with less efficient methods.[7] Fossil fuel companies are also salivating over the opening of more public lands for energy development, and the potential that sea ice melting offers for access to new oil in the Arctic.[8] Globally, combined consumption of all fossil fuels has nearly doubled in the past forty years, with sustained growth though the first two decades of the twenty-first century.[9]

The total amounts of coal, oil, and gas that have already been burned are well-known (Table 3.1), but estimating how much of each remains (the total *resource*) is difficult for at least three reasons. First, while some parts of Earth's subsurface are thoroughly mapped, other areas, like the Arctic, remain relatively unexplored. Second, the extents of some oil and gas deposits are unclear because they are trapped in geologic formations from which extraction is difficult. The US oil shale is an example of a formerly trapped large resource that was liberated by the new technology of fracking. Finally, some resources are known to exist in large quantities but are poorly described because current prices of fossil fuel commodities make them uneconomical to investigate. For example, *methane hydrates*, a form of crystallized methane, are available in enormous quantities on seafloors but mining them would not yield a profit at current prices.[10]

While the total coal, oil, and gas resources available on Earth are thus difficult to estimate, the amounts in the *proven reserves* are reasonably well-determined. Proven reserves

Box 3.1: Isotopes and atmospheric carbon dioxide

A common question about global warming concerns where all the new atmospheric carbon dioxide is coming from. Climate scientists say that the driver is human fossil fuel burning, but how do they know for sure? Volcanoes and hot springs deliver some carbon dioxide directly from the deep underground—couldn't that be the explanation? Or what if the physicists measuring sunlight are getting it wrong, and the extra heat from the Sun is causing more carbon dioxide to outgas from the oceans?

The proof comes from the chemistry of carbon *isotopes*. All atoms are made the same way—they have a nucleus made of positively charged protons and other particles called neutrons, all surrounded by a cloud of electrons. The identity of an atom depends on how many protons are in the nucleus. For carbon, that number is 6 (if it were 7 or 8, then the atom would be nitrogen or oxygen, respectively). When atoms form bonds with each other (like the bonds between the carbon and the oxygens of carbon dioxide) the carbon and oxygen electron clouds overlap and become shared (see the Figure).

Isotopes of an atom have the same number of protons and electrons but a different number of neutrons. A vast majority of carbon atoms have six neutrons, which when added to the six protons, totals 12 particles in the nucleus. These carbons are designated ^{12}C (see the Figure). A rare form of carbon has 7 neutrons and is designated ^{13}C.

Here's the punch line. All the fixed carbon in living cells has more of the lighter ^{12}C isotope as compared to carbon embedded in minerals—like limestone—or dissolved in the oceans, both of which have relatively more ^{13}C. Why? The proteins that fix carbon dioxide during photosynthesis work poorly with ^{13}C and exclude more of the carbon dioxide molecules with that isotope. So, ^{12}C gets concentrated in photosynthetic cells and is then distributed when animals eat the plants. Chemical analysis of atmospheric carbon dioxide shows that its ^{12}C abundance is increasing, consistent with its origin from living material—the plants and marine microorganisms that were the source of the fossil fuels. This proves that the accumulating atmospheric carbon dioxide comes from fossil fuels, not from volcanoes or from ocean outgassing.[74]

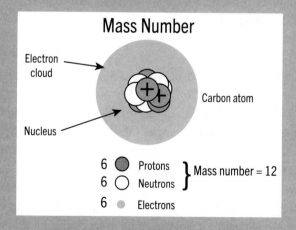

Depiction of the most common carbon isotope, Carbon-12 (^{12}C).

are the subset of total resources that exist in known locations and are economical to develop given current prices. In the energy industry, the amounts of fossil fuels remaining are indicated by a number called the *reserves-to-production ratio*, or R/P. This ratio is just a single number that indicates how many years of production are left, given two pieces of data: the present estimate of proven reserves, and the amount that was used in the most recent year. For example, the R/P for the world supply of oil was 50 years at the end of 2018. This means that, if no new proven reserves are found, and if in each successive year we continue to burn the same amount of oil as in 2018, the world's supply would run out in the year 2068. Similarly, the 2018 world R/P ratios for natural gas and coal are 50 years and 132 years, respectively (Table 3.1).[11]

A key fact about these R/P ratios is that they have not changed much in the past 10–30 years.[12] In other words, as fossil fuels are burned and R/P starts to decrease, some previously unknown or uneconomical resources become available and are transferred to the proven reserves. For example, when fracking the oil shale in the US became technically possible and economical in 2009, *proven reserves* increased and the R/P for oil rose. US energy companies with financial stakes in oil shale happily watched R/P go up, envisioning many more years of oil burning made possible by the fracking revolution. The entire history of fossil fuel development is a steady story of expansion into areas that were once technologically or economically off limits.[13] Despite the access to new oil and gas resources, however, values of R/P for US reserves today are nonetheless quite low (Table 3.1).[14]

The key question now is how much of the proven reserves of coal, oil, and gas can still be burned while keeping Earth's climate in a recognizable state? Table 3.1 shows that the proven reserves of all fossil fuels still substantially exceed the total amounts burned since 1900, and this is not even considering all the resources that are not economical to develop today. We will develop the argument in more detail in the next section, but given the substantial climate change we've experienced just from the amounts burned thus far, this data clearly indicates that most of these profitable reserves are going to have to be left in the ground. Notwithstanding the hopes of some "peak oil" advocates in the 2000s, we are not going to be saved from climate change because the fossil fuels run out.

2. Climate Models and Carbon Budgets

The swift reduction in carbon dioxide emissions is the main goal driving ambitious policies geared toward mitigating climate change. Of course, the atmospheric carbon dioxide level can only be stabilized over a transition period, in which fossil fuels are phased out and replaced by carbon-free energy sources. At the end of this period carbon dioxide emissions from human activities must be reduced to near zero. This is because of the long residence time of carbon dioxide in the atmosphere, which means that a good deal of what we're emitting now will remain in the atmosphere over time frames extending through centuries and millennia.[15]

As we will see, humanity is already on a path to creating a fossil fuel-free economy, a process called *decarbonization*. We will get there eventually, but the severity of climate change depends on how much carbon dioxide and other greenhouse gases are emitted in the meantime. As advocates our job is to build urgency for a rapid transition, leaving as much fossil fuel in the ground as possible. We will be more effective if we have a solid understanding of what well-established climate science says about this.

Nothing in climate diplomacy directly specifies how long the energy transition must take. However, according to a key 1992 international treaty, the *United Nations Framework Convention on Climate Change*, global warming must be kept within limits that minimize dangerous anthropogenic interference with the climate system.[16] In the following decade, climate scientists and policymakers reached a consensus: to accomplish this goal, warming should not exceed 2°C above the preindustrial temperature of 14°C.[17] In another international treaty, the *Paris Agreement* of 2015, it was further agreed that all efforts should be made to keep warming below the more ambitious target of 1.5°C.[18] As we will see, all major climate change impacts on the natural and human worlds become more severe if temperatures are allowed to reach 2°C of warming as compared with 1.5°C.[19]

The question of how fast we must accomplish the energy transition then largely comes down to how much coal, oil, and gas can still be burned while staying below the 1.5°C or 2°C targets. Climate scientists discuss this key issue in terms of a *carbon budget*. Like a financial statement, the carbon budget lays out how much we have to "spend"—that is, how much more carbon dioxide we can afford to still emit.[20] Blowing the budget would invite potentially catastrophic climate change, so it is important to develop new policies with the upper carbon limit clearly in view.

Calculating a carbon budget is not a simple matter because the climate system does not respond linearly when perturbed; it is not appropriate to simply double the known emissions that have led to the 1°C warming so far and call that good for the 2°C budget. One way to see the nonlinearity is to think about the ice albedo positive feedback loop that we discussed in the last chapter. It has a runaway quality to it: melting of white snow exposes dark ocean, which absorbs more heat, leading to more melting. That is certainly not a linear trend. Heat transfer into the deep ocean is also nonlinear because it depends on large-scale ocean circulation and ocean-atmosphere interactions, which are each subject to multiple feedbacks. The effects of clouds on Earth's radiation balance and the strong year-by-year variability of carbon uptake by the land add further complexity. For all these reasons, carbon budgets defy attempts at simplification.[21]

Basics of Climate Modeling

Climate scientists address the complexity of the climate system, including carbon budgets, through *climate models*.[22] A fully developed climate model, known as a *general circulation model*, is a computer simulation that incorporates interactions among the five physical parts of the climate system (Figure 3.1).[23] All such simulations feature two key components. First, they include equations from physics and chemistry that embody the basic laws about how mass and energy are conserved and moved around the climate system. Second, they incorporate experimental measurements, such as the composition of the atmosphere, the topography

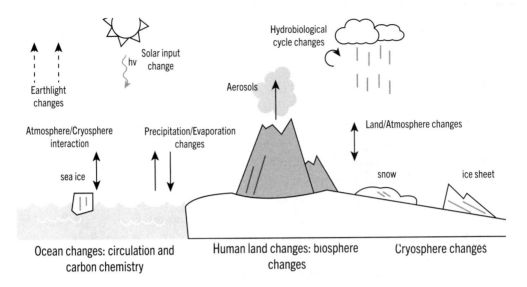

Figure 3.1: Aspects of Earth's climate system that are typically included in models. Perturbations to the system that are studied in climate models include addition of greenhouse gases, changes in land use, volcanic eruptions, and variations in solar radiation.

of the Earth's land surface, and maps of the ocean floor. We will gain a better appreciation for carbon budgets if we understand something about the models that make them possible.

Let's get one basic but super important distinction out of the way: right now, we are talking about climate, not *weather*. Climate is the weather averaged over long spans of time (30 years is a typical period), while weather describes the conditions outside today—perhaps extending into the next two or three weeks—but no longer.[24] Climate models project what average temperatures will be like, say, 30 years from now, and are run many times with different assumptions about what humans will do between now and then to curtail greenhouse gas emissions. Weather forecasts, on the other hand, predict what conditions will be like at the beach next weekend. This distinction may seem very straightforward, but it is easy to misuse the terms.[25] Climate change denialists conflate weather and climate when they claim that cold weather at a given time or place is evidence that the climate is not warming. One such occurrence took place on the US Senate floor in February of 2015, when Senator James Inhofe, Republican of Oklahoma, held up a snowball and solemnly proclaimed that global warming was obviously not a problem.[26]

Weather predictions have limited scope because even small, local disturbances to the atmosphere yield unpredictable consequences elsewhere. There is an intrinsic variability, sometimes called *chaos*, in the system.[27] This inherent, physical limit means that it will never be possible to forecast the weather, say, at the beach at the end of next month. How, then, can climate projections decades ahead be worthwhile? Unlike weather predictions, climate models use our understanding of the Earth system to project general future trends, not the details of what will happen at specific times and places. These projections are of great value in formulating policies for the renewable energy transition.

Climate models work by defining a starting point—for example, the Earth in the pre-industrial year 1800. In the computer simulation, the climate system is changed over time, within constraints imposed by known scientific laws and the way these laws apply to the components of the climate system. It is then possible to test the effect of changes that humans are creating. For example, the atmospheric concentration of methane could be increased by a specified amount over time, and the effects on the climate system then calculated. The results of this methane-change simulation would be compared with an identical simulation in which the methane concentration is held constant.[28]

Climate simulation "experiments" such as this are necessary because the actual climate system is so complex that it is not possible to hold everything constant while varying the tested factor, as can be done in many other areas of science. Certainly, limited experiments are useful; for example, the results of studying how the temperature of a gas mixture affects its ability to hold water are relevant and incorporated into climate models. There are also some "natural" experiments that can be looked at, like the 1992 Mount Pinatubo eruption that caused global cooling. But of course, there is no "control Earth" that would allow us to examine what would have happened had that eruption not occurred.[29] The only way to get insight into that is with a simulation.

Climate Models: Simple and Complex

To get a better feel for climate models, let's revisit the bare rock model of the Earth from Chapter 1. That model is so simple that it includes only a few equations that describe how the Sun's radiation affects the temperature of the Earth. And unsurprisingly, it yields a poor result, predicting that Earth's average surface temperature should be -15°C. Continuing from this, a second model was made by adding an idealized atmosphere to the bare Earth. The new model performed better, predicting a warmer temperature much closer to what our instruments actually measure.[30]

Unfortunately, that second model is still very crude. For example, the only way energy can move around is by radiation, while in actuality, on Earth much heat is transferred in several ways: the evaporation of water near the surface (absorbing heat), rising of the water vapor up into the atmosphere, and then condensation back into liquid (releasing heat). Circulation of the atmosphere and oceans also transports heat from the equator to the poles. To account for all these factors, today's climate models incorporate the major atmospheric and ocean circulation patterns, and the way that these circulations are coupled with each other (Color Plates 7 and 8).

To further account for the complexity of the Earth climate system, climate models also incorporate the many operating feedback loops. The ice albedo positive feedback loop gets a lot of attention because it is causing the melting in Alaska,[31] but a different positive loop called the *water vapor feedback* has even larger effects on overall warming. The atmosphere heats up as greenhouse gases accumulate, and warmer air can hold more water vapor than cooler air. Water vapor is a powerful greenhouse gas, so the warmer the atmosphere gets, the more heat-trapping water vapor can be held within it.

There are also *negative feedback loops* that have dampening effects, which keep the Earth system from spinning out of control quite so badly. One of these is the *blackbody radiation*

Table 3.2: Some climate feedbacks

Feedback	Type	Characteristics
Ice albedo	Positive	Warming causes bright, reflective ice to melt to dark nonreflecting ocean; dark ocean absorbs more heat, leading to more ice melt
Water vapor	Positive	Warming atmosphere holds more water vapor; water is a potent greenhouse gas, leading to more warming
Methane release	Positive	Warming causes methane release from Arctic permafrost thaw, wetlands, and shallow hydrates; increased atmospheric methane traps more Earthlight, leading to warming
Sea level rise	Positive	Warming causes ice melt and sea level rise; continued sea level rise decreases ice shelf stability, leading to more ice melt
Blackbody radiation	Negative	Warming increases surface temperature; hotter Earth radiates more Earthlight into space, leading to cooling
CO_2 fertilization	Negative	Warming from higher CO_2 levels increases the efficiency of photosynthetic carbon fixation, leading to cooling
Weathering	Negative	Higher CO_2 levels increase the rate of mineralization, leading to cooling (over long timeframes)
Ocean solubility pump	Negative	Higher CO_2 levels increase the rate of carbon penetration into the deep ocean via marine photosynthesis and mixing, leading to cooling (also depends on ocean circulation)

feedback: as the Earth's surface warms, it radiates out more Earthlight, some of which is not re-absorbed by greenhouse gases and thus escapes the planet. This process isn't enough to stop the warming, but rather provides a counteracting force that reduces the amount of warming than would otherwise occur. Many more feedbacks, both positive and negative, are also part of the climate system (Table 3.2).[32]

Climate models are able to make reliable projections because they include representations of these feedback loops and many other processes. The built-in complexity of the models then allows estimation of other important quantities besides just overall surface temperature. Such estimations include temperature and pressure variations across Earth's surface and at different altitudes, changes in wind speeds and the large-scale pattern of atmosphere circulation, how ocean and atmosphere circulations influence each other, and alterations to the carbon and nitrogen cycles.[33]

To do all this, comprehensive climate models include the cryosphere, land surface, and biosphere in addition to the atmosphere and ocean (Figure 3.1). This is where carbon budgets come in. By including equations that describe the way carbon moves around the climate system, it is possible to run simulations in which specific amounts of fossil fuels are excavated and quantities of carbon dioxide emitted to the atmosphere are calculated. The models can tell us how much coal, oil, and gas must be burned to add a given amount of carbon to the atmosphere over some period of time, which then gives rise to a projected amount of additional warming. This is the basis for the carbon budgets shown in Table 3.3.

The amount of warming that occurs in response to the addition of a given amount of carbon dioxide to the atmosphere is known as the *climate sensitivity*, which measures how susceptible the Earth system is to human influence. This is most often expressed as the amount of warming associated with doubling of the preindustrial carbon dioxide level (and given the symbol $\Delta T2x$). Because of the challenges associated with incorporating all relevant feedbacks, past climate models returned broad estimates of $\Delta T2x$ in the range of 1.5–4.5°C. However, a comprehensive study incorporating decades of advances in modeling has now narrowed this range to 2.6–3.9°C.[34] This is the temperature increase we can expect if Earth comes back into energy balance when carbon dioxide levels are 560 ppm. With carbon dioxide levels today nearing 420 ppm, we are about halfway there already.[35]

Reliability of Climate Models

Before we look at what climate models tell us about the carbon budget, it is important to learn something about how well they work. Perhaps unsurprisingly, attacks on the reliability of climate models are a mainstay of climate change denialists. A common argument is simply that climate policies should be grounded in hard experimental data and not—as claimed—on speculative models full of fudge factors.[36] Another denialist strategy is to play up the uncertainties inevitably present when modeling something so complex. With that foundation, the arguments can easily slide into implications that the projections are so inaccurate that they aren't even worth paying attention to. Or even if there is merit to a few projections, it would be unfathomable that there could be enough confidence to drive massive policy change, given the economic restructuring and huge price tag required.

Table 3.3: Estimated carbon budget[a]			
Approximate warming (°C)[b]	Remaining carbon budget (GtCO$_2$ starting Jan 1, 2018)		
	33%[c]	50%	67%
-	1080	770	570
1.5	1080	770	570
2.0	2270	1690	1320

[a]Data are from IPCC, 2018.

[b]Warming is compared to the average of Earth's surface temperature in the years 1850-1900.

[c]Probabilities of remaining below 1.5°C or 2.0°C

The denialist argument comes from failure to understand why experimental measurements are not enough to project the future climate. No matter how comprehensive the data sets are, they cannot incorporate the positive and negative feedback loops, which have major influence on the climate system. Feedback loops simply can't be measured directly—they can only be understood at a whole systems level. There is no way to accurately project future climate without incorporating the feedbacks, and there is no way to incorporate feedbacks except through models.[37]

Of course, all the measurements are still important because they provide reality checks to test the models. Computer simulation outputs are compared with the immense work carried out by scientific teams fanned out across the entire planet. Their work includes measurements of ice thickness in the Arctic, humidity in the Western US, wind speeds of Pacific cyclones, sea level rise along the US Atlantic and Gulf coasts, and thousands of other sets of observations. The models then also provide a way to integrate and make sense of all this data.[38]

One way to convince doubters about the reliability of climate models is to point out that they are not just needed to predict the far future, but also to make sense of what is happening now. Anyone who relies at all on a weather forecast is actually depending on a climate model. The same equations, data, and computer networks used to simulate the climate in 2050 are also used to predict the weather tomorrow.

Virtually everybody relies on weather forecasts. Generally speaking, people who use forecasts know that they are based on probabilities, and certainly don't expect perfection, even if they grumble about things that aren't quite accurate. And the forecasts are much better than they used to be. For example, it was remarkable to watch the predicted track of category 5 Hurricane Dorian in September 2019. The storm headed straight to Florida but did not hit land there, and instead headed north while remaining offshore—exactly as predicted. How did the forecasters know enough to confidently make these predictions? That is the power of the *climate* models.[39]

Another reason to place confidence in climate models is that it is possible to carefully calibrate them to match the complexity of the Earth's climate system due to the fact that the past is a known quantity. Color Plate 9 shows simulations combined with measured data on average global surface temperature from 1880 to 2020.[40] The experimental data are shown in black and are exactly the same in all three panels, while the yellow, red, and orange hatched traces show the results from a general circulation climate model that started with the Earth climate system as it was in 1880 and then ran for 140 years. The thicknesses of these lines provides a measure of the uncertainties in the computer simulations, which are obtained by running them many times.

Panel A of Color Plate 9 is a simulation with human influence left out—it simulates the Earth's climate without including input from fossil fuel burning or other human activities. In contrast, Panel B excludes the natural forcings—variations in solar intensity and volcanoes—but instead displays the influence of human fossil fuel consumption. Panel C includes both natural and anthropogenic contributions. Note that the sharp, known temperature increase beginning in 1960 is reproduced only when human contributions are included, and that natural drivers yield no warming at all. Including natural forcings in the model (Panel C) yields only a slightly improved match with the experimental data, compared to anthropogenic influences alone (Panel B). Panel B also shows that the effects of ozone, land use, and aerosols have offset some of the warming from greenhouse gases.[41]

These simulations, and many more like them, demonstrate that the attribution of recent temperature rise to human activity is reliable. This builds confidence that use of that model to predict a particular future scenario—say, one in which fossil fuel burning accelerates—would also yield reliable insights. Model results have also been compared to paleoclimatology data from Earth's past, such as during the Ice Ages, when the distribution of sunlight around the planet was different than it is today (see Chapter 2). They are also tested against contemporary shocks to the system, like the Mount Pinatubo explosion that caused global cooling. In each case, the models performed well, building additional confidence in their predictive abilities.[42]

But are the models full of fudge factors, as critics claim? There are, of course, some significant sources of error. Climate model outputs of temperature, wind speed, and other variables are calculated on a three-dimensional grid, but the resolution (or spacing between grid points) is limited by computer power. Typical high-resolution simulations report data on grid points spaced about 60 miles apart on the Earth's surface, and at perhaps 20–30 altitude layers in the atmosphere (though models on regional scales can be much finer). This means that smaller features that could be important to climate, such as clouds or small islands, are often missed.

To account for this, models calculate average cloud covers at grid points based on factors such as temperature and humidity. Small features on land, such as local variations in elevation and tree cover, are also estimated and averaged based on observational data. These approximate procedures—necessary because of the complexity of the system—are what denialists deride as fudge factors. However, what critics often leave out is that the models are run many times, varying the tuning parameters to optimize the fits to experimental data. Ambiguities are reduced as the structure of the models is fine-tuned, computer power increases, and more experimental data are gathered.

Moreover, all uncertainties in climate models are fully acknowledged by the climate science community. Perhaps the best expressions of this are in the IPCC reports, where syntheses of climate science findings are accompanied by qualitative statements, which indicate the level of confidence held by the expert consensus on each point.[43] Because the IPCC reports also require approval by governments, they are very circumspect about overstating conclusions. In fact, there are numerous examples showing that, if anything, climate scientists are underestimating the climate response.[44] There is little or no concern that findings are overstated.

Worldwide, the climate science community continues to make extraordinary efforts to improve the reliability of the models, and there is good evidence showing that climate projections are gradually becoming better at reproducing experimental observations.[45] It is also important to recognize that it is impossible to eliminate all uncertainty because there is simply too much inherent natural variation. This means that it will always be the case that the signal of human activity must be discerned above the background noise. But at this point a very large number of simulations all point to the reality of climate change and its anthropogenic cause, and an overwhelming consensus exists in the climate science community on both points.[46] Models don't need to be perfect to be right about the big picture. And we will see soon that the impacts of climate change—already well in view—are more than enough to drive home the urgency of the situation.

Carbon Budget Estimates

With this background, we can now look at what climate models say about the carbon budget. The main component of any carbon budget is carbon dioxide emissions from fossil fuel burning. Usually, the effects of other greenhouse gases are incorporated as well, through their carbon dioxide equivalents (CO_2(e); see Chapter 2). The most

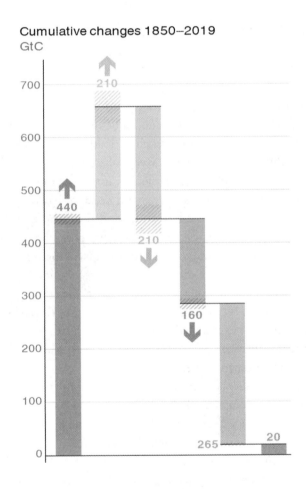

Figure 3.2: Cumulative changes in the anthropogenic portion of the carbon budget from 1850–2019. On the left side, cumulative emissions are shown for fossil fuel burning (+440 GtC) and land use change (+210 GtC). The next two bars depict the amount of anthropogenic carbon absorbed by both the land and ocean (210 GtC and 160 GtC, respectively). The remaining 265 GtC (second bar from right) remains in the atmosphere. The small bar on the far right is a budget imbalance arising from error analysis. The hatched lines on each bar show error estimates. Credit: Friedlingstein, "Global Carbon Budget 2020", Figure 9.

comprehensive budgets also include changes to land use that affect carbon uptake and its return to the atmosphere. Land use change contributed a great deal to emissions in the early-to-mid-twentieth century, and its contribution is crucial for estimating the total amount of carbon dioxide equivalent emitted so far.[47]

Figure 3.2 shows the cumulative human influence on the Earth's carbon budget from 1850-2019. It includes carbon dioxide emissions from fossil fuel burning, industry operations, and cement manufacturing, as well as net emissions from land use change—mostly deforestation.[48] The total amount of carbon dioxide emitted from fossil fuels and industry operations is 440 GtC, with an additional 210 GtC from land use change (see Box 1.3 for definitions of carbon units).[49] Thus, the total anthropogenic carbon load to date has been 650 GtC (2383 GtCO$_2$).[50]

Figure 3.2 also shows what has happened to these carbon dioxide emissions over time. A little over half the carbon dioxide has been incorporated into the land and ocean, with a somewhat higher proportion going to the land. The remaining half is still in the atmosphere and continues to warm the planet. This clearly illustrates how a major portion of carbon dioxide emissions remains in the atmosphere for long periods of time.[51] We should also recognize that the ability of the land and ocean sinks to continue to sequester carbon will likely change as warming continues. For example, when the ocean surface is warmer, there is a smaller temperature difference with respect to deeper waters, making it more difficult for winds to stir up the water column and promote sinking of the carbon dioxide-rich surface water. However, the ocean circulation varies significantly over decades, and sometimes this increases carbon dioxide uptake despite the warmer temperatures. Future trends are thus uncertain.[52]

How much more carbon can be emitted while staying within 1.5°C or 2.0°C worlds? Carbon budgets were first reported in 2013 for the fifth IPCC report, and several other groups have approached the problem since then. Here we will use the carbon budget estimates from a special 2018 IPCC report that assessed the comparative impacts of global warming at 1.5°C and 2°C above preindustrial temperatures (Table 3.3).[53] These estimates are derived from exhaustive modeling efforts and reflect the important distinction between projected climate impacts at these two temperatures, which we will discuss in the next chapter.

The IPCC's carbon budget is expressed in probabilities to account for uncertainties in the modeling calculations. For example, Table 3.3 indicates that humans can only emit an additional 570 gigatons of carbon dioxide to secure a two-thirds probability of staying in a 1.5-degree world. Emitting 770 gigatons would decrease the odds to only 50 percent. For context, estimated global emissions of carbon dioxide in 2019, before the coronavirus pandemic, were 37 gigatons (Figure 3.3).[54] If we continue to emit at this level, the 67 percent budget for a 1.5°C world would be fully depleted by 2033, and the 50 percent budget by 2039.[55] These estimates account for the cooling influence of aerosols from fossil fuel burning, which will be eliminated as fossil fuels are phased out.[56] The calculations assume that emissions of the other greenhouse gases will decrease in tandem with carbon dioxide.

The IPCC budget shown in Table 3.3 does not fully account for the reduced heat uptake by the oceans as the Earth system slowly comes back into equilibrium once fossil fuel burning is stopped. Full equilibration would lead to about 0.3°C of additional

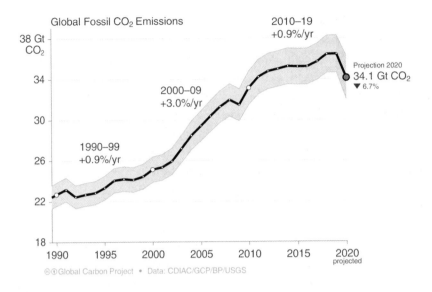

Figure 3.3: Global carbon dioxide emissions from fossil fuel sources, from 1990 to 2020. Units are given in giga-tons of carbon dioxide emitted and do not include other greenhouse gas equivalents. The effects of the COVID-19 pandemic are evident in the projection for 2020. Used by permission of the Global Carbon Project.

warming after fossil fuel burning is halted, but there is considerable uncertainty in this estimate because it is unclear how much atmospheric carbon dioxide might also outgas from the oceans when this occurs.[57] The estimates also don't consider carbon cycle feedbacks, such as melting of the Arctic permafrost. This could release significant amounts of both carbon dioxide and methane, but the temperature threshold that might trigger this is unknown.[58]

Comparing the data in Tables 3.1 and 3.3 allows a rough estimate of how much of the remaining proven fossil fuel reserves are burnable without blowing the carbon budgets. The strictest carbon budget shown in Table 3.3—required for a two-thirds chance of staying within 1.5°C—would allow us to burn just 16 percent of the proven reserves.[59] Even a low stringency budget, such as that required to achieve 50 percent odds of staying within 2.0°C, would still leave slightly over half the proven reserves in the ground. There is no escaping it: preserving the stability of Earth's climate requires that much of the known, economically extractable fossil fuel resource be left in the ground, forever.[60]

There is good geological evidence that about 55 million years ago, a large natural release of about 2500 to 7000 billion tons of carbon occurred. Earth's surface warmed by 5°C at the equator and 9°C at the poles, and the high temperatures persisted for about 200,000 years.[61] This episode is called the *Paleocene Eocene Thermal Maximum* (PETM). The size of this natural release is on the order of ten times more than the 440 billion tons humans are responsible for so far—but our *rate* of emissions is much higher. There is no known historical analog to today's emissions rate; we are in uncharted territory here.[62] And the proven fossil fuel reserves, together with some of the much larger resource, may well be enough for us to create our very own reenactment of the PETM, should we be foolish enough to let that happen.

3. Pathways for Decarbonization

Carbon budgets are important because they frame the possibilities for how we must act. For example, if emissions were to be maintained at the pre-COVID 2019 level, the IPCC climate models predict that we would reach 1.5°C of warming by 2040.[63] This would spend the entire budget in only twenty years: the 2019 emissions of 37 $GtCO_2$ (Figure 3.3) multiplied by 20 years gives 740 $GtCO_2$, putting us at the 50 percent confidence level for a 1.5°C world (Table 3.3).[64] At that point, it would probably no longer be possible to stay even within a 2.0°C world. But if we begin emissions reductions right away, then the fixed budget can be spent over a longer period, enabling a less disruptive transition to a carbon-free energy economy.

In addition to calculating carbon budgets, climate models are also used to project what will happen to the Earth climate system in a variety of hypothetical futures where emissions are curtailed rapidly, slowly, or not at all. In the more favorable futures, less greenhouse gas is emitted, which leads to smaller overall radiative forcings and milder temperature increases. In contrast, significantly greater warming is projected when emissions continue to rise over the next few decades. These different scenarios are called *representative concentration pathways* (RCPs).[65] They are used to estimate the impacts of emissions on global temperatures, ice melting, extreme storms, and other characteristics of the natural and human worlds.[66] Acquiring some feel for the RCPs is helpful for appreciating the range of possible twenty-first century climates, which we will look at in the next chapter.

In 2013 the IPCC developed four distinct RCPs, which it designated RCP2.6, RCP4.5, RCP6.0, and RCP8.5 (Color Plate 10). The numerical parts of these designations represent approximate radiative forcings, in units of watts per square meter (W/m^2), which correspond to each pathway.[67] The original best-case scenario—RCP2.6—peaks at 3 W/m^2 and then declines to 2.6 W/m^2 by 2100. This corresponds roughly with a 2.0°C world. Later, a new RCP1.9 pathway was developed with even lower emissions, consistent with a 1.5°C world.[68] The RCP1.9 and RCP2.6 pathways yield the carbon budget estimates for 1.5°C and 2.0°C worlds, respectively (Table 3.3).

RCP4.5 and RCP6.0 are intermediate pathways, in which carbon emissions peak and decline before 2100.[69] These scenarios are closest to what our future may be like if we continue with business as usual—steady, but slow and uneven, progress toward decarbonization. Finally, in the most extreme RCP8.5 projection, carbon dioxide concentrations rise sharply for the entire twenty-first century before finally plateauing at well over 1000 ppm, sometime after 2100. The rise in global temperature here is about 5°C, a truly catastrophic future, and fortunately not one that we are presently on track for.[70]

What do the RCP1.9 and RCP2.6 pathways tell us about how fast we must act to preserve 1.5°C or 2.0°C worlds? For a 1.5°C world, annual global carbon dioxide emissions must sharply decrease from the present 37 $GtCO_2$ to between 13–20 $GtCO_2$ by 2030, further decreasing to zero by 2050. The 2.0°C target allows for more time and a shallower decrease: 23–30 $GtCO_2$ by 2030, and zero emissions by 2075.[71] Comparable decreases in the other greenhouse gases are also required to meet either temperature target.

Even in modeled scenarios with very sharp emissions reductions, a small fraction of emissions is still hard to eliminate. For example, some industries are difficult to electrify (and thus to power on a zero-carbon electricity grid), and emissions of methane and nitrous oxide from agriculture are also challenging to fully eliminate. To compensate, natural land management methods to improve carbon sequestration in forests and agriculture are invoked to make up the difference—an approach termed *carbon negative*.[72] In essence, this means that some emissions are still allowed as long as they are compensated by removal of as much or more carbon dioxide from the atmosphere. However, if we fail to rapidly cut emissions and too much atmospheric carbon dioxide accumulates, then it will likely be necessary to implement industrial approaches involving carbon capture and permanent sequestration of the carbon in underground geological formations. These approaches will be needed if emissions are not reduced quickly because the natural carbon sequestration capacity of the land is limited.[73]

Many new initiatives will be needed to accomplish the rapid emissions reductions mandated by these strict carbon budgets. These include economy-wide policies for electricity generation, energy efficiency, transportation, urban development, agriculture, and forestry. Fortunately, detailed roadmaps have been developed that describe plausible pathways for the US renewable energy transition over the next few decades. They include descriptions of what needs to happen in all areas of the economy and set priorities for which sectors should take the lead. These roadmaps are outlined in the Interlude, and specific policy approaches to accomplish the goals are detailed in the second half of the book.

Chapter 4

Impacts of Climate Change

We have seen that atmospheric greenhouse gas levels determine Earth's temperature, and that the amount of future warming will depend on how soon fossil fuel burning is stopped. Our purpose now is to explore what the consequences of further warming will be for the natural and human worlds. Many of these downstream impacts are familiar, as they are already becoming a reality. They include accelerating ice melt—with consequences for sea level rise and coastal flooding—increasing rates of biodiversity loss, along with severe hurricanes, droughts, and forest fires. In the human world, they take the form of adverse effects on food production, freshwater availability, and health. The costs of all this will be steep, and almost certain to fall most heavily on vulnerable, low-income populations least responsible for causing the problem.[1]

Climate advocates can make the case for action by highlighting the connection between human activities and the severity of climate impacts. Taking action with a clearheaded understanding of the problem is the antidote to *doomism*, an alarmist response that often appears in public discussion of climate change. Unchecked, apocalyptic thinking about supposedly inevitable disasters can lead to despair and inaction.[2] The problem of climate change is clearly urgent, but as is detailed in the Interlude following this chapter, the best science and policy thinking is unambiguous: solutions are available, and the main barrier lies in generating political will. Advocates can contribute by highlighting the impacts that have already occurred, offering (if needed) evidence of the human cause, and pointing out that how much things worsen is entirely up to us.[3]

1. Evitable and Inevitable Impacts

We will begin with an overall look at the impacts that have already appeared and are expected in the next few decades. In late 2018, the Intergovernmental Panel on Climate

Change released a special report in response to an urgent request from the parties that attended the United Nations' 2015 Paris Conference. The question posed to the IPCC was simple: how much worse will climate change impacts be if we allow the increase in Earth's average surface temperature to increase 2.0°C, compared to 1.5°C? The answer returned was unambiguous—quite a lot worse.[4] Across the board, nearly all aspects of natural ecology and the human world are projected to exhibit a significantly stronger response from just that additional half degree of warming (Color Plate 11).

The report marks a watershed in our understanding of Earth's climate. Prior to having such a clearly focused study in hand it was not well-appreciated that substantial additional damage could really be caused from such a small temperature increase. Recall that, if we continue with business as usual, we will reach 1.5°C of warming within less than two decades.[5] That timeframe is short enough to be on the radar of many elected officials. The simple, central message of this IPCC report can therefore be a powerful way to communicate the urgency of what is at stake.

Let's look at a few of the most important findings from a broader, global perspective, all of which also apply to the US. The IPCC analyzes climate impacts by combining them into five categories termed "reasons for concern" (RFCs), as shown in Color Plate 11.[6] This is sometimes called the "burning embers" depiction of climate change impacts. The horizontal gray bar in the figure indicates the approximate temperature increase of just under 1°C over the period from 2006–2015, as compared with preindustrial times. The severity of the impact is color-coded with the increase in temperature. This depiction is valuable because it reveals that distinct aspects of the natural and human worlds show different susceptibilities to climate change. In addition to the RFC's, the IPCC also analyzed impacts for ten specific aspects of the human and natural worlds (Color Plate 11). We will come back to some of these later in the chapter.

Color Plate 11 shows that "unique and threatened systems" (RFC1) are already impacted. These systems include fragile ecologies such as coral reefs and Arctic environments, where impacts are projected to become severe and eventually irreversible as warming reaches 2.0°C above preindustrial temperatures. This is the worst-case category, since even a 1.5°C world poses extremely serious threats. The hard reality here is that some of these ecosystems are likely to be irreversibly lost no matter how fast action is taken.

The IPCC projections of "extreme weather" (RFC2) show very substantial worsening during the transition from today's Earth to a 1.5°C world. In the US, this means that coastal flooding, wildfires, and extreme storms will all likely become more severe over the next two decades. These findings highlight the importance of investing in adaptation, especially in high-risk areas such as heavily populated shorelines and the urban-wilderness boundaries of western US cities, which are so vulnerable to wildfires.

The "distribution of impacts" (RFC3) includes worldwide impacts and risks that disproportionately affect particularly vulnerable groups based on geography, income, wealth, or other socioeconomic characteristics. One underlying reason for these impacts is the effect warming has on reducing agricultural crop yields and freshwater supplies.[7] Developing countries outside the US (such as South Asian countries) that rely heavily on specific crops or freshwater sources are at a moderate risk of these impacts. In the US, this category includes impacts on low-income, inner-city neighborhoods, exemplified by the vulnerability that many New Orleans residents had to flooding during Hurricane Katrina in 2005. Many of these impacts are preventable but will require inclusion of social equity goals within the framework of climate change policy.[8]

"Global aggregate impacts" (RFC4) reflect impacts that can be looked at on a global scale. This includes loss of certain ecosystems, species extinctions, and financial damages. Color Plate 11 shows that climate change has not (yet) been severe enough to discernibly impact these areas. For example, the global economy is so large that it has yet to show much response to the accumulating damages. However, impacts are projected to appear as the world warms to 1.5°C and beyond. Many of these impacts are evitable.

Finally, "large scale singular events" (RFC5) include large and abrupt changes to ecological or human systems, many of which may be irreversible. This could include the rapid transformation of much of the Amazon rainforest into grasslands, or the collapse of Arctic biomes. Large and abrupt die-offs of corals and degradation of some Arctic ecosystems are already detectable, although other impacts—such as rapid sea level rise from ice sheet melting—may only reach very high-risk levels at much higher temperatures. Except for some of the most fragile ecosystems, the impacts in this category are evitable. Note that Color Plate 11 does not depict the impacts of further warming to 3.0°C or beyond. At those temperatures, extremely high and potentially irreversible impacts occur in all five of the RFC categories.

Climate Tipping Points

Large-scale singular events encompass the notion of climate *tipping points*.[9] These are thresholds of energy imbalance that, if surpassed, could cause dramatic changes in the state of the climate system. The idea is that global climate may exhibit an abrupt shift at certain specific junctures, with large changes occurring within a few decades or less, rather than a longer, smooth transition. The basic notion is shown in Color Plate 12.[10] At the top of the figure, the "glacial-interglacial limit cycle" refers to two stable Earth states corresponding to Ice Ages (far left) and interglacial periods. In this depiction, anthropogenic forcings have already moved the Earth away from Ice Age cycling and into the center of the drawing, placing us at a point where our actions may soon determine whether we will inhabit a recognizable (if warmer) Earth, or a hothouse planet (see the dotted gray arrows). *Anthropocene* refers to the proposed name of the most recent period in Earth's geological history. This name suggests that human influences on the planet have become so pervasive as to merit renaming the present period, which, so far, is still called the *Holocene*.[11]

One reason that the idea of tipping points seems credible is that the climate has undergone extremely large natural fluctuations in the past. These are best documented during recent Ice Age glaciations and are called *Dansgaard-Oeschger (D-O) events*.[12] These events show up in the Greenland ice core record as a very fast warming of up to 10°C within a few years, followed by a slower period of cooling back to the original temperature.[13] These repeated events differ from tipping points because they don't represent long-term, irreversible climate shifts. But they are relevant because they demonstrate that, at least regionally, climate has the capacity to rapidly undergo large changes.

We do not know when these tipping points might be reached, but many climate scientists set a precautionary boundary at 2°C of warming. Surpassing this limit would make the transition into a stable, hotter Earth more likely. And after the transition occurs it could be difficult—or impossible—to return to the original climate state. Examples of individual tipping points that might contribute to the flip into hothouse Earth include the collapse of all or part of the Greenland or West Antarctic ice sheets, rapid release of greenhouse gases from thawing Arctic permafrost, and fast transformation of tropical rainforests into grasslands (Table 4.1).[14] A recent study suggests that some large vulnerable ecosystems, like the Amazon rainforest, may collapse within just a few decades of reaching a tipping point.[15] We should note that the idea of tipping points is not without its critics within the climate science community, as it stands in fairly sharp contrast to the conventional notion that impacts will worsen more incrementally as forcings increase.

Table 4.1: Climate change tipping points[a]

Tipping point	Potential causes and effects
Arctic permafrost loss	Melting permafrost leads to large scale releases of methane and carbon dioxide
Greenland ice sheet collapse	Retreat of ice sheet in response to warming leads to sea level rise
West Antarctic ice sheet disintegration	Cascading collapse of ice sheet sections leads to sea level rise
Amazon rainforest dieback	Warming and deforestation causes biome shift to savannah, accompanied by biodiversity loss and decreased rainfall
Northern boreal forest shift	Forest expansion into the far northern Arctic and dieback of southern sections, leading to regional warming
Coral reef dieoff	Warming waters cause bleaching and dieback, loss of unique ecosystem and fisheries
Breakdown of Atlantic meridional overturning circulation[b]	Melting Arctic ice spills into Northern Atlantic, decreasing density of salty seawater and blocking return flow to South Atlantic; leads to regional cooling and ecosystem change
West African monsoon shift	Rainfall changes in the African Sahel region lead to disruption of agriculture and ecosystem change
Indian monsoon shift	Increases in carbon dioxide emissions or aerosol levels trigger rainfall extremes and disruption to agriculture

[a]Information from Robert McSweeney, Explainer: Nine 'tipping points' that could be triggered by climate change. CarbonBrief, February 10, 2020. (Note 14).

[b]See Color Plate 8 for a depiction of ocean circulation patterns.

In the remainder of this chapter, we will look at the effects of global warming on the US and the rest of the world. The next three sections examine the consequences for nature: ice melting, weather extremes, and degradation of the biosphere. The last section describes the impacts that have already affected human society—many of which have come about from the damage to the natural systems we depend on.

2. Ice Loss and Sea Level Rise

There is nothing complicated about sea level rise, which is occurring today for two reasons. First, warming causes the oceans to expand, shrinking beaches and pushing coastlines inward.[16] Second, much of the meltwater from mountain glaciers and the large ice sheets in Greenland and Antarctica eventually reaches the oceans, causing them to rise. Between 1900 and 2016, the sea level rose 7–8 inches. The biggest contribution came from the thermal expansion, while melting of the large polar ice sheets contributed least.[17]

Some parts of the cryosphere melt without causing sea level rise. Like ice cubes in a glass, the melting of ocean ice in the Arctic sea does not raise sea level. Similarly, melting *ice shelves*, extensions of ice sheets which are attached to land but float on the ocean, make no contribution either. When warming temperatures cause an ice shelf to break off, however, the adjacent glaciers and land ice sheets become more prone to melting because their ice has an easier path to the sea.[18] That is why climate scientists watched with some concern when, in mid-2017, a Delaware-sized chunk of the large Antarctic Larsen C ice shelf broke off and floated away, threatening to destabilize the nearby land ice (Figure 4.1b).[19]

We noted earlier that warming is accelerating, so it should be no surprise that the rate of cryosphere melting is increasing as well. For example, the previously slow-melting Greenland ice sheet is now contributing about 25 percent of sea level rise—the largest, single contributing factor.[20] The European and Arctic heatwaves of summer 2019 caused melting in Greenland at altitudes above 10,000 feet, a rare occurrence in this "dry snow zone".[21] And the thickness and extent of Arctic Ocean sea ice is also decreasing rapidly. Fortunately, modeling suggests that restraining warming to 1.5°C is enough to slow or even halt further melting. But if we allow warming to increase above this, we may see a fully ice-free Arctic Ocean in late summers by as soon as the 2040s (Figures 4.1a and 4.1b).[22]

Melting is also accelerating in Antarctica, especially in the highly vulnerable West Antarctic Ice Sheet (WAIS), as seen in Figure 1. Ice shelves from the WAIS extend over the surrounding ocean, and the boundary between these shelves and the land portion of the sheet is retreating inland, destabilizing successive sections of the sheet.[23] From 2009–2017, Antarctic ice in the WAIS melted about six times faster than it did in the 1980s.[24]

Glaciers around the world are losing ice at increasing rates. Perhaps most troubling is the melting of extensive Himalayan glaciers known as "Earth's third pole," which is threatening the supply of freshwater runoff to Chinese, Indian, and other Asian populations numbering in the billions.[25] In the US, over a fourth of the named glaciers disappeared from Glacier National Park in Montana between the mid-1960s and 2015, and nearly all that remain are thinner glaciers that extend over smaller areas.[26] Glaciers are the most

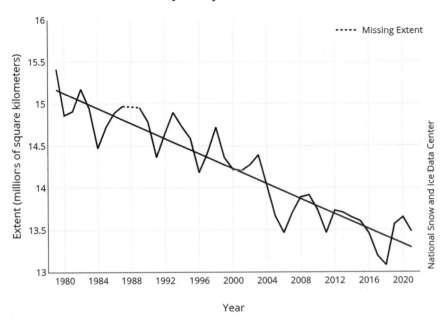

Average Monthly Arctic Sea Ice Extent
January 1979 - 2021

Figure 4.1a: Shrinkage of January Arctic sea ice cover from 1979 to 2021. About one-eighth of the area covered by winter ice has been lost over the past four decades. This corresponds to a shrinkage of 3.1% per decade. Credit: "A lopsided January", National Snow & Ice Data Center, February 2, 2021, http://nsidc.org/arcticseaicenews/2021/02/a-lopsided-january/

vulnerable part of the cryosphere, and large, further losses are inevitable even with a rapid response to the climate challenge.

Although we will have to come to terms with the loss of many glaciers, much of the rest of the cryosphere can still be preserved. The current trajectory will generate an additional 6–14 inches of sea level rise by 2050.[27] In the US, this will certainly require expensive adaptation or relocation of vulnerable communities, especially on the Southeast and Gulf coasts. But if warming can be held to 1.5° or 2.0°C, the cost and effort of adaptation will still be manageable, and the large ice sheets could restabilize. In contrast, continuing with business as usual may lead to as much as eight feet of sea level rise by 2100, a truly catastrophic outcome that may even surpass irreversible tipping points (Table 4.1).[28] As we ponder the situation, it is helpful to remember that the scary pictures of a submerged future Florida only represent the consequences of scenarios where we fail to act for a 2.0°C world, not the scenarios where we act with resolve.[29]

Nuisance flooding is a further consequence of sea level rise. These are "sunny day" floods that occur at high tides along susceptible coastlines causing road closures, deterioration of infrastructure, and other impacts. In 2020, the US National Oceanic and Atmospheric

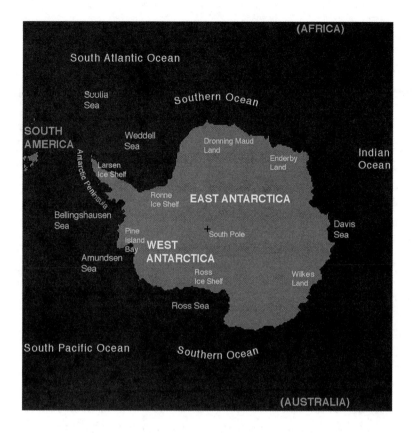

Figure 4.1b: Map of Antarctica showing the positions of the East and West Antarctic ice sheets. The collapsing Larsen ice shelves are indicated on the Antarctic Peninsula at the upper left. Credit: National Snow and Ice Data Center.

Association (NOAA) released a comprehensive report detailing the large increase of this type of flooding in recent years (Color Plates 13 and 14).[30] For example, Charleston, South Carolina, experienced nuisance flooding on 13 days in 2019, the same number of incidents that it recorded in the previous 50 years combined. Rising seas also intensify storm surges when hurricanes and other strong storms make landfall. These high-water levels, which can reach 20 feet or more above calm sea level, occur when strong winds over the ocean combine with low atmospheric pressure to force water over the coastland.[31]

Sea level rise, nuisance flooding, and storm surges don't impact all US coasts equally, because geological factors also contribute. One of these factors is the still-ongoing rebound of the land since the Ice Age glaciers retreated. This makes land subside in the Northeast, worsening sea level rise in that region compared to the West Coast and Alaska, where the issue is of less immediate concern in most places. Severe land subsidence along the Louisiana and Texas shores of the Gulf Coast and in Virginia has another explanation—groundwater or fossil fuel excavation, which causes the earth to sink in response. At Grand Isle, Louisiana, the sea has risen by a remarkable eight inches in just the past ten years, mostly from subsidence of the land. Yet oil and gas mining in the Gulf of Mexico continue apace.[32] In the last section of this chapter, I describe the massive and costly adaptation project planned to deal with this.

Are the melting cryosphere and its consequences caused by human activities? To address this question, climate scientists asked whether the observed sea level rise could be simulated using models that either included or omitted anthropogenic forcings.[33] Only the models that included fossil fuel burning and land use change could match the present data, with a confidence level above 90 percent. Similar results were found when the models were used to evaluate data on sea ice loss, glacier mass loss, and Northern hemisphere snow cover.[34] This is an important finding, which follows up the 2013 IPCC assessment that it is extremely likely (over 95 percent probability) that human activities cause global warming.[35] Although it may seem obvious that the human role in warming implies a similar role in ice melting, actually demonstrating this required many years of painstaking analysis to separate the human signal from the background noise.

3. Extreme Weather

Climate change is increasing the intensity of extreme weather events. Unlike gradual sea level rise, for example, the destructive consequences of severe weather are immediate and inescapable. They bring home the impacts of climate change and can impress lawmakers and community leaders with the urgency of the problem and the need to act. Extreme weather events have an outsized influence in shaping public perceptions of climate change.[36]

Questions often arise about the extent to which human activity is responsible for specific extreme weather events. The best general answer is to emphasize that Earth's entire climate system has been fundamentally altered by increases in heat-trapping greenhouse gases and changes in land use. We are in completely new territory, having now changed the chemical composition of the atmosphere. In the new climate system, it is more likely that hurricanes will be stronger, heatwaves longer, that wildfires will extend over larger areas, and so on. The extra heat in the system means that, in climate scientist Jim Hansen's potent phrase, "the dice have been loaded" in favor of greater intensity whenever a heat-related, extreme weather event occurs.[37]

Recently, climate scientists have gone beyond this general description using the methods of *attribution science*, which is able to tease out the size of the human contribution to specific weather events.[38] Climate scientists who work in this field compare the frequencies, durations, and intensities of recent extreme weather events with the available past climate record, when greenhouse gas concentrations were lower. Many studies in this field also use climate models to simulate conditions with and without human influence on greenhouse gas levels. Table 4.2 lists six recent events where the analysis reveals that human-induced climate change played a part in making the event both more extreme and more likely to have occurred.

Some of the clearest applications of attribution science are in the analysis of heatwaves. The increase in Earth's average surface temperature has shifted the bell curve of temperature distribution so that extreme periods are now hotter (Color Plate 15). The US temperature record has several features that made identifying a human fingerprint on heatwaves

Table 4.2: Attribution of specific extreme weather events to anthropogenic climate change[a]

Year	Event	Human contribution
2020	Siberian heat wave	>600 times more likely
2020	Australian heat wave (contribution to bushfires)	10 times more likely
2019	Tropical storm Imelda (Texas)	2.6 times more likely
2019	European heat wave (results for France)	At least 5 times more likely
2018	Capetown drought	3 times more likely
2017	Hurricane Harvey rainfall (Texas)	3 times more likely

[a]Attribution studies conducted by the World Weather Attribution initiative, https://www.worldweatherattribution.org

challenging. First, the extraordinary heat that accompanied the Dust Bowl in the 1930s was a natural variation that required later temperature increases to be stronger before they could be confidently assigned a human origin. Second, unusually low temperature increases in the US Southeast compared to other parts of the country also added to the complexity of the problem. However, a clear connection from anthropogenic forcings to global warming is now clear in several heatwaves that occurred between 2011 and 2014 in some Eastern states, Texas, and California.[39] Heatwaves elsewhere in the world are also due to human activity. The 2020 Siberian heatwave was off the charts—it is virtually certain that an event of this magnitude could not have occurred without anthropogenic influence (Table 4.2).[40]

Heatwaves also occur in the oceans. An international research team recently found that, compared to the 30-year time period from 1925–1954, the number of annual marine heatwave days had increased by more than 50 percent by 2016.[41] There will almost certainly be more frequent and more intense events like the sustained 2016 marine heatwave off the Alaskan coasts, which killed large numbers of shellfish and seabirds. Like the Siberian heatwave, not only is the human contribution clear, but the event was so severe that it could not have occurred in a world unperturbed by humans.[42] This higher level of attribution to human activities has only emerged for a few extreme weather events since 2016 but will almost surely increase as we move farther away from the climate of the pre-industrial world.

In the US, decreases in the length and intensity of coldwaves are detectable beginning as early as the first decade of the twentieth century.[43] Notwithstanding this overall trend, extreme cold has been in the news because of the devastating winters that have occurred recently in New England and the upper Midwest. The extreme cold happens when the *polar vortex*—a frigid, naturally occurring counterclockwise circulation of air around the North Pole—unexpectedly dips southward for some period of time, freezing out the Northeast and North Central US.[44] The unusual polar vortex flow is causing extreme US coldwaves and is also responsible for Arctic sea ice loss.[45] We don't yet know what is driving the altered circulation, but many climate scientists suspect that anthropogenic forcings play at least some role.

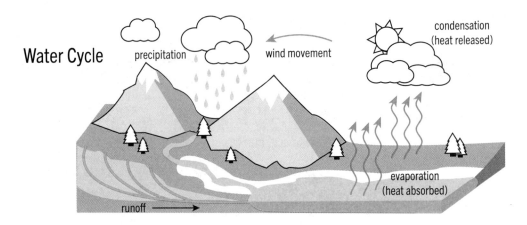

Figure 4.2: Depiction of Earth's water cycle, showing the warming of the atmosphere as part of the evaporation, condensation, and precipitation cycle. The movements of water vapor through the atmosphere and liquid water across the surface complete the cycle.

Earth's Hydrological Cycle

The *hydrological cycle* is the key to understanding how global warming affects hurricanes, floods, droughts, and wildfires (Figure 4.2). This cycle describes the way water circulates through Earth's climate system, especially the atmosphere and oceans. One simple idea goes a long way toward understanding extreme weather: a warmer atmosphere can hold more water vapor before it condenses and falls out as rain. The overall temperature increase of just over 1°C has already increased the atmosphere's water holding capacity by 6.5 percent, and precipitation in turn has risen by about 3 percent.[46]

This is only part of the story, however. The condensation of water vapor to form clouds changes Earth's radiation balance in complicated ways, producing either heating or cooling in different regions.[47] Another issue is that water at Earth's surface absorbs heat from its surroundings when it evaporates, causing the area to cool while the water warms. After the warmed water vapor rises, it releases that heat into the atmosphere when it condenses back into liquid droplets. So, the hydrological cycle is like a pump that moves heat upwards from the surface (Figure 4.2). This process also drives stronger winds, as the air moves from higher to lower pressure regions. Winds in turn drive the ocean circulation, which is already affected by the melting cryosphere and greenhouse heating of surface layers. Through this chain of events, all parts of the hydrological cycle are intensified when the temperature is higher.

Sunlight is most intense in the tropics, and the greater heat drives evaporation of more water than at the poles or midlatitudes. The warm, moist tropical air rises until the temperature of the surrounding atmosphere is cold enough for it to condense and rain out. More water being held in tropical air because of global warming means that rainfall near the equator is more intense than in a cooler world. When the rising air column around the equator reaches the top of the troposphere, it turns north and south and then descends in the subtropics of each hemisphere. That air is now very dry, and as it descends, it creates dry, high pressure regions centered around 30° latitude in both hemispheres. This is where the Earth's major deserts are found. The air then circulates back to the equator near ground level, completing a cycle called a *Hadley cell* (Color Plate 7).[48]

Climate models consistently predict that Hadley cells will extend farther toward the poles as the planet warms, which means that desert regions will expand towards midaltitudes. This *desertification* is occurring already in the Mediterranean, Southern Africa, and the Southwestern US.[49] In the US, the impact of desertification on the freshwater supply is worsened by the regions' dependence on the melting Rocky Mountain snowpack.[50] Attribution studies linking droughts to human activity are now advanced enough to establish the connection in some cases. For example, the extensive California drought that lasted from 2012-2016 clearly has a human fingerprint.[51] An otherwise moderate drought that has captured the North American Southwest since 2000 has also been pushed into "megadrought" status by anthropogenic influences. The drought in this region is now the most severe since the late 1500s.[52]

Linking extreme weather to climate change is intuitive—and more likely to be attributable in part to anthropogenic influence—when the event is clearly connected to increased moisture in a warmer atmosphere. Very heavy rainfalls are in this category, including those that occur during hurricanes. Hurricanes also gain strength from warmer surface ocean temperatures, which further contributes to increased intensity. For example, an anthropogenic fingerprint has been connected to Hurricane Harvey, which inundated the Texas coast in August, 2017.[53] Other massive floods, such as the 2019 deluges in the upper Mississippi valley, are a consequence of very heavy rains, but their link to anthropogenic climate change is not yet certain because the record of such events is still too sparse.[54]

It is intuitive that wildfires will increase with warmer temperatures and dryer air in the Western US.[55] Indeed, since the 1980s there have been many more large forest fires and much more acreage burned in the American West.[56] Detailed study of the forest fire record, together with climate modeling, has shown that since 1984 the cumulative area burned in the Western US is twice what would have occurred in the absence of human activities.[57] Moving to specific events, it has been shown that the devastating Australian wildfires that occurred during the 2019–2020 Southern hemisphere summer months have a strong anthropogenic fingerprint (Table 4.2).[58] The extreme breadth and intensity of the wildfires that ravaged the Western US in the summer and fall of 2020 will likely also be attributed to human activity.

One important contributing factor to wildfires is that, as temperatures warm, plants open up pores in their leaves and stems to absorb carbon dioxide, and this also releases water vapor to the atmosphere. This process is known as *transpiration*.[59] Plants in warmer climates then have to take up more water to make up for the increased loss, drying the soil and contributing to wildfire spread. The anthropogenic fingerprint for soil moisture loss with planetary warming is very strong.[60] Wildfires are even burning in the boreal forests of Greenland, Canada, and Alaska, where the eventual consequences could be particularly severe because these lands hold about 50 percent of the world's soil carbon.[61] The boreal fires create an albedo-based, positive feedback loop because melting accelerates as dark soot released from the wildfires settles on the white snow and ice.[62]

4. Ecology and Biodiversity

The natural world sustains and enriches us, and many of the services it provides are irreplaceable. For most of human history, small populations and preindustrial technologies limited how much we could take from nature, and terrestrial and marine ecosystems remained close to their natural states. However, the human population has grown nearly eightfold in the past two centuries and we now depend heavily on extracting Earth's mineral and biological resources for food, fuel, medicines, and materials.

The cumulative depletion of many natural resources has reached critical limits. One-third of Earth's land is now dedicated to agriculture and animal husbandry, diminishing the total stock of plant biomass by 35–40 percent when compared with prehistoric times.[63] More damage to our natural ecology comes from direct harvesting of the living biosphere, environmental pollution of every kind, and infiltration by invasive species—carried either directly or inadvertently—by human expansion.

All these insults diminish the health of forests, wetlands, grasslands, and both marine and freshwater ecologies.[64] Indeed, outside of Antarctica and parts of the northern boreal forests, little remains in the state that existed before *Homo sapiens* came to dominate. As demands for energy and materials continue to grow, the imperative to use nature sustainably only increases.[65] A comprehensive 2019 UN report on biodiversity and ecosystem services offers the expert conclusion: the damage already done seriously threatens societal goals to achieve human well-being, and urgent and concerted efforts fostering transformative change are needed to restore and re-balance the natural environment.[66]

In the context of ecosystems, climate change is a threat multiplier—it worsens the impacts already caused by land use change and harvesting of the biosphere. The effective habitats and ranges of millions of species are already constricted by human activities, but now climate change further stresses organisms adapted to a cooler world with fewer extremes of heat, rainfall, drought, and fire. Many plants and some animals cannot migrate fast enough to keep up with the rapid changes to their local ecosystems and so are increasingly unmatched with their environments, as well as more likely to succumb to pests and diseases.[67]

Endangered Species

The combined impacts of land conversion, biosphere harvesting, and climate change have increased the rate of species extinction tens to hundreds of times above the average for the past 10 million years.[68] Today no fewer than one million plant and animal species are threatened with extinction within the next few decades. We are now in the sixth major extinction of Earth's history—the first five were due to natural events, but this time there is no question that humans are responsible.[69]

The previously mentioned UN report unflinchingly catalogs the devastation: about half a million terrestrial species are committed to extinction unless their habitats are rapidly restored, and at least 680 vertebrate species have already gone extinct since the year 1500 (Figure 4.3).[70] There have also been declines in insect abundance and sharply reduced

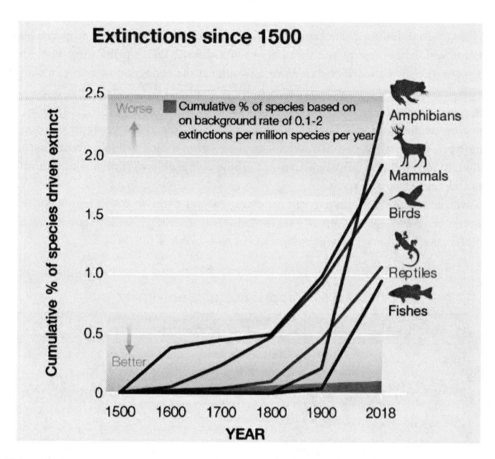

Figure 4.3: Sharp increases in extinctions in vertebrate groups, from the year 1500 to the present. Credit: Intergovernmental Science-Policy Platform on Biodiversity and Ecosystem Services

numbers of local varieties of domestic plants and animals, reducing the pool of genetic variation that is important for food security.[71] In marine environments, there are similar concerns: 40 percent of amphibians, one-third of sharks and their relatives, and one-third of marine mammals are all threatened.[72]

It is clear enough how ecosystem disruption from climate change can drive species to extinction, but perhaps less evident why biodiversity is important to begin with. Those with a natural affinity or reverence for the natural world need no convincing, but not everyone shares this perspective. What makes biodiversity so critical?[73]

A central difficulty to understanding the importance of biodiversity is the failure to recognize how interconnected life is with its natural environment. It is sometimes easy to think that the animate and inanimate worlds are separate, and that (for example) the local freshwater resource would be perfectly fine whether it is populated with organisms or not. This is a serious misconception because in every natural resource that humans depend on, there is a full spectrum of life ranging from the smallest microorganisms all the way to large apex species such as fish and mammals. Each has a critical role in contributing to the overall ecology. When the natural hierarchy of life begins to break down, so does the linked health and the integrity of the resource.

Conservation biologists Paul and Anne Ehrlich discuss the importance of species preservation with what they call the rivet-popper hypothesis.[74] Like a plane wing with a few rivets popped off, a natural resource that has suffered the extinction of a few individual species can still usually provide the services we depend on. But if too many "rivets" are lost, the system collapses. And with climate change now threatening the system through changing temperature and rainfall patterns, preserving the rivets becomes even more important. For instance, in a freshwater lake preservation means avoiding overfishing and pollution and confining human activities to those compatible with maintaining healthy function at all ecological levels.

Swift action toward reducing carbon emissions and effective conservation measures can still preserve a large measure of Earth's biodiversity, securing its many contributions to both the sacred and practical dimensions of our lives (Table 4.3).[75]

Table 4.3: Importance of biodiversity	
Principle[a]	**Reason why important**
Healthy ecosystems	Heavy dependence on the natural world for productive soils, fresh water, and medicinal products
Human health	Disease vectors emerge from degraded ecosystems
Carbon sequestration	Forests, farms, grasslands and other natural resources play a crucial role in the climate system
Economic benefits	Food, fiber, and fuel extracted from the natural world play a major role in the global economy
Cultural identity	Religions, indigenous cultures, and contemporary developed societies all take symbolism from living Nature

[a]From Julie Shaw, Why is Biodiversity Important? Conservation International, Nov. 15, 2018. https://www.conservation.org/blog/why-is-biodiversity-important

Fragile Marine Ecosystems

Some ecological systems are threatened by climate change because they are inherently fragile (Color Plate 11). For example, the fragility of the Arctic ecosystem comes from its dependence on sea ice, which is susceptible to melting. And because of the cold temperatures, Arctic plants and microorganisms—the foundations of the ecosystem—have had to adopt very frugal, low-energy lifestyles that in turn, are easily disrupted by climate change.[76]

Among all ecosystems, the coral reefs are perhaps most at risk due to climate change. Hard shell corals are invertebrate animals that cooperate to build enormous living structures, like the 1500–mile Great Barrier Reef north of Australia. All coral reefs anchor rich ecosystems that sustain more than a quarter of the ocean's biodiversity. The larger ecosystem includes salt-tolerant mangrove trees and seagrasses, which protect shorelines from storms and sustain the food web that supports many unique species of fish, sea turtles, and other marine life. The coral-based marine ecosystem is also of great economic value for fisheries and tourism and is a major source of natural compounds in medicine.[77]

The rainbow colors of reefs come from chlorophyll pigments in algae that live inside coral tissues. The coral exoskeletons are built by scavenging calcium and carbonate ions from the seawater, generating a form of solid calcium carbonate, or limestone.[78] These features of coral biology make them highly susceptible to climate change. Because corals are adapted to an optimal ambient temperature, their life processes degrade sharply upon warming. Warmer temperatures bleach the corals by expelling the colorful algae, turning them white and weakening their skeletons.

What happens next is a matter of chemistry. When carbon dioxide dissolves in seawater, it initiates a set of reactions that cause *ocean acidification*, a large-scale process that changes fundamental physical properties of the oceans. In turn, these reactions trigger yet more chemistry that starts to dissolve the solid calcium carbonate coral shells and the shells of other important shellfish (Box 4.1). The Earth will likely lose 70–90 percent of its coral reefs in a 1.5°C world, and 99 percent at warming of 2°C.[79] Coral reefs are present in seven US states and territories—including Florida and Hawaii—where they have economic value for fisheries and tourism, and, additionally, an estimated $1.8 billion in flood risk reduction potential. That number is dwarfed by the full worldwide impact of corals on goods and services, which is estimated at $375 billion.[80]

The only way to prevent the oceans from becoming more acidic is to stop burning fossil fuels, and the only way to reverse acidification is to actively remove carbon dioxide from the atmosphere.[81] If this can be accomplished, some of the ocean's carbon dioxide would outgas back into the atmosphere, rebalancing ocean chemistry closer to its natural state. But if we don't take these actions, future damage could go far beyond coral reefs and the shellfish industry. We know this because extreme ocean acidification from volcano-derived carbon dioxide emissions helped drive Earth's largest past extinction at the end of the Permian period 252 million years ago.[82] At that time, the ocean pH dropped by 0.7 units over a period of 10,000 years, wiping out nearly all marine species alive then. The meteor impact at the end of the Cretaceous period, which killed off the dinosaurs 66 million years ago, also acidified the oceans and caused wide extinction of marine life.[83]

Forests and Terrestrial Ecosystems

Among all ecosystems, any threats to forests are especially serious because they store so much carbon. Thus, it came as very good news—and a surprise to many—when a comprehensive satellite-based study showed that, globally, tree cover increased by 7 percent between 1982–2016.[84] A significant part of this gain was in the US, where forested land has been growing by about one-tenth of 1 percent per year (nearly 700,000 new acres per year).[85] Despite this, there are still major concerns for the large boreal forest belt in the extreme north and for the tropical rainforests, especially in Brazil. Alaskan, Canadian,

Box 4.1: Ocean acidification and its consequences

How do increased atmospheric carbon dioxide (CO_2) levels make the oceans more acidic? To understand this, the first thing to recognize is that CO_2 is always dissolving in and evaporating from the ocean surface. Given this two-way movement, adding more CO_2 to the whole system inevitably causes some of the new amount to find a home in the oceans.

What happens when CO_2 dissolves in water? Let's first recall that all atoms have nuclei containing neutrons and protons, surrounded by an electron cloud (see Box 1 in Chapter 3). In a process called ionization, water molecules are easily able to shed one of their two hydrogens, which each have only one proton (and zero neutrons) in their nuclei. When the hydrogen lets go, it does not take any electrons with it and becomes a free proton, written H^+ (see the figure). What remains of the water is now called a hydroxyl ion, written OH^-. The superscript indicates a negative charge because the OH^- kept all the water's electrons. The proton has a positive charge because it has no electrons at all.

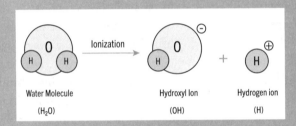

Water Molecule	Hydroxyl Ion	Hydrogen ion
(H_2O)	(OH)	(H)

Here is why carbon dioxide acidifies water. When the CO_2 dissolves a number of reactions occur, with a net effect acting as though the CO_2 reacted with the OH^- from water, removing it from the solution. The product is a molecule called bicarbonate, written HCO_3^-, while the proton remains free. The net reaction can be written $CO_2 + H_2O -> HCO_3^- + H^+$.

Protons now exceed hydroxyl ions, and this is what is meant by an acidic solution. As more CO_2 dissolves, protons continue to be produced in excess of hydroxyl ions. The pH value of the ocean, calculated from the proton concentration, decreases as the proton excess gets larger and larger. Today the ocean pH value is 8.1, a decrease of about 0.1 pH unit since the beginning of the industrial revolution. This amounts to a 30 percent increase in acidity.[140]

The damage to coral reefs and shellfish comes from a side effect of this chemistry. When hydroxyl ions are taken up by the extra CO_2, a rebalancing reaction occurs to partly replenish them. Carbonate ions (CO_3^{2-}) from the hard calcium carbonate shells of corals and other shellfish dissolve in the seawater to produce hydroxyl ions and more bicarbonate, by the following reaction: $CO_3^{2-} + H_2O -> HCO_3^- + OH^-$.[141] The released hydroxyl ion compensates for part of the proton excess, so the ocean is somewhat less acidic than it otherwise would have been. But this is coming at an extraordinarily high price. The figure shows a pristine shell at left. At right is a shell dissolving in our acidic ocean, with excess carbon dioxide.

Credit: NOAA

and Siberian boreal forests have been impacted mainly by wildfires, but researchers are worried about the possibility of diebacks from rapid temperature increases and accompanying insect infestations.[86]

Deforestation in the Amazon has sharply increased since the time of the satellite study. About 15 percent of the forest has been lost since the 1970s, and some scientists fear that a tipping point is approaching (Table 4.1).[87] Climate change adds to these impacts because it produces higher temperatures and an increase in droughts. This limits tree growth and increases arboreal mortality, leading to a long-term decline in the forest's ability to serve as a carbon sink. Amazonian, African, and Indonesian rainforests create their own wet weather by transpiration of enormous amounts of water through their leaves. More deforestation thus leads to less rainfall, injuring the unique rainforest ecology that depends on the high humidity. These factors are now outweighing gains from carbon dioxide fertilization.[88] At some stage, a fast transformation to grasslands might occur, which would be accompanied by a large release of much of the forest carbon that is presently sequestered.[89]

These threats to forests are severe, but in general terrestrial ecosystems are slightly less vulnerable than their marine counterparts.[90] The risks are less immediate, with substantial increases in climate risk and impact coming in the transition from a 1.5°C to a 2.0°C world (Color Plate 11). Therefore, a fast energy transition preserving a 1.5°C world has potential to substantially lessen damage. In particular, extreme wildfires and the spread of invasive species and diseases are projected to be reduced by limiting warming to 1.5°C. Globally, climate models suggest that the fraction of land susceptible to ecosystem transformation doubles at 2.0°C compared to 1.5°C.[91]

5. The Human World

For a long time, climate change has been low on the radar screen for most US citizens. This is changing as more of us recognize that the issue is not just about faraway ice caps and tropical rainforests. Recognition is dawning that extreme weather is not going to stop. More systemic effects are coming next, including impacts to crop yields and freshwater resources, direct threats to health, and real consequences for the economic bottom line. Questions of social equity will also more clearly emerge. Low-income communities and communities of color have been more greatly affected by air and water pollution. These and other environmental impacts are only worsening as the world warms.[92] And since low income US households produce smaller carbon footprints they have also contributed less to the problem.[93]

While we focus attention on the US for advocacy purposes, we must keep in mind that greater impacts are already occurring elsewhere. In many developing nations with fewer resources and capacities to adapt, the threats are much more tangible and immediate than they are here. It is a moral imperative that first-world nations dedicate funding to international mitigation and adaptation efforts, such as those set out in the United Nations 2030 Agenda for Sustainable Development. Like low-income communities in the US, most citizens of developing nations have contributed a small fraction of the emissions and must not be left to bear the brunt of their effects.[94]

Food and Water

Climate change impacts agriculture by changing the length and timing of the growing seasons and disrupting seasonal patterns of rainfall. Historical weather patterns no longer offer good guidance to farmers, who are working hard to respond effectively but have no clear roadmap to follow.[95] Other challenges include extreme storms that threaten soils and rural infrastructure, intense heatwaves (which are particularly damaging to livestock), increased droughts that impact irrigation water supply, and greater threats to crops and livestock from pests and diseases.[96] Of course, there is a great deal of regional variation in susceptibility to each of these impacts. The economy of the Midwest, which is the most dependent of US regions on productive agriculture, has already experienced these climate shocks.[97]

The health of the rural economy is tied heavily to agriculture, and these impacts will be front and center in any discussions about climate. A major portion of US agricultural output is locked up in large, industrialized enterprises, especially the major commodity crops such as wheat, corn, rice, cotton, and soybeans. These operations are heavily managed and, while obviously not impervious to the influence of climate change, have so far been able to adjust without large losses. Industrialized agriculture is also a major source of greenhouse gas emissions, and this problem can be addressed through new technologies and policy approaches to incentivize climate-friendly practices.[98]

It is a different story for smaller, family-run farms. Here, resources are scarcer and local communities have been obliged to recognize the climate problem and imagine innovative ways to adapt.[99] Rural farming communities are both heavily affected by the climate problem and have particular challenges in responding to it because of greater poverty, heavy year-to-year economic dependence on crop revenue, and more limited access to the internet and other technology resources.[100] Cohesive social networks can help combat these challenges. In the Northwest, the Pacific Northwest Tribal Climate Change Network offers ways for indigenous tribes to work together and access government and university resources to build resilience.[101] Another example is the Great Lakes Climate Adaptation Network, a network of local government staff who collaborate to act on climate challenges that are unique to the region.[102]

Water resources are also more precarious in the wake of climate change. Freshwater supply is tied to the hydrological cycle, and as this cycle intensifies in a warming world, both floods and droughts will become more severe in areas already prone to these events. There is a strong consensus that the Southwestern US is at severe risk of longer, more intense droughts as the twenty-first century unfolds. As mentioned above, changes in global atmospheric circulation will drive desertification by bringing dryer air over more US territory. Higher temperatures will decrease soil moisture, lowering crop yields in drier parts of the country. Water supplies for all uses will be further threatened by reduced Rocky Mountain snowmelt and depletion of groundwater in the Ogallala Aquifer, located in the nearby Southern Great

Plains.[103] Given rapid population growth in this region, even a very fast renewable energy transition will certainly demand increased attention to water management.

The potential for prolonged drought is most severe in the Southwest, but every region of the US is threatened by water-related impacts. Extreme rainfall drives flooding along the banks of the Mississippi and other large rivers, while extreme precipitation and other climate impacts on aging water infrastructure in the Eastern US will require multibillion-dollar responses. In the Midwest, Lake Erie has been overwhelmed by toxic algal blooms that are favored by warmer temperature and extreme rainfall. And warmer streams and oceans combined with decreased streamflow from snowmelt threaten salmon habitats in the Pacific Northwest and Alaska.[104]

Resiliency in the face of droughts and other impacts to water supply can take many forms. Water utilities in stressed regions typically adopt conservation plans and connect these to efforts by local and regional governments in order to enact municipal ordinances and pricing plans that promote savings. For example, owners of water rights in Colorado are encouraged to work with market-based tools to facilitate transactions between sellers and buyers that take local conditions into account and encourage water to be used where it is most needed.[105] Agriculture accounts for a huge portion of water consumption, so urban-rural partnerships in which towns and cities engage with farmers to maximize allocation efficiencies are crucial. Reuse of wastewater and desalination are also viable strategies. There are many successful examples of communities across the country that have been implementing all these approaches.[106]

Health

Climate change puts human health at risk, from more intense heatwaves and worsened air and water quality, to increased spread of infectious agents and displaced populations. There are strong social and geographic dimensions to these increased risks in the US today, which fall disproportionately on children, older adults, low-income communities, indigenous communities, communities of color, and those living in the Southern and Southeastern coastal regions and Caribbean US territories.[107]

Increased illness and death from heat extremes are among the most well-publicized health impacts from climate change. The European heatwaves that occurred in summer 2003 were responsible for 70,000 deaths—although the toll would have been greatly reduced had adaptive measures, such as neighborhood cooling centers, been in place. While this impact was extreme, US emergency room data consistently documents upward spikes in illness and mortality from excessive heat.[108] This is expected to worsen as warming continues—particularly under conditions of both high heat and humidity—which can make it impossible for the body to shed enough heat.[109] Sometimes it is argued that decreases in death and illness from fewer cold extremes will compensate for the detrimental effects of extreme heat, but most studies project that reductions in premature death from cold temperatures will be smaller than the increases in heat-related deaths.[110]

Sweeping new research on the health of newborn infants shows that premature births and low birth weights are both strongly correlated with a warming climate, and that Black mothers and their children are harmed at a much higher rate.[111] This is just one aspect of the disproportionately greater climate change impacts felt by communities with large minority populations. These communities are often located in the least desirable neighborhoods—such as floodplains—and in close proximity to manufacturing and electric power plants, which generate high levels of local pollution. Several US governmental agencies confirm that minority groups are greatly overrepresented in the ten US counties most vulnerable to weather-related disasters and extreme heat.[112] These are some of the latest data to underline the importance of connecting climate change solutions with the movement for environmental justice.

In the US, warmer temperatures increase the prevalence of ticks and mosquitoes that carry bacteria or viruses. The 2014–2016 Zika virus epidemic in Texas, Florida, and offshore islands may have been among the consequences.[113] Projections show that continued warming will make it more likely that Lyme-disease carrying ticks will expand both geographically and seasonally, increasing risks of human exposure.[114] Harmful bacteria and parasites are also projected to degrade water quality, especially in regions prone to coastal flooding and its detrimental impacts on the function of sewage systems.

Warming also increases risks from influenza and could multiply the threat of reemergence of COVID-19 or related viruses. Research has shown that warmer winters are followed by more severe flu seasons—perhaps because fewer people become infected when the winter is mild—increasing the odds of early and more severe epidemics in the following year.[115] With enough warming, influenza may no longer be limited seasonally but could instead become a year-round risk, increasing the odds that strains posing more severe health risks will emerge.[116]

A warmer climate will also increase air pollution from ground-level ozone. Ozone is formed during smog episodes from the combination of sunlight and chemicals in vehicle exhaust. Higher temperatures promote more ozone formation from both human-caused smog and natural emissions. Atmospheric methane also leads to ozone formation, so decreasing methane emissions would simultaneously mitigate global warming and improve health.

The levels of particulate matter in the atmosphere will also increase in a warmer world. A major cause of this is the increased extent of wildfires, especially in the West. Both ozone and particulate matter increase the incidence of respiratory illnesses and have greater impact on vulnerable groups such as children and the elderly.[117]

It is reasonable to expect that these health impacts will be mitigated to the extent that warming can be limited to 1.5°C or 2.0°C. Recent research backs this expectation, finding that limiting warming to 2.0°C would save 1.4 million lives in the US over the next 20 years, and amass over $700 billion in savings per year from avoided health care costs and losses in labor productivity—an amount substantially greater than the cost of the energy transition.[118]

Climate effects on human health also have important advocacy dimensions. Concern for health transcends political boundaries, and common ground can be found in discussions of how climate change amplifies threats to well-being. Studies have shown that most Americans are unaware of the many ways climate change harms health. Because physicians are highly trusted messengers, this work suggests that they could play an important role in public climate change education.[119]

Economy

The financial costs of climate change are most obvious in the direct damages caused by extreme weather, and in the increased nuisance flooding and storm surges caused by sea level rise. But this is only one part of the picture. Adaptation to climate change is also costly, since it often involves a lengthy process of study, planning, implementation, and evaluation. Mitigation imposes heavy financial burdens as well, as it often requires the development of entirely new technologies or an overhaul of existing infrastructure, like the electricity grid.[120] There is a reason why some folks who accept the findings of climate science nonetheless resist the changes that must be made: they are going to cost an awful lot of money.

Damages from storms, severe weather, droughts, and wildfires are carefully tracked by the National Centers for Environmental Information (NCEI). Its statistics show that the number of events exceeding $1 billion in costs has been increasing steadily since 1980, with a cumulative price tag approaching $2 trillion.[121] The six most costly events—and over half the total financial damage—has come from hurricanes, with the 2005 devastation of New Orleans by Hurricane Katrina leading with an estimated cost of $170 billion.

These costs should really be thought of as the tip of the iceberg because they reflect easily quantified damages, including losses paid on insurance claims. Other losses can't be readily enumerated in dollars, like systemic effects on health and personal welfare from long-lasting events such as droughts and floods, or the displacement of communities. Expansion of the population into vulnerable areas, such as beachfront communities and the urban/rural interface in fire-prone regions, also drives damages higher. Natural resource degradation includes economic costs from loss of coral reefs in Florida and Hawaii, salmon habitat in the Pacific Northwest, grassland productivity in the Great Plains, and maple syrup production in New England among many other examples.[122]

While impacts from all these factors are severe, they are still small compared to the size of the US economy, with GDP of $20 trillion. Even the $307 billion damages of quantifiable costs in 2017—which were 50 percent higher than in any other year—still amounted to only 1.5% of GDP, or about a third of the GDP loss from the Great Recession of 2007–2009. That recession had crippling impact for many people because it was accompanied by a large increase in unemployment and a 20 percent decrease in the combined net worth of all households and nonprofits.[123] Climate change impacts do not yet possess nearly this reach, but the potential for economy-wide effects grows each year. Regional economic impacts in the Southeast and Gulf Coast may appear early since climate models project a substantial worsening of extreme weather in the next few decades, and hurricanes are by far the costliest extreme events (Table 4.4).

A recent analysis from the Center for Climate Integrity offers a striking illustration for one category of impacts. This study estimated that the cost of seawall construction—to protect coastal communities in 22 states from sea level rise, storm surges and nuisance flooding—will exceed $400 billion within the next two decades alone (Box 4.2).[124] These results were obtained using conservative estimates of sea level rise, favorable IPCC emissions scenarios (RCP2.6 and RCP4.5), and calculations of storm surges typical for historically average years. The costs are so high that they imply a strong likelihood that many beachfront communities will have to be abandoned. Recognizing this, the Government Accountability Office—a watchdog agency that investigates federal spending practices—is calling on Congress to establish a pilot program for domestic climate migration.[125] Of course, depending on how much warming occurs, migrations may also be spurred by intolerable heatwaves, wildfires, or other manifestations of climate change.[126]

The Center for Climate Integrity also poses the essential question: Who will pay for the massive adaptation and relocation costs? This will not be an easy issue to resolve. One set of culprits are the fossil fuel companies that have been funding climate change denialism for many years.[127] But the US government also bears responsibility through its National Flood Insurance Program (NFIP), which covers most property owners in flood zones. NFIP heavily subsidizes flood insurance compared to what it should cost given the risks, thus encouraging coastal settlement.[128] Private mortgage lenders also bear some responsibility since they have begun to offload risky mortgages in flood-prone areas to the US government-sponsored lenders, Fannie Mae and Freddie Mac. According to existing regulations, these lenders cannot charge additional premiums to account for disaster risk.[129]

Table 4.4: Cumulative US climate change damages, 1980-2020

Event class[a]	Cost ($ billions)[b]
Tropical cyclones	$954.4
Severe storms	$268.4
Droughts	$252.7
Flooding	$150.4
Wildfires	$85.4
Winter storms	$49.6
Freezes	$30.6

[a]NOAA National Centers for Environmental Information (NCEI) U.S. Billion-Dollar Weather and Climate Disasters (2020). https://www.ncdc.noaa.gov/billions/summary-stats

[b]Incorporates individual events with damages of over $1 billion. Data are as of July 6, 2020.

Box 4.2: Adaptation to sea level rise in Louisiana

Southern Louisiana is losing land faster than almost anywhere else on Earth (see the Figure). This is happening because of sea level rise, although oil and gas drilling is making matters worse by causing the land to subside. Another issue is that the damming and encasement of the Mississippi River for flood control blocks much of its flow into the delta. This prevents land buildup that would otherwise occur from the accumulation of sediments, which are now deposited far offshore instead. With the ecology of the delta, oil and gas interests, the fishing industry, and the river export of 40 percent of US agricultural output all at risk, the state of Louisiana has embarked on a $50 billion Coastal Master Plan, perhaps the most ambitious climate change adaptation enterprise ever attempted.[142]

In its 2017 Master Plan, the State of Louisiana described the completion (or funding) of 135 separate projects, including construction of new levees and barrier reefs, land restorations, and shoreline protection.[143] It also projected that 2250 additional square miles of land could be lost in the next fifty years, on top of the over 2000 square miles that has disappeared since the 1930s. Going forward, the state envisions many more projects to save this land, with multiple lines of defense—essentially the use of natural and manmade features to impede storm surges and reduce damage. This effort will be coupled with environmental habitat protection. Funding for much of this effort has yet to materialize.

Although the Master Plan enjoys nearly unanimous support in the Louisiana state legislature, it is not without its critics, not the least of whom are inhabitants of some small fishing communities that could be sacrificed for the sake of saving the greater whole. The problem is simply that building flood protection for some areas, including parts of New Orleans, exposes other regions to inundation. Inevitably, this plan sacrifices some interests for others, although it is not forthright about this.

Given the central role of oil and gas in Louisiana's economy, it is no surprise that the Master Plan also falls short of recognizing that these industries have themselves caused a major part of the problem. Given the funding challenges, it would not be unreasonable for large oil and gas firms to invest much more money—through judicial orders, if need be.[144] The necessary rapid demise of oil and gas need not damage either employment opportunity or the health of the local economy. In fact, the Gulf can provide clean, renewable energy through offshore wind power, a huge undeveloped resource that it has in abundance.[145]

 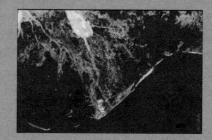

Land loss from 1932 (left) to 2011 (right) of a section of Louisiana's coast.
Credit: NOAA.

Do future generations have the right to inherit a stable climate system that is not threatened by potentially catastrophic tipping points? This is a crucial question that determines how aggressively climate change must be tackled today. However, while almost nobody argues that we do not owe a healthy climate to posterity, this agreement is not enough to generate consensus on a course of action. The problem is that the costs of the energy transition have to be paid today, but the damages are in the future and of uncertain scope. Estimates of costs to future generations vary widely because neither the sensitivity of the Earth system to increased emissions nor the likely pace of technological development for mitigating and adapting to damages is well understood.

To estimate future damages, climate scientists employ *integrated assessment models*, which are models that include the physical laws that govern Earth's climate system (as discussed in Chapter 3) together with representations of the global economy.[130] Because these models combine natural and human systems, they are extremely complex. Specifically, the need to model future socioeconomic behavior produces a great deal of uncertainty in the findings. For example, the IPCC estimates that damages will reach 4.5 percent of global GDP by 2100 if we do not enact new climate policies.[131] However, another group projects much higher damages of 15–25 percent of global GDP by 2100, and this is for a 2.5–3.0°C world in which new emissions reductions policies are included in the models.[132]

One reason why these projections vary so much is that some modelers do not incorporate catastrophic climate change tipping points (Table 4.1) or difficult-to-value resources (like biodiversity) in their calculations, because these factors are difficult to quantify. This leads to smaller damage estimates. However, those events do have potentially enormous costs, and leaving them out of the calculations is equivalent to assuming that they can never occur, which is certainly incorrect. Difficulty in estimating costs does not justify ignoring them.[133]

Another major point of controversy among models that estimate low versus high climate damages is a factor known as the *discount rate*. This is a bit like a financial interest rate, except it is applied over multiple human generations. Money is worth less in the future than it is today, but the future is when climate change damages will mostly accrue. The controversy is then over how much to *discount* the future. Many argue that we should optimize present-day investments rather than paying the high costs to rapidly build a new energy economy. Doing so will build wealth for future generations, making them richer. That is the equivalent of assigning a high discount rate, which corresponds to a low *social cost of carbon* (SCC), a price assigned per ton of carbon dioxide emitted. A low SCC implies modest future climate damages which, for example, might come about if technological improvements greatly enhance the capacity to adapt at low cost.[134]

Anyone familiar with financial planning will know about the magic of compound interest—how small amounts of money today can grow to large sums in a few decades—at a rate that one might hope for in the stock markets, like 5 percent. Discounting climate change costs works the same way, except in reverse. If the cost of climate change 100 years from now is $1 trillion, and we adopt a high discount rate of 5 percent, then we would only be willing to spend about $8 billion today to avert it—less than 1/100 of the projected cost.[135] But at a discount rate of 1 percent, the implication is that we are willing to pay $370 billion. Such a low discount rate implies a very high SCC, which means that carbon pollution is so damaging that we should invest a great deal to eliminate it.[136]

The whole issue of how aggressively to tackle climate change can be boiled down to this debate about what the discount rate should be. But because we cannot estimate future damages with any accuracy, the debate comes down to pure politics—with the key issue simply the extent to which we are willing to invest large sums now to protect future generations. This is the question of *intergenerational equity*. Politically, the problem is that discounting gives theoretical ammunition to those who would rather wait and see, a position unsurprisingly embraced by those with heavy financial stakes in the fossil fuel industry. President Obama embraced a relatively low discount rate, resulting in an SCC estimate of $50 per ton of emitted carbon dioxide in 2020, but under the Trump administration, the SCC estimate plummeted to $7 per ton.[137]

Many argue on ethical grounds that the discount rate must be kept very low and the SCC commensurately high, because it is wrong to value future generations less than the present.[138] This implies a strong motivation to pay what it takes to preserve a 1.5°C world. By investing in the problem now, a low discount rate would incorporate the potential for catastrophic climate change tipping points and would not ignore hard-to-value resources, like biodiversity. A high discount rate, on the other hand, assumes a healthy growing economy with many good investments to be had, and a future much richer than the past. Unfortunately, climate change itself increasingly threatens to degrade natural resources to the point that this richer future will be impossible.[139] That may be the best reason why a low discount rate is more appropriate.

This discussion of discount rates provides an argument for rapid action on climate change based on the analogy of buying insurance. Young people buy health insurance and West Coast homeowners buy earthquake insurance; these are rational decisions even though individual risks are low. Similarly, green energy investments are nothing less than the insurance policy for maintaining a habitable Earth. The policy must be bought whether the damages would have materialized with or without it. Illnesses and earthquakes are familiar to us, but the increasingly visible financial damages from climate chaos reflect a cry for help from nature for us to get our house in order. It is our job to help lawmakers learn to hear it.

Interlude: The Coming Energy Transition

The transition to a net-zero carbon economy is one of humanity's greatest challenges. There is no overarching international roadmap, but a broad consensus exists about which elements should be present in the plans for any country. In this short interlude we will look at these elements, presented in six energy transition roadmaps proposed for the US (Table i.1).[1] This will set the stage for a more detailed exploration of the ideas and policies in the rest of the book.

The US energy transition is taking place together with similar efforts going on all over the world. At the 2015 Paris meeting of the United Nations Framework Convention on Climate Change (UNFCCC), almost all countries submitted voluntary plans for nationally determined contributions to the global effort. Each set of plans includes pathways for low emissions development and climate resiliency that recognize the unique cultures and social norms of a particular nation.[2] Judging by the near-universal participation, this has been a successful formula. The Paris Agreement replaced the 1997 Kyoto Protocol, an international treaty that required emissions cuts by developed countries with industrialized economies but did not make similar demands of developing nations. This distinction recognized that developed countries are responsible for the lion's share of emissions, but it also created tensions that proved irresolvable.[3]

The key elements found in roadmaps to a net-zero carbon US economy consistent with a 1.5°C world are:[4]

- Convert all electric power generation from fossil fuels to carbon-free energy sources, and create a nationally unified, smart power grid.
- Electrify the transportation, industry, commerce, and residential sectors, and use green fuels where electrification is difficult or less desirable.
- Improve energy and materials efficiency in all economic sectors.

- Reduce emissions of climate pollutants such as methane, nitrous oxide, black carbon and halocarbons as much as possible.
- Remove excess atmospheric carbon dioxide by natural land management and industrial approaches.

Although some of the roadmaps to net zero emissions deemphasize one or more of these five elements, the most comprehensive plans recognize the central importance of all of them.

The Present Situation

The US submitted ambitious plans for its energy transition at the Paris Agreement, but the federal government has not yet enacted them. Nonetheless, US carbon dioxide emissions have been falling fairly consistently since the mid–2000s, and both US and global economic growth are starting to become decoupled from these emissions (Figure i.1). The ability of the US and some other countries to increase growth while decreasing emissions is good evidence that carbon-free energy sources can efficiently power the economy.[5] This decoupling is modest, but it does offer a response to the argument that ever-increasing fossil fuel production is necessary to maintain prosperity.

Worldwide, however, emissions are still increasing, and the modest progress so far is not enough to stabilize atmospheric greenhouse gas concentrations at levels consistent with either

Table i.1: Some roadmaps for the US renewable energy transition		
Roadmap	Year	Characteristics
US Deep Decarbonization Pathways Project	2014/2015	80% reduction in emissions by 2050 Primary focus on energy and its use in the commercial, industrial and residential sectors
Evolved Energy Research	2019	Builds on 2014 Deep Decarbonization report Includes pathways to 350 ppm CO_2 by 2100 Emphasis on circular carbon economy
C2ES/Climate Innovation	2019	Alternate sociopolitical scenarios for 80% emissions reduction by 2050: international competition, state initiatives, and low-carbon lifestyles
US House Select Committee	2020	Net zero emissions economy wide before 2050 Includes national clean energy mandate, efficiency standards, promotion of electric vehicles, natural land management
Rewiring America	2020	70-80% reduction in carbon emissions by 2035 "Electrify everything" Maximum feasible transition with focus on power grid
Princeton University Net Zero America	2020/2021	Five comprehensive pathways for net-zero emissions by 2050 Emphasis on cost-effective, existing technologies Granular detail at state and county levels

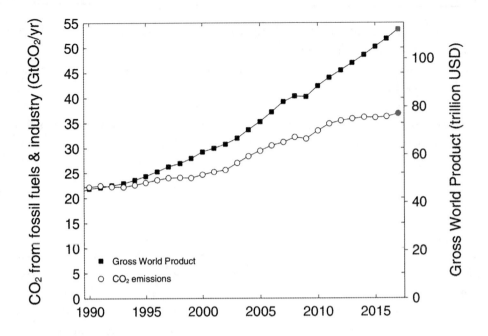

Figure i.1:

Top: Increased decoupling of global carbon dioxide emissions from economic growth, as measured by the Gross World Product, tracked from 1990 to 2019. The data includes emissions from fossil fuels and industry.

Bottom: Carbon dioxide emissions from top emitting countries and the remainder of the world, including projected 2020 decreases from the pandemic. These data only include emissions from fossil fuel burning. Both graphs report emissions in units of gigatons (Gt), or billions of metric tons. Used by permission of the Global Carbon Project.

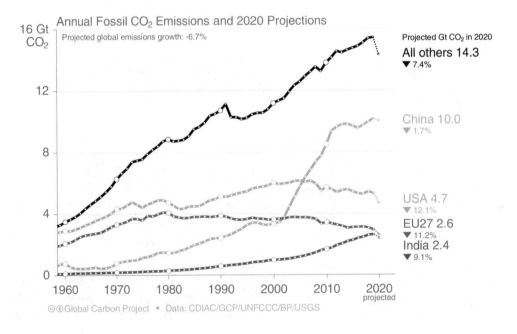

1.5°C or 2.0°C worlds.[6] Had sustained emissions reductions begun several decades ago, the 1.5°C target would have been reachable with cuts of just a few percent per year. However, in 2019 the United Nations Emissions Program (UNEP) concluded that a 1.5°C world now requires carbon emissions to decrease by 7.6 percent per year,[7] while other experts estimate that cuts of 15 percent per year are required.[8] These estimates are comparable to the approximately 7 percent drop in global carbon dioxide emissions in 2020, caused by the coronavirus pandemic.[9]

The hard reality is that the world has waited so long that meeting the ambitious 1.5°C target is now extraordinarily difficult, but perhaps not yet impossible.[10] Fortunately, warming can still be stabilized at this level by removing and safely sequestering excess carbon dioxide from the atmosphere.[11] Carbon removal and sequestration is a crucial element of decarbonization plans and will almost certainly be needed even if the 1.5°C target is met without overshoot. Although many climate scientists regard a 1.5°C world as reasonably safe from climate tipping points, this is not certain. We should also note that sea levels in the last interglacial period were far higher than today, even though atmospheric carbon dioxide levels did not exceed 300 ppm. These data suggest that stabilizing some parts of the large ice sheets from eventual irreversible melting may require large-scale atmospheric carbon removal.[12]

Since the Paris Agreement is now enshrined as the international mechanism of choice to accomplish the carbon-free energy transition, it is relevant to ask whether it sufficiently meets climate targets. Figure i.2 shows International Energy Agency (IEA) projections of emissions in global models where the Paris pledges are met (Pledged policies), showing that the improvement compared with current policies is modest and not sufficient to meet the required steep emissions cuts for a 2°C world.

The news is not all bad, however. Present policies are a significant improvement compared to those in place in 2005, and this is enough to make worst-case scenarios of 4–5°C warming increasingly unlikely (Figure i.2). The IEA projects that 3°C of warming is the most probable outcome given present trends.[13] This is a more optimistic view than offered by the United Nations Environment Program (UNEP), which instead estimates that meeting all Paris pledges would create a 3.2°C world.[14]

Of course, our examination of climate change impacts in the last chapter showed that a 3°C world is hardly an acceptable outcome, and the possibility that progress will be slowed or reversed cannot be discounted. Indeed, while some countries are making good progress on their Paris pledges,[15] others are faltering, and the US became the worst actor of all when the Trump administration withdrew its support from the Agreement (although President Biden has now reengaged the process).[16] So finding that the worst-case scenario appears unlikely is not a reason for complacency.

Social and Cultural Dimensions

The energy transition is challenging because there is more to it than finding effective technical solutions, which is difficult enough. Rather, the problem also has crucial cultural and political dimensions. This is shown vividly by the Coronavirus pandemic, which illustrates how an event with global impact can fundamentally change behavior and alter

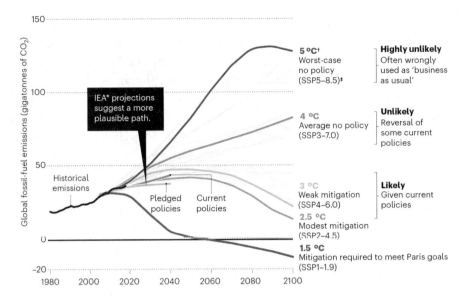

Figure i.2: Five projections of future fossil fuel emissions using the Shared Socioeconomic Pathway (SSP) framework, introduced in 2021 in the sixth IPCC report (see the text for details). The projections report emissions trajectories needed to constrain warming to 1.5°C, 2.5°C, 3°C, 4°C, or 5°C. Superimposed are projections from the International Energy Agency (IEA) corresponding to the future emissions that would occur with active policies (Current policies), as well as with current policy intentions and targets (Pledged policies). Given current policies, likely total global warming is projected to be approximately 3°C. Reproduced by permission from Z. Hausfather and G. Peters, "Emissions—The "Business as Usual" Story is Misleading," *Nature 577,* 618-620 (2020).

climate trajectories.[17] As tragic as it has been, the pandemic has also altered consumption and travel habits worldwide and has clearly shown how individual lifestyles and societal norms affect climate pollution. Some of these changes, such as acceleration of the trend toward telecommuting and electronic conferences, may lead to significant emissions reduction over the long term. Even more important, directing stimulus and recovery efforts to clean energy investments offers an unexpected opportunity to catalyze the structural changes in the economy that are needed for decarbonization.[18] This is crucial for further accelerating the decoupling of economic growth from carbon emissions (Figure i.1).

A broad research program is examining how social and cultural norms may influence the effectiveness of policies for the energy transition. This is done through a framework known as *shared socioeconomic pathways,* or SSPs (Table i.2).[19] The SSPs model five distinct ways in which the world might develop without any new climate policy—not even the Paris pledges. These possible baseline futures vary in population, economic growth, technology, and international relations, and lead to projected warmings of 3–5°C depending on the pathway. The wide range of outcomes shows the importance of social cohesion and other values in determining the future climate, apart from technological solutions.

A key factor examined in the SSPs is the extent to which the future worlds remain reliant on fossil fuels. In all envisioned futures, continued economic growth drives increasing energy demand, and this could spur an "all of the above" approach to energy development where both carbon-free and fossil fuel sources grow (for example, in the SSP5 future). But in a world where the value of environmental sustainability comes to the fore, much more of the increased demand would be met by carbon-free energy and better energy efficiency (the SSP1 future), even without any specific policy.

Critical insights emerge when the SSPs are combined with *relative concentration pathways* (RCPs), the scenarios in which climate policies are explicitly modeled to produce the emissions reductions needed for 1.5°C, 2°C, or warmer worlds.[20] For example, RCP2.6 is a favorable future pathway that models the steep emissions reductions needed for a 2.0°C world. That scenario can be projected onto a green and more equitable baseline SSP1 world, an SSP4 world that has high socioeconomic inequality, or any of the other SSPs. All combinations of RCPs and SSPs can be modeled, producing dozens of possible futures to examine in detail.

This modeling suggests that not every SSP is compatible with each RCP pathway. Most notably, the outcomes reveal that it may not be possible to achieve either 1.5°C (RCP1.9) or 2.0°C (RCP2.6) worlds in the context of the SSP3 socioeconomic pathway, in which resurgent nationalism and security concerns lead nations to "bunker down" and focus on domestic issues. This shows the danger associated with elections of nationalist governments in the US, Brazil, and elsewhere. Reaching these ambitious temperature targets is difficult within SSP5 as well.

Early results also offer a lesson about the importance of linking climate change mitigation with reducing socioeconomic inequality. Six modeling teams combined each of

SSP	Moniker	Warming in 2100	Characteristics
SSP1	Green Road	3 - 3.5°C	Gradual but pervasive shift to a sustainable trajectory, attention to social equity; less carbon-intensive fuels; recognition of ecological limits.
SSP2	Middle Road	3.8 - 4.2°C	Current sociological trends continue. Decreased intensity of energy and resource use, uneven global growth, slow progress toward sustainable development
SSP3	Rocky Road	3.9 - 4.6°C	Resurgent nationalism and regional conflicts; little focus on global sustainability, environmental degradation. Material- intensive consumption persists, social inequalities worsen
SSP4	Divided Road	3.5 - 3.8°C	Increasing inequality, loss of social cohesion, increasing unrest. Deeper fragmentation into rich and poor countries. Strong development in rich countries, all-of-the-above energy use
SSP5	Fossil Fuel Road	4.7 - 5.1°C	Integrated global markets, rapid technological progress driven by carbon-intensive fossil fuel use. Investment in health and education, faith in technology, geoengineering

Table i.2: Shared socioeconomic pathways[a]

[a]SSP narratives are excerpted from: Hausfather, Z. Explainer: How "Shared Socioeconomic Pathways" explore future climate change, CarbonBrief, 19 April 2018, https://www.carbonbrief.org/explainer-how-shared-socioeconomic-pathways-explore-future-climate-change

the five SSP pathways with six distinct RCP frameworks, creating a matrix of 30 possible futures.[21] Among these, the modeling teams all found that the SSP1 pathway, which features inclusive development and reduced inequality, is consistent with reaching a 1.5°C world. In contrast, only two-thirds of the teams found a way to stabilize temperatures at 1.5°C within an SSP2 context (the socioeconomic framework most similar to today's world). The fraction of models succeeding in this effort within SSP5, SSP4, and SSP3 worlds were all even smaller. These are compelling findings for progressive climate advocates who want solutions that also address social justice concerns.[22]

Roadmaps for the US

To get a feel for the US energy transition, we will first look briefly at the most important technologies for reaching net-zero emissions by 2050, and the prospects for bringing these to fruition. In the following section, we will consider some of the most immediate ways in which advocates can contribute in the next few years. We will also look at the first full legislative roadmap from the US Congress, released by the House of Representatives in the summer of 2020.

The first comprehensive US energy transition roadmaps for deep decarbonization were prepared in anticipation of the Paris Agreement. The plans featured 80 percent emissions reduction by 2050—the "80 by 50" goal —as compared with 1990 levels and used modeling to lay out a variety of different pathways to get there.[23] This was a significant pledge, consistent with the objective to limit warming to 2.0°C.[24] The most important findings are that the "80 by 50" goal is technically achievable with presently available or near-commercial technology. Further, the costs of the transition are modest, amounting to only about 1 percent of the US GDP. Four distinct scenarios were able to meet the 2.0°C target. For electricity generation, these pathways to 2050 feature strong growth of renewables, nuclear power, or fossil fuels with carbon capture and storage (CCS), respectively, while the fourth scenario is a mixed case.[25]

Among the options for decarbonizing the power sector, solar photovoltaic and wind energies are the least expensive options presently and are poised for rapid growth. Rapid buildout of solar and wind, including aggressive exploitation of the offshore wind resource, must be accompanied by accelerating work on energy storage technologies, including advanced batteries and renewably produced hydrogen. This is necessary to accommodate the intermittency of the PV solar and wind resources. During this buildout, zero-carbon nuclear power should be maintained because it offers a reliable backstop that ensures the robustness of the grid during the transition period. Complementary goals for the electric grid include creating a nationwide architecture that fully accommodates home solar power. This will enhance the reliability of the grid and allow it to operate more efficiently.[26]

The buildout of the carbon-free grid has to be accompanied by a reduction in the use of fossil fuels, so that any increases in energy demand are not met by an "all of the above" approach, as they have been to date.[27] Coal power is already in irreversible decline because it is no longer cost-competitive, but the same cannot be said of natural gas, which is presently the leading power source. Industry rhetoric notwithstanding, natural gas is not needed as a "bridge" to a carbon-free future; solar photovoltaics and wind turbines are already better investments and energy storage technology is developing rapidly.

How fast can the all-important transformation of the power grid happen? A maximal outcome is given in the roadmap proposed by the Rewiring America group in 2020, which combines an intensive analysis of the US energy sector with modeling of the transition to carbon-free sources.[28] The findings are that 70–80 percent of all US emissions can be eliminated by 2035 with a crash program that swiftly converts the power grid while also electrifying many end uses, including much of the transportation sector. Remarkably, the crash program relies only on existing solar, wind, battery storage, electric vehicle, and heat pump technologies. It would increase the capacity of the grid by three-to-fourfold, including doubling nuclear power output. And because of the better performance of electric motors, deep cuts in fossil fuel mining, and other efficiency gains, total energy demand is projected to be sharply cut.[29] This may be the only proposal capable of meeting the US share of emissions reductions needed to stay within the global 1.5°C climate target.

A slightly less ambitious but still very aggressive plan is offered by a Princeton University proposal called Net Zero America (Table i.1). Five distinct roadmaps are proposed, each consistent with net-zero carbon emissions by 2050. This is ambitious enough for the US to meet its share toward the global 2°C target, although it likely will not suffice for the 1.5°C target. The five roadmaps vary in their emphases on the extent of electrification and the contributions of carbon capture and sequestration, nuclear power, and biomass energy sources. Like the Rewiring America proposal, all would require massive and unprecedented buildout of new infrastructure—including 60 percent expansion of the electricity transmission grid, complete shutdown of coal-fired power plants, doubling the number of homes heated by electricity, and 50 percent of new car sales in battery-powered models—all by 2030. Benefits would include huge decreases in health care costs from the lowering of fossil fuel air pollution, and the creation of millions of new green energy jobs to compensate for losses in the fossil fuel industry. Remarkably, the projected cost of the transition is within the historic range of US energy spending, which is about 4–6 percent of GDP each year.[30]

All of the Net Zero America pathways clearly represent very aggressive technology takeovers compared to what has been historically realized. This is the basis for pessimistic views about the prospects for any rapid energy transition. For example, the influential energy historian and analyst Vaclav Smil is convinced that fossil fuels will still be the dominant energy source by the mid-twenty-first century.[31] If Smil turns out to be right, it will clearly not be because key replacement technologies have not been invented. Rather, their rapid deployment depends on both replacing entrenched fossil fuel-based approaches and bypassing the more exhaustive demonstrations of feasibility typically required before investors are willing to commit to something new.[32] In the worlds of finance and politics, pessimism—abetted by corruption—unfortunately becomes a self-fulfilling prophesy.[33] But it is important to recall that no laws of physics prevent the US (and the world) from executing the carbon-free energy transition to the maximum limit that technology allows, as Net Zero America, Rewiring America and other groups advocate.

Although both Rewiring America and Net Zero America emphasize buildout of commercially ready technology, there is little doubt that minimizing costs and maximizing long-term efficiencies will also require scaling up other approaches – especially renewable hydrogen, smart electricity grids, and carbon capture and sequestration. The International Energy Agency projects that about half the emissions reductions needed to reach net-zero carbon by 2050 come from technologies that are not yet on the market.[34] Regardless of the extent to which new technology is required, however, there is little doubt that the US Department of Energy and other agencies must massively increase the budgets for energy research and development. This will require large increases in the budgets for the US Department of Energy and other agencies, with a focus on carbon-free technology and on bridging the gap from basic findings to large-scale commercialization.[35] To enable the latter, the push for new technologies can be paired with policies that provide incentives for private industries to invest, such as a national carbon pricing plan or aggressive clean energy mandates for the industry, power generation, and transportation sectors.

The ultimate destination of the energy transition could be the *circular carbon economy*, a vision emphasized in the Evolved Energy Research roadmap (Table i.1).[36] Although many decades from fruition, this vision sheds reliance on both fossil fuels and biomass as carbon sources, instead substituting carbon that is captured from the atmosphere. The carbon dioxide is then transformed into fuels and commodities by combining it with energy-rich, renewable hydrogen produced directly from carbon-free electricity.[37] While we do not need to fully implement the circular carbon model to reach a net carbon-free US economy by 2050, its key component technologies—carbon sequestration and renewable hydrogen—are each nonetheless crucial for achieving climate goals.

An Advocacy Agenda for the 2020s

Let's pivot from this high-level analysis to get a better understanding of what citizen advocates can do to help make some of this happen. Table i.3 offers some suggestions for top policy priorities in the present and succeeding decades, all of which are described in more detail later in the book. Reaching Rewiring America's vision would require a World War II-like command economy to direct ten-fold increases in both solar and wind generation within three to five years.[38] Because such an economic restructuring is not politically plausible, however, an aggressive, escalating national carbon price should become the top priority. This policy will drive transformative change throughout the economy and is projected to meet net-zero emissions by 2050. By forcing fossil fuel manufacturers to pay for their climate pollution, a national carbon price will shift the entire energy playing field by favoring the lowest-emission technologies while simultaneously disfavoring coal, oil, and gas. A national clean energy mandate could also accomplish much of this, although it would likely not reach as broadly into every segment of the economy. There are grass roots advocacy opportunities for both implementing a national carbon tax and for expanding state and regional carbon pricing programs, which have already had some success in California and the Northeast.[39]

A carbon price can be effectively paired with direct actions to limit the scope of future fossil fuel development. There are many avenues for advocates to oppose new infrastructure projects such as pipelines, fossil fuel export terminals, and the like. *Divestiture*, the cutting off of financing for fossil fuel operations, is another crucial area for advocacy work. And while less known, the role of state-regulated public utility companies in driving construction of new solar and wind power plants for the grid is well worth looking into. These groups are also influential in determining early retirement of coal and natural gas-fired facilities, establishing the infrastructure for electric charging, and weighing in on state policies for accelerating home solar power. Of course, direct lobbying of state legislatures to accelerate carbon-free energy mandates for electricity generation is also an option.[40]

Specific advocacy efforts should target both rural and urban climate policies. *Afforestation*, the planting of new forests, is the most effective and ready-to-go approach to enhance carbon

Table i.3: Suggested key actions by decade	
Year	**Actions**
2020s	• Implement aggressive economywide carbon price or national clean energy mandate • Accelerate research and development for carbon capture, energy storage, renewable hydrogen and synthetic methane • Accelerate buildout of solar photovoltaics and wind energy, including offshore wind development • Begin large scale afforestation and carbon sequestration practices in agriculture • Phase out construction of new fossil fuel infrastructure and accelerate divestiture from fossil fuel operations • Create national and state green banks to fund equitable and climate-friendly initiatives, especially in urban development • Plan and initiate restructuring of the electricity grid, including railroad electrification corridors and accommodation of electric vehicles • Maintain zero-carbon nuclear power generation
2030s	• Fund public-private partnerships to drive commercialization of carbon capture and key energy technologies, especially renewable hydrogen • Maintain large scale afforestation and carbon sequestration practices in agriculture • Establish unified national electricity grid that accommodates distributed generation • Identify and find solutions for difficult decarbonization bottlenecks in industry, including expansion of carbon capture for essential industries • Maintain zero-carbon nuclear power generation by facilitating replacement of aging reactors with small modular reactors • Phase out production of corn ethanol in tandem with the accelerated electrification of the light vehicle fleet • Fully eliminate coal power and accelerate retirement of natural gas power plants
2040s	• Begin implementation of direct air capture and/or enhanced weathering technologies for large-scale atmospheric drawdown of carbon dioxide • Maintain large scale afforestation and carbon sequestration practices in agriculture • Begin large-scale implementation of renewable hydrogen production and the circular carbon economy • Modify natural gas pipelines and/or construct new pipeline infrastructure for renewable hydrogen

stores on the land. This natural approach to atmospheric carbon removal can offset emissions that are difficult to reduce in some industries and modes of transportation. There are also many opportunities to advocate for expanding carbon storage on agricultural lands, approaches that are also able to constrain releases of methane and nitrous oxide. In the context of a fast energy transition, like what is advocated by Rewiring America, the relatively small overshoot of emissions beyond the 1.5°C target might be offset by these methods alone. This would allow more time for developing industrial scale removal to reduce atmospheric carbon levels, and to redirect that carbon to the synthesis of renewable fuels and commodities.[41]

In urban areas, advocacy efforts can focus on developing and implementing city climate plans, which set goals for energy-efficient buildings, urban density, mass transit, and other policies that help reduce vehicle use and the like. Electrification of as many end uses as possible is key, including promoting the use of electric vehicles, heat pumps, induction stovetops, and electric furnaces and water heaters.[42] The urban environment is also an ideal setting for working to incorporate equity into climate policy, since cities are often homes for frontline and disadvantaged communities that have already been more greatly impacted by climate change.

Almost all aspects of the advocacy agenda are incorporated into the first full-fledged energy transition plan developed by the US Congress: the Climate Crisis Action Plan introduced by the US House Select Committee on the Climate Crisis in June of 2020 (Table 1.1).[43] This plan was inspired by the Green New Deal, a vision that called for linking solutions to the climate crisis with promoting social justice and redressing structural inequalities in American society.[44] The Climate Crisis Action Plan plan is not legislation, but rather a fully articulated roadmap to build a clean energy economy that values workers, advances environmental justice, and meets the challenges of the climate crisis.[45]

The Climate Crisis Action Plan invokes the 1.5°C target and calls for aggressive emissions reductions for a net-zero carbon economy by 2050. It includes cleaner and more resilient infrastructure, investments in research and development for critical technologies, a national climate bank, stringent climate pollution standards, carbon sequestration by natural approaches, and a price on greenhouse gas emissions. It also places environmental justice at the center of climate and environmental policy, with special attention to the impacts of climate change on health and resiliency of historically disadvantaged communities.[46] Modeling of the key recommendations in the report suggests that implementing them would reduce carbon dioxide emissions by 88 percent compared to 2010 levels, and that net-negative emissions from carbon sequestration on the land would suffice to offset the remaining 12 percent, enabling a net-zero carbon economy in 2050 (Color Plate 16).[47] The modeling also provides key details about the scope of the required transition, setting out requirements for how rapidly solar and wind power, electric vehicles, battery storage, and other technologies need to be deployed.

The federal government plays a major role in the Climate Crisis Action Plan, but it is far from the only actor, and much effective advocacy can take place at state and local levels. In the next chapters, we will see how progressive states are driving the energy transition, and how the work of grassroots advocates nationwide is making a difference. We will begin by exploring the resources and strategies that can be mobilized to make our advocacy work effective.

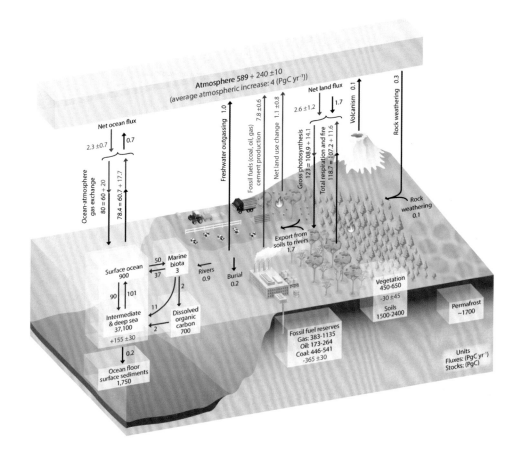

Color Plate 1: Black numbers indicate estimated amounts of carbon in Earth reservoirs in preindustrial times in units of petagrams (10^{15} grams), which is equivalent to billions of metric tons. Black numbers adjacent to arrows indicate movement of carbon among reservoirs in preindustrial times. Red numbers are carbon movements from human activities averaged for the years 2000–2009. Ciais, "Carbon and Other Biogeochemical Cycles," Figure 6.1, p. 471. Used with permission.

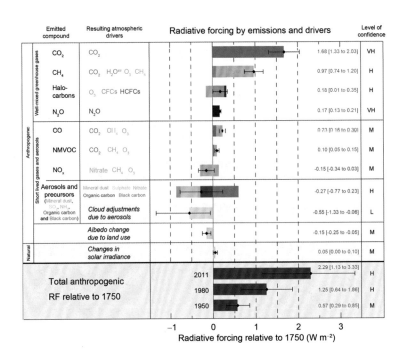

Color Plate 2: Drivers of climate change. See Chapter 2 for detailed explanations. From IPCC, 2013, Summary for Policymakers, Figure SPM.5, p 14. Used with permission.

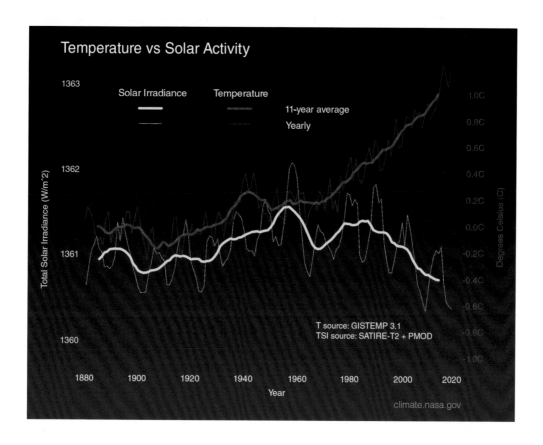

Color Plate 3: Comparison of global surface temperature change with variation in the rate of solar radiation energy reaching Earth. The units of solar irradiance are given in watts per square meter of surface. Credit: NASA.

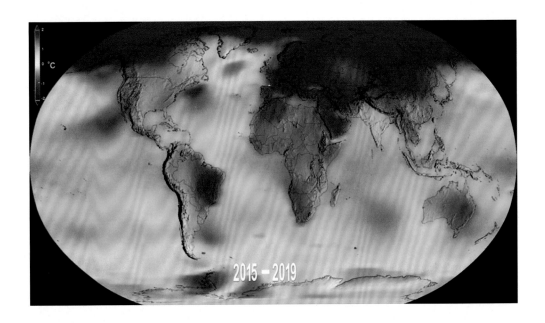

Color Plate 4: Representation of global surface temperature changes mapped on Earth's surface. Red indicates temperature increases of 2°C or greater, yellow depicts increases of approximately 1°C, and blue shows regions that have cooled. Data are an average of temperatures over 2015–2019, compared to a 30-year baseline period from 1951 to 1980. Credit: NASA.

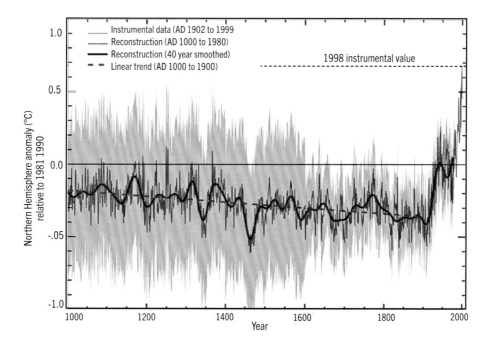

Color Plate 5: Temperature reconstruction from climate proxy data (blue) combined with modern instrumental data (red). The gray background shows error estimates. See Chapter 2 for details. From IPCC, Observed Climate Variability and Change. In: Climate Change 2001: The Scientific Basis, Chapter 2, p. 134. Used with permission.

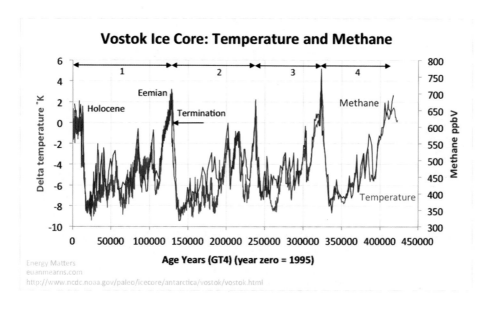

Color Plate 6: Ice core data from the Antarctic Vostok site, which shows the close relationship between temperature and atmospheric carbon dioxide levels over multiple Ice Age glaciations. Credit: NOAA.

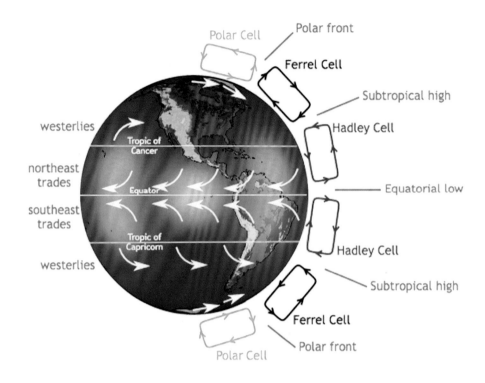

Color Plate 7: Depiction of the major atmospheric flow patterns that climate scientists seek to incorporate into climate models. Credit: NASA.

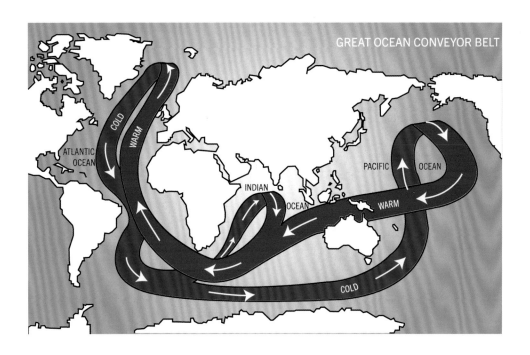

Color Plate 8: Global ocean circulation patterns, distinguishing between surface and deep flows. In the North Atlantic, sinking of cold seawater is driven by its high salt content, which makes it denser. The melting cryosphere may affect this flow; see Chapter 4, Table 1. Credit: NOAA.

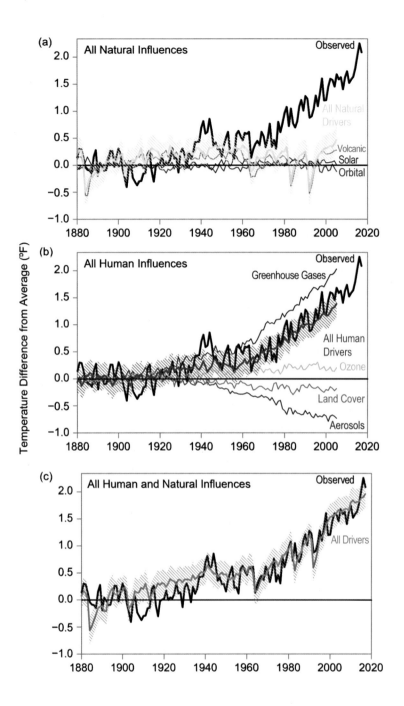

Color Plate 9: Observed temperature changes (the identical heavy black line in all three panels) are compared to the projections of climate models, when the models include (a) natural influences only, (b) human influences only, or (c) both human and natural influences. See Chapter 3 for details. Credit: Fourth National Climate Assessment, Volume II, Figure 2.1.

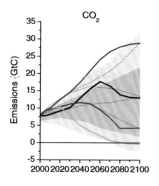
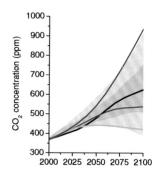

Color Plate 10: Projections of greenhouse gas emissions (left) and atmospheric carbon dioxide concentrations (right) for the four representative concentration pathways: RCP2.6 (green), RCP4.5 (red), RCP6.0 (black), and RCP8.5 (blue). From DP van Vuuren et al., "The Representative Concentration Pathways: an Overview," Climatic Change, no. 109 (2011).

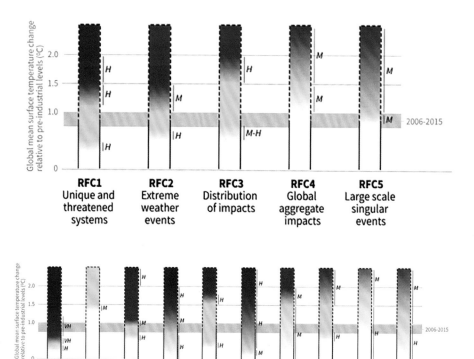

Color Plate 11: The impacts and risks associated with warming of up to 2.5°C above preindustrial temperatures. The meanings of the colors are given at right. Confidence levels for the transitions are indicated as low (L), medium (M), high (H), or very high (VH). From IPCC, Global Warming of 1.5°C, Summary for Policymakers. p. SPM-13. Used by permission.

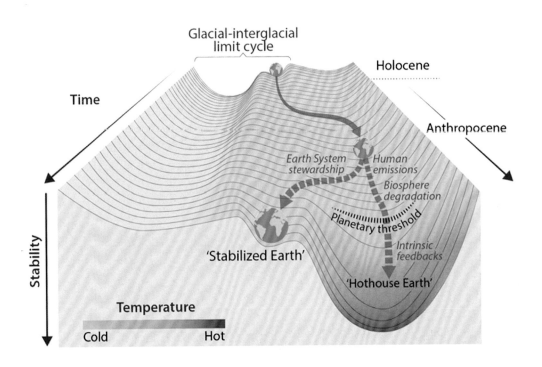

Color Plate 12: Depiction of climate tipping points. Two potential alternative trajectories for Earth's climate future are shown. See Chapter 4 for details. W. Steffen et al., "Trajectories of the Earth System in the Anthropocene," Proceedings of the National Academy of Sciences in the United States of America, no. 115 (2018): 8252-8259, https://www.pnas.org/content/pnas/115/33/8252.full.pdf. Used by permission.

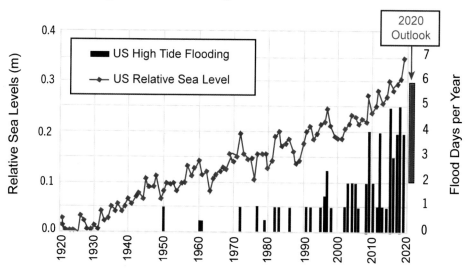

U.S. High Trade Flooding and Coastal Sea Levels

Color Plate 13: Depiction of high tide flooding (HTF) in the US over the past 100 years, shown as black bars indicating the number of flood days per year, and as blue points showing increasing relative sea level (RSL). The values of HTF and RSL for 2019 are shown in red.

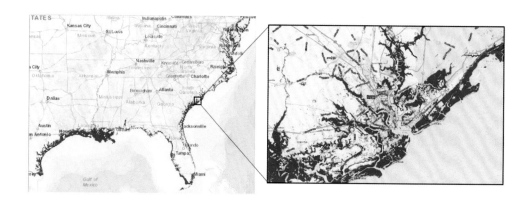

Color Plate 14: Depiction of regions of the US coastline that are vulnerable to high-tide flooding. The close-up on the right—of a section of the South Carolina coast—shows that the impacts are not only confined to beachfronts.

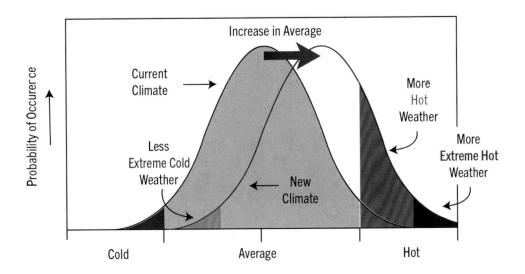

Color Plate 15: Global temperature shifts from climate change. As warming increases, the overall distribution of temperatures shifts to the right, signifying more extreme heat events and fewer extreme cold events. Credit: EPA.

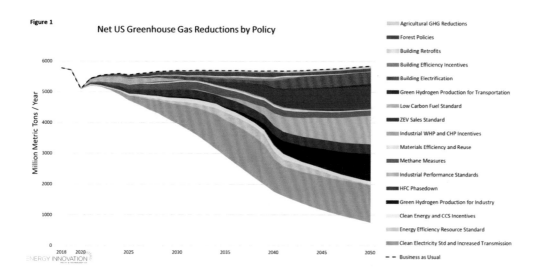

Figure 1

Net US Greenhouse Gas Reductions by Policy

Legend:
- Agricultural GHG Reductions
- Forest Policies
- Building Retrofits
- Building Efficiency Incentives
- Building Electrification
- Green Hydrogen Production for Transportation
- Low Carbon Fuel Standard
- ZEV Sales Standard
- Industrial WHP and CHP Incentives
- Materials Efficiency and Reuse
- Methane Measures
- Industrial Performance Standards
- HFC Phasedown
- Green Hydrogen Production for Industry
- Clean Energy and CCS Incentives
- Energy Efficiency Resource Standard
- Clean Electricity Std and Increased Transmission
- – – Business as Usual

Color Plate 16: Modeling of a subset of the recommendations offered by the House Climate Crisis Committee in its US energy transition roadmap, carried out by the modeling group Energy Innovation. The contributions of the various policies are color coded at right. Used by permission.

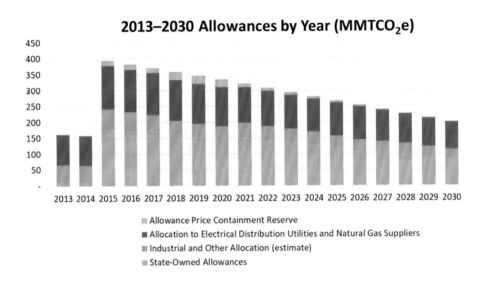

2013–2030 Allowances by Year (MMTCO₂e)

■ Allowance Price Containment Reserve
■ Allocation to Electrical Distribution Utilities and Natural Gas Suppliers
■ Industrial and Other Allocation (estimate)
■ State-Owned Allowances

Color Plate 17: Allocation of emissions allowances in the California cap and trade program. Free allocations to utilities and industry are shown in purple and blue, respectively. Allocations made by the state agency and offered at auctions are shown in orange. Green bars represent an initial reserve. See Chapter 7 for explanation of the California program.

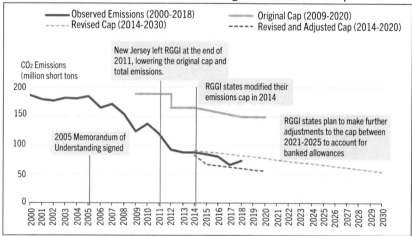

Observed Emissions and the Original and Revised Caps

Observed Emissions (2000-2018)   Original Cap (2009-2020)
Revised Cap (2014-2030)   Revised and Adjusted Cap (2014-2020)

CO_2 Emissions (million short tons

New Jersey left RGGI at the end of 2011, lowering the original cap and total emissions.

RGGI states modified their emissions cap in 2014

2005 Memorandum of Understanding signed

RGGI states plan to make further adjustments to the cap between 2021-2025 to account for banked allowances

Source: Prepared by CRS; observed state emission data (2000-2018) provided by RGGI

Color Plate 18: Observed CO_2 emissions (heavy blue line) and projected future emissions (dotted light blue line) from large electric power plants located in states participating in the Northeast Regional Greenhouse Gas Initiative (RGGI). Source: Congressional Research Service.

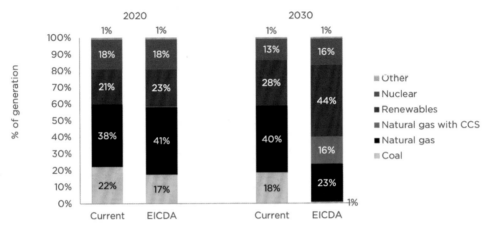

Source: Rhodium Group analysis. Note: Only central-energy cost results are shown.

Color Plate 19: Modeled increases in energy generation from zero carbon and low carbon sources by 2030, had the federal Energy Innovation and Carbon Dividend Act been enacted in 2020. Used by permission of the Columbia | SIPA Center on Global Energy Policy, and Rhodium Group.

Chapter 5
Climate Advocacy

We've taken a look at where we need to go, and at some roadmaps for getting there, so it is time to get into the nuts and bolts of what must be done. We begin this effort by looking at the strategies and resources available for climate advocacy. This will allow us to get a handle on the approaches that are most effective for persuading lawmakers to enact the advocacy agenda—or at least, the parts of that agenda that are unfolding in the communities where we live.

The biggest challenge for advocates is that climate change is a wicked problem: a problem that human nature is noticeably ill-suited to solve.[1] Many factors contribute to the wickedness—the frightening nature of the problem, the lack of a clear best approach that is obvious to the public, the need for a great many people to fundamentally change aspects of their worldview, and the fact that costs have to be borne today for the sake of future generations.[2] Human psychology also contributes to the challenge because the problem of climate change can appear so overwhelming that it becomes paralyzing to confront directly. It is then all too easy to carry out half-measures and turn our attention to something else. This is a form of denial—possibly hard-wired into the human brain—and operating below a conscious level.[3]

This sounds pretty bleak, yet many individuals, organizations, and governments have been able to take effective action anyway. As we have seen, the future arc of climate change is already bent away from the worst-case scenario outcomes that were imagined twenty years ago, progress we should not overlook.[4] The key to the past success and what it will take to keep up the momentum lies in adopting powerful climate narratives that offer win-win scenarios in spite of the *free-rider problem*—the unwillingness of some groups to pay the upfront costs of investing in solutions, even though they benefit from the emissions reductions made by others.[5] The problem is overcome when the necessary actions are seen as beneficial rather than burdensome. Storytelling is everything.

Figure 5.1 Visual equation illustrating IPAT formula.

1. Is Technology Necessary?

The roadmaps to climate stability all rely heavily on technologies for clean power as well as innovation, all while maintaining economic growth. However, some Americans reject this approach in favor of a back-to-nature paradigm, which addresses climate change by spurning the growth model and returning to small, local economies of scale. The classic work in this area is E.F. Schumacher's *Small is Beautiful*—which is still influential today.[6] Before we jump into the politics and strategies for advocacy, it is relevant to take a look at why Schumacher's approach, while well-intentioned and inspiring, is not able to solve the problem.

To do this, we can look at the impacts of climate change in terms of societal forces: population, affluence, and technology. This approach has been applied generally to many types of environmental impact, such as air and water pollution, and it is known as the *Kaya identity* in the context of global warming (Figure 5.1).[7] Although the Kaya identity can be written as an equation (I = P x A x T) it is more useful to think of it as a list of important human factors that contribute to the severity of climate change.

With all else equal, a higher population will produce heightened climate impact. Yet efforts to limit population growth need not be front and center in climate advocacy because this problem is largely already solved.[8] This is counterintuitive, given that the population has more than doubled since 1970. But for the past three decades, plunging fertility has brought about continuous decreases in the population *growth rate*.[9] In earlier centuries, women would have six babies so that two could survive to adulthood. Today, on average, women have two babies, and both survive. As public health improved in the twentieth century, there was a period when fertility was still high, but many more children survived. In the next few generations, however, the resulting great abundance of young versus old people will even out—by the late twenty-first century, the world population structure will be flat, with roughly the same number of people in each age group.[10] At that point, the world's population will no longer be increasing.

The United Nations projects that by 2100 the population will have leveled off between 10–12 billion.[11] Over half the growth in the next three decades is expected to be in Africa, where fertility rates are the highest.[12] To be sure, depending on the trajectory of a transition to renewable energy in that region, the extent of African population growth could either worsen global warming or make it less severe. Certainly, African development that bypasses fossil fuels and moves directly to renewable energy would ease global warming, while persistent higher fertility in fossil fuel-dominated economies would worsen it. If you do wish to concentrate your climate advocacy on population growth, Africa is the place to focus on. But in the US, projected future emissions are quite insensitive to how fast the population grows.[13]

What about the affluence and technology portions of the Kaya identity? Affluence is usually measured as *gross domestic product* (GDP) per capita. GDP is defined as the value of all the goods and services produced in an economy over a specified period of time. It is far from perfect as a measure of total economic value since many intangibles are left out. But these factors, which include children's health and education, are reasonably correlated with the transactions that GDP does measure.[14]

The technology term in the Kaya identity can be approximated by how much carbon dioxide is emitted for a given amount of economic growth. This is also called the *carbon intensity* of the economy. An economy dominated by fossil fuels would make this term large, so the climate impact would be high. Conversely, carbon intensity, and thus, climate impact, would be much lower in an economy based on renewable energy. Carbon intensity is decreasing in the US and many other countries because it saves money to be more energy-efficient—accomplishing the same job while burning less fossil fuel. Decreasing carbon intensity is another way of saying that economic growth is becoming decoupled from fossil fuel burning.[15]

Climate modeling studies can separate out the importance of the population, affluence, and technology aspects of the Kaya identity.[16] In models of potential future climate where we continue with business as usual (a trajectory for a 3–3.5°C world), increases in GDP per capita are closely balanced by decreases in carbon intensity. This means that a largely unregulated, growing world economy generates high carbon emissions, but they would be even higher were it not for the decreasing carbon intensity. The models also showed that population growth has much less of an influence on climate change than either affluence or technology.

It looks as if both sides are correct—either decreasing wealth per capita or drastic lowering of the carbon intensity (or both together) can, in theory, cut carbon dioxide emissions and lower climate impact. But we can look to the economic effects of the COVID-19 pandemic to understand immediately why the former approach is not viable. US GDP dropped by annualized rates of 5 percent and 34 percent in the first two quarters of 2020,[17] and while emissions did indeed drop, the tremendous financial hardships can hardly be sustained. Even advocates of *degrowth*, a conscious shrinking of the economy to decrease GDP per capita, recognize that no sizable political constituency exists to advance that agenda.[18]

It seems clear that climate change can only be solved by working within the pro-growth economic system that we have. The decoupling of emissions from economic growth in two-thirds of the states at least encourages the notion that this approach is viable.[19] And while it may be scant consolation to adherents of Schumacher's vision, the energy transition roadmaps do not neglect careful stewardship of natural resources but rather recognize it as an integral part of the solution.

2. The Climate Advocacy Landscape

The scope of climate advocacy in the US is very large and includes organizations from across much of the political spectrum (Table 5.1). Some of the groups are entirely dedicated to climate work, but many also engage other issues such as air and water pollution, environmental justice, and land and wildlife conservation. For the new advocate, it is especially valuable to connect with nearby chapters of organizations that also have national scope. It is then possible to become immediately immersed in the local advocacy community while simultaneously gaining access to the resources of the larger group. Four highly accessible grassroots organizations with cross-country local chapters stand out:

The Sunrise Movement is a youth-led initiative that emerged in the past few years to exert a decisive influence on the Democratic climate platform.[20] Sunrise adopts the values of the Green New Deal, holding that solutions to the climate crisis must be coupled with social equity policies.[21] The group embraces a strategy of direct action, joining with other organizations in large-scale climate strikes and organizing protests in the US Congress. Sunrise has rapidly become a highly influential player in climate politics, so much so that the 2020 Democratic Presidential candidates were unable to avoid the issue of climate change, instead making it prominent in their campaigns and debates.[22]

350.org, the organization founded by the author and activist Bill McKibben, is a grassroots movement looking to end fossil fuel use and facilitate the renewable energy transition.[23] It places a strong emphasis on linking climate change solutions with social justice. Individual chapters engage state, county, and city politicians and other leaders on the most relevant, climate-related local issues. Stopping the proliferation of local fossil fuel infrastructure is a common theme among many of the chapters.

The Climate Reality Project, initiated by former Vice President Al Gore, is a progressive organization built around a unique event that all volunteers participate in—the Climate Reality Leadership Corps Training.[24] The three-day training includes lectures and question-and-answer sessions on climate science and policy, as well as participatory exercises that build skills in climate communication. Graduates connect to a large global network while working in local chapters to catalyze change close to home. They also get access to the Vice President's signature slideshow on climate science and are encouraged to build the skills necessary to give their own presentations.

Citizens' Climate Lobby (CCL), now 200,000 strong in the US and internationally, is a non-partisan group organized around a single issue: developing the political will for Congress to enact an aggressive, economy-wide fee on carbon emissions.[25] Citizen lobbyists convene in Washington, D.C. twice a year for conferences, followed by intensive lobbying in the Senate and House. Over 500 active chapters around the country plug in for monthly calls with the national office in Washington, D.C. Like volunteers with the

Table 5.1: Climate advocacy groups

Progressive	Big Green	Climate Law	Non-partisan	EcoRight
350.org	Sierra Club	Earthjustice	Citizens' Climate Lobby	Climate Leadership Council
Climate Reality Project	Defenders of Wildlife	Our Children's Trust	Environmental Defense Fund	RepublicEn
Sunrise	National Wildlife Federation			Niskanen Center
League of Conservation Voters	National Resources Defense Council (NRDC)			American Conservation Coalition
Center for Biological Diversity	Nature Conservancy			ConservAmerica
Friends of the Earth	Wilderness Society			ClearPath
Extinction Rebellion	World Wildlife Fund			
Fridays for Future				
Zero Hour				

Climate Reality Project, CCL members also benefit from training on lobbying skills and communications strategies.[26]

The first three of these groups are progressive, emphasizing linkage of climate change policies with social equity. The other progressive groups—Extinction Rebellion, Zero Hour, and Fridays for Future (the school climate strike movement founded by Swedish youth activist Greta Thunberg)—also have this emphasis (Table 5.1). They are distinguished by their focus on direct, highly visible actions to galvanize public opinion.[27] In particular, the symbolism of youth walking out of classrooms to demand action for a livable Earth has inspired climate advocates worldwide, earning Thunberg recognition as Time Magazine's 2019 Person of the Year.[28] There are many other progressive climate groups with national scope who are working through policy advocacy, legal action, voter mobilization, and other venues. However, the scope of grassroots citizen involvement with these organizations is more limited.

The more traditional face of environmental advocacy in the US is represented by the groups listed under the moniker Big Green (Table 5.1). Many of these nonprofit organizations were traditionally focused on stewardship of the land and wildlife but have now also developed programs on climate and renewable energy. Big Green organizations are distinguished by large staffs, yearly budgets in the tens of millions of dollars, and well-developed relationships with business and political leaders. Although all of these organizations have many donors, they are not grassroots groups, but rather insiders working to effect change from within the political system. This is an important role but has also exposed some of the organizations to criticism from grassroots progressives, who accuse them of cutting too many deals with business interests.[29] The influential progressive author and activist Naomi Klein has even suggested that, through their willingness to accept business-friendly approaches, which she sees as flawed, Big Green groups have caused more damage to the climate than right-wing denialism.[30]

Two Faces of Climate Advocacy

On November 13, 2018, following the blue-wave midterm elections that returned the House of Representatives to Democratic control, two advocacy groups converged on Capitol Hill. Sunrise Movement protestors joined by just-elected Representative Alexandria Ocasio-Cortez of New York staged a sit-in at future Speaker Nancy Pelosi's office to demand more aggressive action on climate change.[31] At the same moment, 500 impeccably dressed Citizens' Climate Lobby volunteers hustled to their lobbying appointments to push a new bipartisan carbon pricing bill with the potential to sharply cut fossil fuel production.[32] Sunrise and CCL did not coordinate actions on that day, but their distinctive approaches illustrate the strength and diversity of the climate advocacy movement, and speak to its potential to catalyze change.

The Sunrise Movement promotes an agenda of direct action with a compelling, easily understood message: climate change threatens our air, water, and communities, but solutions are available. And while the greedy few are driving us toward catastrophe, we are indivisible, and we will win. Sunrise envisions a five-step strategy culminating in mass noncooperation to force enactment of the Green New Deal.[33] The contrast with conventional advocacy could not be more stark: this is not an attempt to work within the system, but a movement to transform it.[34]

In contrast, CCL has worked for over a decade to gain credibility and insider status in Congress while simultaneously cultivating and expanding a grassroots network of citizen volunteers from both political parties. Its volunteers are now well-known on Capitol Hill for their knowledge, diligence, and respectful style.[35] CCL held lobby meetings in November 2018 that represented a culmination of many years of effort but yielded little in the way of tangible success. In retrospect, however, these meetings laid crucial groundwork for the new bill.[36]

CCL volunteers do not lack sincerity and diligence, yet their approach has not been enough to win the day. The bipartisan introduction of an aggressive carbon pricing bill in both houses of Congress was certainly noteworthy, but the bill could not garner further Republican support in the following session. This means that more pressure from Sunrise and its allies is needed to open a larger political window, so that all climate advocates can be more successful in their push for durable solutions. With that combined effort, CCL's insider credibility may allow it to play a fruitful, mediating role.

Whether CCL's carbon pricing vision becomes part of an overarching federal climate policy or not, further growth and especially solidarity of the climate advocacy movement is crucial. A large and well-founded countermovement with several coalitions is challenging the legitimacy of climate science, and advocates against policies to mitigate climate change.[37] This movement is not limited to a few rogue groups, but includes virtually all carbon-intensive US industries, as well as the organizations that represent them in state and federal government.[38] To combat this, greater coordination among climate advocacy groups is needed. Disagreement over style and substance among progressives, Big Green groups, and other advocates limits the growth of the pro-climate movement by feeding the doubts of many Americans who—while alarmed and potentially recruitable as advocates—also believe that no solution is possible.[39]

The EcoRight

The EcoRight is committed to the idea that climate change is solvable through the application of conservative principles: limited government, free enterprise, competition, property rights, and entrepreneurship (Table 5.1). These organizations accept the validity of climate science. However, some waver at the edges of how much humans are actually responsible and on whether the impacts are likely to be as severe as the IPCC consensus suggests. Nonetheless, the groups are putting forth solutions and are clearly distinguishable from the many denialist organizations who actively sow doubt and fund misinformation campaigns.[40] In fact, the center-right Niskanen Center has an active program of climate education that is worth consulting, and its conservative credentials in many other policy areas allow it to play a constructive role in building Republican support for meaningful solutions.[41] RepublicEn similarly connects its supporters with important findings in climate science.[42]

EcoRight groups are heavily invested in market-based mechanisms for promoting clean energy, new technology development, and land stewardship. The Climate Leadership Council (CLC), led by past Republican luminaries James Baker and George Schultz, has a leading role in advancing a carbon pricing plan, which resembles the bill promoted by CCL, although it is less aggressive.[43] The CLC has spawned other advocacy groups, including Young Conservatives for Carbon Dividends and Students for Carbon Dividends, who each promote the CLC policy.[44]

Other policies promoted by many groups on the EcoRight include the expansion of carbon-free nuclear power and the development of carbon capture and storage technologies. Although these approaches are sometimes shunned by progressives, they are well-supported in most of the scientific community. EcoRight organizations are also proponents of carbon sequestration on the land—especially the expansion of tree cover—which echoes grassroots Republican support for conservation and use of the land for hunting, fishing, and other recreation. This is certainly an area where common ground with progressives has been found.

The EcoRight policies, however, typically fall short in at least two ways. First, at least some of the groups do not recognize the importance of rapidly ending fossil fuel production. For example, the American Conservation Coalition, which offers one of the more thoroughly articulated statements of EcoRight policy, endorses American energy dominance, promotes the use of natural gas as a "transition fuel," and contextualizes renewable energy development as an addition to traditional fuels.[45]

The main shortfall of the EcoRight is its failure to confront the full scope of necessary solutions and situate its policies within the context of the energy transition roadmaps. The CLC promotes carbon pricing, for example, but does not acknowledge that the tax rates it proposes are likely inconsistent with 1.5°C or 2°C worlds.[46] EcoRight groups support carbon capture and storage not only in industry settings, where it is beneficial, but also in natural gas-fired electricity generation plants, where it takes market share away from wind and solar power.[47] By developing a more comprehensive agenda and linking its policies to the roadmaps, the EcoRight could gain more credibility, even if it continues to oppose (as it surely will) any linkage of climate change solutions with the values of the Green New Deal.

3. Equity and Climate Policy

Equity is now at the forefront of the climate conversation. Progressives are a key constituency in the Democratic party, and their message is amplified by the health, economic, and racial justice crises of 2020, which continue to batter the country. Those events laid bare the structural inequities and injustices in our economy and society, opening possibilities for a new conversation about the kind of country we want America to be.[48]

There are compelling arguments for why climate change and social justice should be tackled together. High-poverty communities within the US have fewer resources for adapting, so a future world in which climate change is not brought under control would be even more inequitable than it is today. And if low-income communities, indigenous peoples, and communities of color are not fully included in climate coalitions, it will be harder to build a resilient, nationwide network that maximizes the potential for mitigation and adaptation.

There is also a strong argument based on simple fairness. Low-income communities have been more substantially affected by air and water pollution, and these and other environmental impacts are worsening as the world warms.[49] Since low-income US households produce smaller carbon footprints, they have contributed less to the problem.[50] These arguments about impact and causation also hold true on the international level in comparing developed and developing nations.

It is best to incorporate equity at the outset of climate negotiations. One way to accomplish this is by following the *Jemez Principles* for democratic organizing: a framework to achieve common understanding among diverse cultural and political groups.[51] The principles include inclusiveness and diversity at the planning table, community outreach, just and respectful mutual relationships among team members, and a process by which goals and values of other groups are incorporated into one's own work.

Incorporating the Jemez Principles into climate discussions among stakeholders of varying political persuasions is a challenge. Progressives might begin by reconciling divisions among their own constituencies. A good example is the split between labor and environmental advocates, exemplified by the refusal of AFL-CIO president Richard Trumka to support the Green New Deal in its initial form.[52] It is understandable that the interests of labor and the environment are not always in harmony; good jobs have traditionally been abundant in fossil fuel industries. Yet at a deeper level, the two communities are aligned against excessive private ownership of energy interests and the outsized role of corporate influence in politics generally. In fact, writers of the Green New Deal resolution included pro-labor language about renewable energy jobs, yet labor officials assert that they did so without consulting them.[53] When a conflict like this makes it into the public square, it signals that the Jemez Principles were not fully embodied in the negotiating process.

Is there a general framework that advocates can agree on to incorporate equity into climate policy? In the international arena, negotiations are approached with a win-win mentality: a context where an effective energy transition and the amelioration of social inequity are mutually compatible.[54] In America, this would mean asking political centrists and conservatives who emphasize economic efficiency and entrepreneurship, to recognize that building inclusiveness into climate policy does not require sacrificing these values. For their part, progressives might better recognize that solving climate change will itself reduce inequality, not least because the heavily polluting fossil fuel operations in low-income communities will be replaced by clean energy facilities. There must be room to find rapid compromise because our centuries-old arguments about the role of government will probably not be resolved quickly enough to avert a climate meltdown. As often stated, there is little point in equitable access to a train wreck.[55]

The Green New Deal

These tradeoffs are crystallized in the politics surrounding the Green New Deal resolution, which reflects the sentiments of the progressive wing of the Democratic party. The big question dividing the two sides of the Democratic tent is whether capitalism in its present form can (or should) be sustained.[56] For Alexandria Ocasio-Cortez and the 94 Green New Deal resolution cosigners in the House of Representatives, the answer appears to be no. In practice, this means that progressive advocates couple their climate advocacy with the redress of systemic injustices to frontline and vulnerable communities.[57] The goal is to achieve net-zero greenhouse gas emissions through a fair and just transition that will also create jobs, promote justice, invest in infrastructure, and secure a sustainable environment.

The critique of capitalism has its origins in an earlier Green New Deal manifesto, which emerged in the United Kingdom (UK) during the global recession of 2008. For UK economist Ann Pettifor and her colleagues, who wrote the manifesto, the environmental crisis cannot be independently solved because it was created by the existing globalized financial system. According to this view, the privatized, deregulated system currently in place provides large amounts of credit to existing industries and creates incentives to accelerate the extraction of all types of natural resources. Only when governments regain public control over the financial system will it really become possible to move in the direction of sustainability.[58] Of course, the present financial system also allows capital to be mobilized towards green energy, but this is not enough for a sharp energy transition and an end to fossil fuel dependence. Rather, it creates an "all of the above" energy world in which fossil fuels continue to be burned, which increases the likelihood of climate chaos.

The US Green New Deal that has been introduced into the House of Representatives has three essential elements that follow the UK blueprint. The first is the decarbonization of the economy to reach net-zero emissions. This part of the proposal calls for meeting 100 percent of US power demand through clean, renewable, and zero-emission power sources within ten years. Next, the proposal insists on a socially just, inclusive economy, which includes a federal jobs guarantee and new, large public investments to put people to work. Finally, the policy calls for specific attention to the frontline and vulnerable communities, which includes low-income communities, communities of color, indigenous peoples, and youth.[59]

The proposed ten-year timeframe to achieve a zero-carbon US power sector is infeasible on practical and technological grounds, but this has not stopped Green New Deal proponents from transforming the climate change conversation in Washington.[60] As mentioned above, EcoRight-affiliated Republicans have now begun to reengage the issue, all of the major 2020 Democratic presidential candidates felt compelled to put forward plans to address the crisis, and President Biden has embedded climate change policy throughout his administration. Another important trend is the increasing number of Democrats who have signed pledges to reject campaign contributions from fossil fuel companies.[61] All of this represents an enormous and very promising shift.

The question remains, of course, as to whether this new energy can be translated into meaningful policy in the crucible of Washington, D.C. politics. The main problem for progressives is that they do not presently command a majority, even among Democrats—just 95 House and 14 Senate Democrats cosponsored the Green New Deal resolution when it was introduced in early 2019.[62] The party's moderate wing also played the dominant role in returning the House to Democratic control in 2018.[63] And, of course, President Biden, a standard-bearer for moderate Democrats, triumphed over progressive candidates in the 2020 Democratic primaries.

Democratic control of both the Senate and House of Representatives in the 117th Congress (2021-2022) certainly boosts the prospects for comprehensive climate policy. Nonetheless, it remains very unlikely that the full Democratic coalition in either branch of Congress will be on board to link climate solutions with major Green New Deal initiatives in non-climate policy, such as a federal jobs guarantee. In the Senate, filibuster rules require 60-vote majorities to overcome procedural objections to debating bills, so given the likely strong Republican opposition, even modestly ambitious climate legislation may need to be passed under the complex rules of budget reconciliation.[64] Under these rules, climate legislation with significant impact on the federal budget could be packaged with the yearly bill to fund the government, which requires only a 51-vote majority. This could be an opportunity to pass far-reaching legislation, such as an aggressive carbon tax or clean energy standard, but its success will depend on progressives' willingness to compromise with Democrats from conservative states.[65]

4. Practical Strategies for Advocacy

A good deal of advocacy work is connected to conversations with lawmakers and community and business leaders, where our job is to be persuasive about the need for solutions. Let's look at some of the resources and strategies that can help maximize effectiveness in these situations.

Resources for Advocates

The job of an advocate is to build the political will to solve climate change (Box 5.1). This is accomplished in many different ways, including letter-writing campaigns, phone banks, participation in local, state, and federal administrative processes, and visible, targeted campaigns against the construction of new fossil fuel infrastructure. Climate advocates are also voices for comprehensive public education on the magnitude of the climate crisis and the range of solutions that are possible.

If all this seems daunting, rest assured that it is not necessary to go it alone. We have already mentioned the benefits of joining local advocacy networks, but many other resources are also available. The Yale Program on Climate Change Communication offers insights into the many ways to take political and personal action—one such resource is a video by climate scientist Katharine Hayhoe on the easy lifestyle changes we can all make to cut personal carbon emissions.[66] A complementary resource provided by the Union of Concerned Scientists offers tips for getting meetings with local leaders, writing effective letters, and tackling a variety of other practical matters.[67] Other materials have been made available by the Climate Advocacy Lab, a project supported by philanthropic foundations in partnership with advocacy leaders.[68] The Green Corps training program is highly recommended for advocates who wish to learn the organizing skills needed for leadership roles. Many other groups provide useful materials and networking opportunities (Box 5.2; Table 5.2).[69]

The development of local climate action plans is often a central part of advocates' work.[70] In these efforts climate modeling groups provide key resources, including estimates of local greenhouse gas emissions and projections of how much they will increase if action is not taken, as well as realistic goals for emissions reductions. For example, the US Deep Decarbonization Pathways Project (USDDPP) assists states and broader US regions by modeling decarbonization pathways, which helps ensure that enacted policies are effective and coordinated. The group has already provided deep decarbonization analysis for California, New York, and Washington states, in addition to the Northeast and Northwest regions, and the RE-AMP network of nonprofits and foundations in the Midwest.[71]

If deep decarbonization analysis has not been completed, citizen advocates can still petition municipal environment and sustainability departments to develop the plans. While it is optimal to connect local work to larger state and regional efforts, substantial guidance is available for local governments to begin development, mainly drawing on experience from California.[72] The nonprofit Rocky Mountain Institute (RMI) also offers research and systems analysis for local governments to develop transition plans.

The receding Rocky Mountain snowpack, melting glaciers, and early spring snowmelts have generated substantial apprehension among skiing enthusiasts, with forecasts for the years ahead looking ever more dire (see the Figure).[94] Many non-skiers also have good reason to be unhappy. These are the folks who hold the nearly 700,000 jobs created by the US winter sports enterprise, an economic sector that generates $72 billion per year in consumer spending.[95]

With their bottom line clearly threatened, you might think that ski resort executives would be all over this—pushing the aggressive policies needed to preserve the winter climate. Indeed, they are making their snowmaking process more energy efficient and promoting sustainability and climate awareness. But they are also adapting and changing business models to emphasize more revenue from summer activities.[96] Given the size of the threat, the scope of the executives' response is small.

The nonprofit advocacy group Protect Our Winters (POW), founded by a professional snowboarder, has stepped aggressively into this breach. POW lobbies Congress to enact serious legislation to slow climate change.[97] Its actions helped inspire the National Ski Areas Association to join with other groups in forming the Outdoor Business Climate Partnership, which pursues similar advocacy.[98] Importantly, outdoor sports are big in many conservative states. In 2018, Montana Senator Jon Tester—who supports climate change action—was reelected with the support of these groups.[99] Additionally, a POW-sponsored letter-writing campaign appears to have catalyzed a turnaround by the president of the International Ski Federation, which has now signed on to a United Nations climate initiative: the UN Sports for Climate Action Framework.[100] The work of POW and the Outdoor Business Climate Partnership illustrates the power of climate advocacy. Their actions help dispel paralysis, create hope, and build momentum.

Jackson Glacier, Glacier National Park, photographed in 1911 (left) and 2009 (right). Credit: National Park Service.

For example, local government and citizen groups in Holyoke, Massachusetts, have enlisted RMI's help with an initiative to completely transition the city's energy grid off fossil fuels.[73]

Climate Narratives

Narratives can play a decisive role in determining action on climate change. For example, in international negotiations the dominant narrative for several decades was a top-down story about high-level agreements binding all countries to specified emissions reductions. At the Paris meeting of 2015, this was replaced by a bottom-up narrative in which countries

Table 5.2: Resources for advocacy	
Group	**Resources**
Yale Program on Climate Change Communication	Climate opinion maps, social science research, public engagement with climate change issues
Climate Advocacy Lab	Intersection of climate engagement, social science and data analytics. Resources organized by state. Guide to communication.
Union of Concerned Scientists	Education on climate science, science/law interface, renewable energy, science and democracy, letter-writing opportunities
National Audubon Society	Tools for reducing personal and community impact, coping skills
350.org	Tools for facilitators, guide to constructing visual aids, using the arts to fight for climate and against injustice
Climate Reality Project	Inspiring videos, eBooks and blogs.
Green Corps	Hands-on training program for environmental organizers
We Are Still In	National network that builds community support for the US Climate Alliance efforts in international negotiations

specified their own plans. Many observers feel that the new narrative is more empowering by allowing nations to chart their own courses. The roadmaps created by individual countries might foster win-win scenarios where the transition to clean energy creates new opportunities and benefits that would not otherwise have been realized. This is the *green growth* narrative, which creates a sense of hope and optimism. It effectively counters the *doom and gloom* narrative—a story that is all about unaffordable costs and skyrocketing adverse impacts.[74]

The COVID-19 pandemic has generated some new narratives about climate change. One storyline is that getting climate change on the political agenda will now be much more difficult because governments are being spent dry fighting this pandemic and everyone is emotionally exhausted. But another, much more inspiring narrative is that Congress can act forcefully when really needed, and if it injects trillions of dollars to boost the economy, the stock market can recover while unemployed folks get the help they need. If we can do this for the pandemic, we can also do it for the climate.[75] Remembering the green growth narrative, we can also frame the climate message in terms of the many exciting growth possibilities associated with renewable energy.[76]

Narratives can also be tailored to specific audiences with particular personal values and political orientations.[77] Recent work by the evolutionary psychologist Jonathan Haidt has identified five moral axes of individual development: compassion/harming, fairness/cheating, in-group loyalty/betrayal, authority/subversion, and purity/degradation.[78] According to Haidt, progressives are motivated primarily by compassion and fairness, while conservatives also value in-group loyalty, authority, and purity. Conservative morality gives roughly equal weight to all five of these values, while progressives are very weakly motivated—if at all—by loyalty, authority, or purity.

Box 5.2: We are still in

The 2017 US decision to exit the Paris Agreement was received with shock and dismay by the international community but was met with strong resistance here at home. In response to the announcement, 24 governors formed a bipartisan coalition of states known as the US Climate Alliance. Cumulatively these states represent 55 percent of the US population and have a larger combined economy than any country, except China. These state leaders pledged to meet the US Paris Agreement commitments within their jurisdictions, even while the US was in the process of withdrawing.[101] Now that the US has rejoined the Agreement, the legacy of work from this coalition provides a powerful platform to move forward with aggressive climate initiatives.

Efforts in these 24 states are demonstrating that transitioning to a carbon-free energy economy is both feasible and profitable. Greenhouse gas emissions in these states have dropped more than twice as much as in the rest of the country, and their economies have grown three times faster.[102] States in the Alliance have tracked their progress and reported it to the global community—like at the yearly United Nations climate negotiations—which continued to implement the Paris Agreement while the US was absent. Alliance states lead the US in green bank financing, solar deployment, grid modernization, and clean transportation, as well as most other elements of the renewable energy transition. An initiative called Fulfilling America's Pledge looks in detail at how well the decentralized state and local efforts are meeting climate goals. Their reports also lay out short-term roadmaps for reaching the United States' Paris Agreement target by 2025.[103]

The We Are Still In movement is an extensive grassroots US network that supports the US Climate Alliance by reassuring the international community that US citizens—not just state governors—are committed to a green future.[104] This is important because it will take time for other nations to regain their faith that America really can provide consistent climate leadership. We Are Still In includes cities, tribes, businesses, universities, faith communities, and healthcare organizations. The movement offers a platform for individuals and local organizations to share resources and build community, and it collects information about successful local actions from across the country.

Progressive audiences motivated primarily by compassion and fairness are usually the most receptive to the climate advocates' message, although it is important to keep in mind that these values are not missing altogether from the conservative worldview. Among the other uniquely conservative values, in-group loyalty and authority do not have a particular resonance with climate change. However, purity and sanctity values do connect and are often expressed as a desire to preserve God's creation. In the right context, this offers a way to bridge differences. Religious leaders and environmentalists share a respect and concern for the Earth even if they use different language to express this value.[79] Pope Francis's encyclical *Laudato sí* is an eloquent example.[80]

Another useful framework has emerged from the Yale Program on Climate Change Communication.[81] This group's polling of the American public revealed separation into six unique audiences according to their attitudes about climate change (Figure 5.2).[82] For each group, the Yale researchers have developed a climate messaging strategy that connects to their cultural values, openness to new information, and other characteristics.[83] A key takeaway is the sharp increase in the fraction of Americans who describe themselves as "alarmed" or "concerned": in 2013 these groups made up 42 percent of the public, increasing to 57 percent by the end of 2019. Correspondingly, the fraction of Americans who are "doubtful" or "dismissive" decreased from 28 percent to 20 percent in that same time frame.[84] This is a significant shift, which was evident in the 2020 race for the Democratic presidential nomination. Climate change was high among voters' concerns and received substantial attention from the candidates. In contrast, there was far less discussion of climate between the two major presidential candidates in the 2016 elections.

Some narratives work because they connect to a concern that the audience has about a particular climate impact. For example, many progressives and conservatives share a passion for biodiversity and natural resources. This explains why public support for the Endangered Species Act has been consistently strong, and why bipartisan land and water conservation bills pass Congress with healthy margins even when high levels of partisanship block other initiatives.[85] Developing a forceful narrative about the linkage between climate change and natural resources can be an effective strategy to engage lawmakers and community leaders across the political spectrum.

It can be tempting to prepare data-laden, heavily science-based narratives, but these can backfire with some crowds. In those situations, a climate advocate may well be met with a similarly urgent yet opposing response, generating arguments that will probably be counterproductive.[86] However, one possible science-based narrative that may even reach resistant audiences is that the pandemic has alerted us to the importance of listening to experts. States that ignored expert advice and reopened too early sustained much higher numbers of infections and deaths in the summer of 2020.[87] A similar dynamic is happening with respect to investments in climate change adaptation. States projected to suffer the most financial damage from climate change are also spending the least to protect against it.[88]

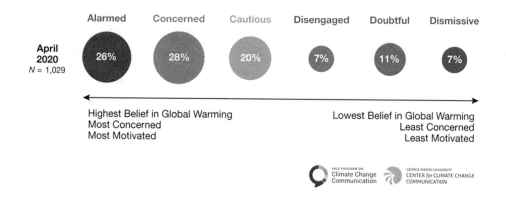

Figure 5.2: Unique audiences in the American public that respond to climate change in distinct ways. This is called Global Warming's Six Americas. Used by permission.

With receptive audiences, however, a science-heavy approach can be persuasive. For example, in these situations it becomes possible to dive more deeply into climate impacts, distinguishing how scientists have been able to pinpoint human influence both broadly and with specific extreme weather events such as heatwaves.[89] The not-for-profit environmental think tank Climate Interactive has developed a variety of tools for science-savvy advocates, which can be used to initiate broad conversations about the energy transition (Box 5.3).[90]

All meetings with important audiences should be carefully planned, drawing from these themes to create a persuasive narrative arc. One approach is to use local climate impacts as a conversation opener, and with some common ground established, then turn the dialogue to the projects that would mitigate greenhouse gas emissions or build community resilience. It can be relevant to emphasize that aggressive mitigation will lower adaptation costs over the long run, though well-executed, proactive adaptation can also lead to substantial savings and even aid mitigation efforts.[91] For example, high-yielding heat – and drought-tolerant crops introduced as an adaptation measure might in turn decrease the need for additional farmland. Given typical agricultural practices, this would likely enhance land carbon stores.[92]

Some individuals are unreachable with any climate change narrative because they are too entrenched in a denialist worldview. The best approach in these cases is to simply drop all mention of climate change and focus exclusively on the economic benefits of the green growth narrative. This has clearly worked in some deeply conservative parts of the country where mere mention of the word "climate" can be a conversation-stopper. Several leading states in wind energy development, including Texas and Oklahoma, are governed primarily by Republicans, and there are many other examples of red states embracing green solutions that provide an economic advantage.[93]

In all advocacy work, the narratives we adopt are of premier importance and deserve careful consideration. As we have seen, this requires knowing the audience and tailoring approaches accordingly. A win-win storyline from a well-prepared advocate with all the key facts in reserve is the basic formula for success.

Box 5.3: The En-ROADS Opportunity

En-ROADS is a free, user-friendly modeling program that educators, policymakers, and citizen advocates can use to understand how the climate system will respond to policy decisions. It is based on the best available science and relies on the same state-of-the-art computational methods that climate scientists use in their research. En-ROADS is valuable because it supports engaging and thoughtful dialogue, which leads to greater understanding of the climate problem, thus serving as a catalyst for action. Advocates can access En-ROADS through a common web server and have the opportunity to become designated En-ROADS Climate Ambassadors by completing a short training program.[105] Tens of thousands of people, including a number of US lawmakers, have already participated in En-ROADS climate workshops.

En-ROADS seminars can be run as either interactive workshops or role-playing games in which different groups assume identities as government and business leaders. Almost anyone who dives into En-ROADS is certain to learn a great deal about how the Earth climate system is affected by human activities in energy production, transportation, agriculture, and all other economic sectors (see the Figure).

A typical workshop begins with a projection of how the climate will evolve through the year 2100 if the world follows business-as-usual trends. Participants are then invited to make suggestions for new policies, and the program returns information about how various aspects of the climate system are altered. After some familiarity with the program has been gained, the facilitator may challenge the audience to see what it takes to limit global temperature increase in 2100 to 2°C. This exercise is particularly useful because it reveals that policies are not independent of each other but generate different outcomes depending on what else is already being done. Participants discover that there is no silver bullet to solve climate change and are able to learn why some interventions produce larger or smaller gains than they had anticipated.

En-ROADS models the global climate and energy system, but the intention is to gain insights in order to put policies into effect locally. Seminars thus often close with reflections and invitations to look at how the workshop experience can inspire new actions or shift priorities for what should be done.

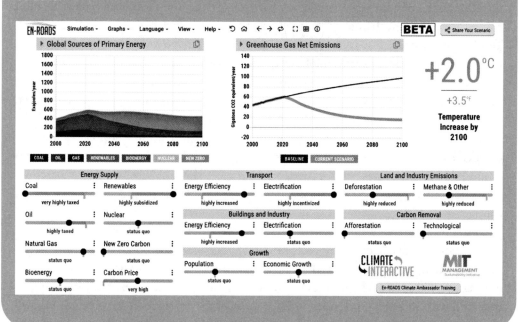

Chapter 6

Fossil Fuels:

Business & Politics

To reach the goal of net-zero greenhouse gas emissions by 2050, it will be necessary to rapidly eliminate almost all new fossil fuel infrastructure construction, retire existing assets early, and halt exploration and drilling in new areas. These actions are necessary because existing fuels do not magically disappear when new energy sources are added to the mix.[1] In the past 40 years, the share of US energy consumption from renewables and nuclear power doubled from 10 to 20 percent, yet fossil fuel use also rose (Figure 6.1). This is some progress, but at nowhere near the pace required to avoid severe damage from warming. Rather, rapid expansion of carbon-free power must replace fossil fuels while also accommodating any increases in demand from a growing economy.

There is no alternative to directly confronting the fossil fuel juggernaut. Between 2016–2019, banks accelerated lending to fossil fuel companies, with 30 percent of the $2.7 trillion in investments coming from the four largest US banks.[2] And over the next decade, fossil energy firms worldwide plan to invest nearly $5 trillion in new capital expenditures,[3] blowing well past carbon budgets for a 2.0°C world. This planned overproduction is termed the *production gap*, a consequence of worldwide failure to enact policies to contract fossil fuel supply.[4]

Fossil fuel energy companies ruthlessly exploit deficiencies in our laws and politics to earn large profits, leaving the costs of climate change to be borne by the public, yet there are many reasons to think that the era of fossil fuels is coming to a close. In this chapter, we will look at the business models and political strategies of fossil fuel firms as they deal with increasing pressures to leave most of the resource in the ground. We will also discuss how climate advocates have been working in this critical area. Fossil fuel

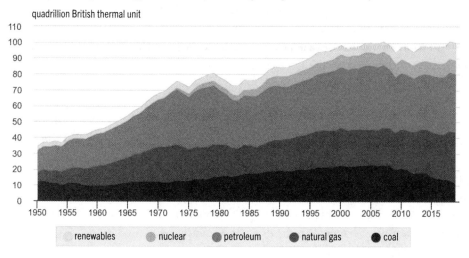

U.S. primary energy consumption by major resources, 1950-2019

quadrillion British thermal unit

renewables · nuclear · petroleum · natural gas · coal

Figure 6.1: Changes in the sources of energy used in the US over time. Despite the recent growth in renewable energy, fossil fuels still supply nearly 80 percent of US energy. Credit: Energy Information Administration.

companies depend on public permits to build new infrastructure and they need investors to fund their operations. These are two of the key leverage points where citizen action has been effective.

1. The Business of Fossil Fuels

The business of fossil fuels is really three businesses: coal, oil, and gas. However, while their end uses are distinct, the economics of oil and gas are linked because they are often produced together. Given the proximity of oil and gas resources, companies such as ExxonMobil, Royal Dutch Shell, and BP find it profitable to produce both. Completely distinct enterprises, such as Peabody Energy and Arch Resources, produce coal. Coal and natural gas are each used to produce electricity, so coal and oil/gas companies compete with each other for that large market.[5]

The Demise of Coal

Coal is a relatively small economic sector that employs about 50,000 workers, with about three-quarters of the mining operations located in Wyoming, and Appalachian and nearby states (Table 6.1). Once dominant in electricity production, the industry has suffered severe, almost certainly irreversible losses in the past decade. As recently as 2008, coal still fueled over 50 percent of US electricity generation; today that fraction is below 20 percent.

The spectacular decline did not come about from specific policies, but almost entirely from market forces. It is expensive to generate electricity from coal, and the industry has simply been unable to compete against cheaper alternatives, mainly natural gas. The rise of solar and wind energy has been a factor as well, but in 2019, these renewables still contributed less than 10 percent to the US electricity mix.[6] It is the abundance of cheap gas from fracking that is driving coal off the grid.

Three-quarters of all coal plants already cost more just to operate, compared to building and operating new solar and wind facilities—whose costs have plummeted as the industries begin to mature.[7] The financial losses have forced the three top-producing coal companies—Peabody Energy, Arch Resources, and Murray Energy—into bankruptcy and restructuring.[8] In fact, most US coal is now produced by companies that have gone through bankruptcy protection. The Unfriend Coal advocacy coalition is tracking the rapid shrinkage of the insurance market for coal operations, a significant driver of its decline.[9] Not a single new coal-fired power plant remains on the drawing board in the US today.

The demise of coal has been further accelerated by the impact of the COVID-19 pandemic. As the contracting economy lowered demand for electricity, utilities' profit margins were squeezed. And as coal is the most expensive source to run, cutting its use was the most effective cost-saving response.[10] In January through June of 2020, US coal use dropped a remarkable 30 percent compared to the first half of 2019, while consumption of both natural gas and renewables increased.[11] Idling of existing coal plants is significant because as electric utilities consider their future plans, it will be more compelling to retire those facilities outright.[12]

Coal used to generate electricity is called *thermal coal*. Because wind and solar power are intermittent, today's coal promoters argue that coal is needed to keep the electricity grid running smoothly. However, a great deal of US electricity is generated by natural gas and nuclear power, which are not dependent on weather conditions. Further, the experience of the past decade is that the sharp drop in coal generation was easily accommodated

Table 6.1: Fossil fuel production and reserve leaders			
Fossil fuel	Country Production[a]	Country Reserves	US state production[b]
Coal	1. China 2. Indonesia 3. US	1. US 2. Russia 3. Australia	1. Wyoming 2. West Virginia 3. Pennsylvania
Oil	1. US 2. Saudi Arabia 3. Russia	1. Venezuela 2. Saudi Arabia 3. Canada	1. Texas 2. North Dakota 3. New Mexico
Natural gas	1. US 2. Russia 3. Iran	1. Russia 2. Iran 3. Qatar	1. Texas 2. Pennsylvania 3. Oklahoma

[a]Country production and reserve data are from the 2020 BP Statistical Review of World Energy

[b]US state production rankings are from the EIA, https://www.eia.gov/state/rankings/#/series/47

without any ill effects on the grid.[13] For these reasons, there is no longer a rational argument for retaining the thermal coal industry. There is also no reason to entertain the use of coal as a feedstock to generate liquid or gas fuels, or to retrofit old plants with expensive carbon capture technology, as once contemplated.[14]

Coal mining cannot be fully eliminated anytime soon, because in the US about 10–15 percent of the product is *metallurgical coal*, an essential ingredient in steelmaking.[15] This high-grade coal is heated to produce *coke*, a nearly pure carbon product. The coke is then combined at high temperature with iron ore to make steel, an alloy of iron and carbon. While a small fraction of the solid carbon in the coke is incorporated into the steel, most of it is converted to carbon dioxide in the hot blast furnaces. Unlike electricity generation, where low-cost zero-carbon options are available, we do not presently have good alternatives to metallurgical coal for steelmaking.[16]

The decline in US coal stands in contrast to its growth elsewhere, especially large recent expansions of the industry in South and East Asia (Table 6.1)[17]. While coal use dropped slightly in 2019, hundreds of new coal plants are currently under construction in China and neighboring countries, with many more planned.[18] This persistence of coal use in other countries is obviously inconsistent with meeting global benchmarks for 1.5°C or 2°C worlds, making it imperative to both cancel new facilities and retire existing ones.[19] The US can exert leverage to hasten this process, but it must first reestablish strong national leadership on climate change and enact policies to accelerate decline in its own coal use.[20]

Advocates seeking to help bring about further decline in thermal coal production can act in many ways. Over 40 percent of coal production is on federal land, and climate advocates can ask federal lawmakers and administrators to stop leasing for this purpose. The Environmental Protection Agency (EPA) could also more strictly regulate coal industry pollution, which would increase industry costs and further stress financial bottom lines. Methane release from coal seams, atmospheric emissions of mercury and other toxins, and contamination of groundwater by coal ash residue are among the most severe environmental impacts of the coal industry (Table 6.2).[21] Finally, since existing market dynamics

Table 6.2: Greenhouse gas emissions from fossil fuel operations in 2018[a]		
Process	**MMT CO_2(eq)**	**Climate Pollutant**
All US emissions	5249	all
Petroleum systems	37	CO_2
Petroleum systems	36	CH_4
Natural gas systems	35	CO_2
Natural gas systems	140	CH_4
Coal mining	59	CH_4
Embedded fossil fuel carbon[b]	135	[CO_2]

[a]Data are from the EPA, Inventory of U.S. Greenhouse Gas Emissions and Sinks, 2018

[b]Refers to carbon that is incorporated into plastic, asphalt and other products, and is stored for some period of time.

are already so unfavorable, even a modest national carbon price would rapidly accelerate retirement of the remaining coal-fired electricity generation.[22]

The emissions from coal-fired electricity power plants can be reduced either by retiring facilities outright or by running them less often—as has taken place during the COVID-19 pandemic.[23] Advocates can engage with this work at the state level by participating in the regulation of electric utility companies. While the approach can be time-consuming, it is an established way to make progress—the accelerated retirement of coal-fired plants has already decreased the capacity of the US coal fleet to generate electricity by over 25 percent in the last decade. Because coal burning generates so much carbon dioxide, accelerating retirements—especially of the larger plants—by even a few years is significant.[24] The governments of coal-producing states could also be directly petitioned to end mining on public or private lands, which could be faster. However, this approach is less likely to succeed because of the dependence of state coffers on industry revenue.

Twenty-six counties in Wyoming, Appalachia, and nearby regions depend on coal for a significant fraction of local government revenue, and up to 10–20 percent of local employment. This means that financial consequences will be severe in these communities as coal shrinks further, with losses perhaps resembling those in US regions where industries, such as automaking, have departed. Policies to shrink coal should thus be accompanied by state and federal investments in hard-hit areas, including worker retraining programs.[25]

Oil and Gas Production

The fossil fuel industry is dominated by oil and natural gas companies, which directly employ nearly two million Americans and provide many more jobs in supporting industries. Natural gas makes the largest contribution of any fuel to powering the US electricity grid and has many other uses in heating and manufacturing. In contrast, crude oil contributes only marginally to producing electricity—instead, it provides the feedstock for a wide variety of transportation fuels, plastics, and other commodities.[26] Because the US leads the world in both oil and gas production, curtailment of our domestic industries will have global consequences (Table 6.1). While production could increase elsewhere to pick up some of the slack, a decisive US shift away from fossil fuels would likely spur broader movement toward carbon-free energy.

Oil and gas extraction contributes about 1.8 percent of US GDP, but this underestimates the industry's economic impact, since there is also a large infrastructure for transport, refining, distribution, and export.[27] The industry is divided into upstream companies that focus on exploration, drilling, and recovery of the hydrocarbons, as well as downstream firms that refine the crude products and deliver them to customers. In between are midstream operations that transport and store the products, including the companies that build oil and gas pipelines. The largest US firms, ExxonMobil and Chevron, are known as integrated companies because they are involved in both upstream and downstream activities.[28]

Box 6.1: The fracking revolution

Hydraulic fracturing (or "fracking") is an approach to oil and gas drilling in which water, sand, and other chemicals are injected into deep rocks at high pressure, causing them to crack (frac ture) and yield much more product as compared to conventional drilling (see the Figure). After the rocks fracture, the sand particles hold the cracks open to allow the oil or gas to flow out of the well. The fracking procedure includes *horizontal drilling*: after the well reaches a depth of 5,000–10,000 feet, the drill turns and then proceeds horizontally for another mile or more. These methods allow access to large amounts of previously unreachable oil and gas.[174] Unsur prisingly, however, fracturing rocks is also linked to a greater incidence of earthquakes.[175]

Fracking is concentrated in a few key regions of the US where oil and gas are found in fine-grained sedimentary rocks called shales. The Marcellus Shale—which runs through Pennsylvania, Ohio, and adjacent states—is a major region where fracking has proliferat ed. Texas, Colorado, Wyoming, and North Dakota are also rich in the shale oil and gas that supports fracking operations.

There is no question that fracking has brought jobs and economic growth to these parts of the US, but it comes with significant health and environmental costs. For example, a major vic tory that helped secure expansion of the fracking industry was a new law dubbed the "Hal liburton loophole", named for one of the major oil and gas field operators. Under this law, injection of the fracking mixture is exempt from regulations under the Safe Drinking Water Act (SDWA).[176] Concerns about polluted groundwater are compounded by further pollution from inadequate treatment and disposal of the fracking mixture, which flows back up the drilling pipe before oil and gas start emerging. Significant human health concerns are also emerging in communities adjacent to fracking operations—especially with respect to pregnancy and birth complications—including lower birth weights and more high-risk pregnancies.[177]

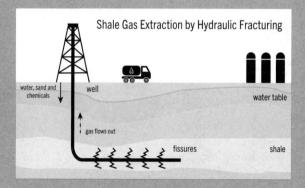

Depiction of the fracking process. The source rocks are located well below the water ta ble, but potential for drinking water contamination remains because the chemical mixtures and extracted gas must be transported across the table.

The recent past has been one of boom and bust for the oil and gas industry. The fracking revolution led to a 70 percent increase in oil and gas production between 2009 and 2019, making the US a net fossil fuel exporter (Box 6.1).[29] By 2015, two-thirds of the natural gas and half the oil produced in the US was already coming from these operations.[30] This created an economic boom in some parts of the country. For example, fracking of the

Bakken shale formation in North Dakota increased oil production by ten-fold, making the state the second largest producer after Texas.[31] In Appalachia's Ohio River region, fracking has enabled plans to develop a new national center for producing *natural gas liquids*. These are hydrocarbons such as ethane and propane that are found mixed with the methane in natural gas and are good feedstocks for plastics, chemicals, and other commodities.[32]

The fracking revolution has also made the US into a leading oil and gas exporter. A longstanding ban on oil exports ended in 2015,[33] and natural gas exports have spawned an entirely new liquified natural gas (LNG) industry. Domestic natural gas is cooled to very low temperatures, compressed to a liquid, and then shipped abroad on pressurized, refrigerated tankers (Figure 6.2).[34] US LNG exports grew from near-zero in 2015 to four billion cubic feet per day three years later—already over half the amount exported by pipeline.[35]

A major downside to fracking is an increase in carbon dioxide and methane emissions from refineries and other facilities that produce oil and gas. These emissions contribute nearly 5 percent of all US greenhouse gases, with the largest damages caused by methane leaks from pipeline connections and other infrastructure (Table 6.2).[36] One major source of methane emissions is the enormous Permian Basin, which stretches from Texas to New Mexico and accounts for significant shares of US oil and natural gas production. Satellite measurements in 2018–2019 showed that 3.7 percent of extracted Permian basin natural gas leaked to the atmosphere, which is more than enough to offset the savings in carbon dioxide emissions gained by substituting natural gas for coal in electricity generation.[37] Other leaking natural gas is flared—or burned at the wellhead—by the drilling companies, contributing to the carbon dioxide emissions from oil and gas operations (Table 6.2).[38]

Although fracking greatly boosted local economies and fossil fuel company profits, from the outset it was built on the accumulation of large amounts of debt. Despite enormous

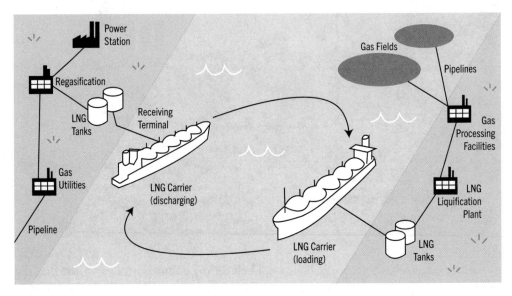

Figure 6.2: Components of the liquified natural gas (LNG) production and transportation chain. After extraction, natural gas is compressed and liquified at very low temperatures, stored in LNG tanks, and shipped in specialized carrier vessels, which maintain the fuel in that state. At the destination port, the fuel is warmed and depressurized before it is distributed by gas utilities and used in electricity power stations.

investor enthusiasm for many years, by 2019 banks and other lenders had begun to retreat from the industry in the face of decreasing prices from the oil glut and a wave of industry bankruptcies.[39] At that unstable moment, the COVID-19 pandemic hit. This caused a sharp decrease in oil consumption, forcing removal of many shale drilling rigs, and generating a number of industry mergers and further bankruptcies.[40] Going forward, the decrease in available capital for the industry, combined with the possibility that the pandemic may produce long-standing decreases in consumption, will likely advance the date of world peak oil demand.[41] Although the pandemic has also hobbled renewable energy development, there is now clearly potential to solidify the energy transition by relocating investment capital away from fossil fuels and toward the green industries of the future.

Business Strategies of Oil and Gas Firms

The global shift to carbon-free energies has pushed oil and gas companies toward developing new business models. For example, the rapid expansion of electric vehicles predicts a much earlier peak in world oil demand than previously thought, perhaps within the next decade.[42] And while natural gas is flourishing today, the rise of cheap wind and solar power threatens its new dominance in electricity generation.

For oil and gas firms, the key element of the climate change threat is uncertainty. One common fear is that the government response to future climate emergencies could be dramatic—perhaps even leading to a rapid, disorderly rush away from fossil fuels and a collapse in prices. Pressure from climate advocates and investors coupled with the plunging costs of renewables could help drive this transformation. Fossil energy firms would then face enormous losses from idle infrastructure and oil and gas reserves left in the ground, consequences of investments that should never have been made.[43] Needless to say, the recent drilling slowdown driven by the COVID-19 pandemic has heightened industry fears of this outcome.

This business dynamic provides an incentive for oil and gas companies to shift their perspectives and engage governments and environmental advocacy groups on policy solutions. Three recent initiatives illustrate this development: a move to greater investment in plastics and other products that are synthesized from crude oil, voluntary efforts to shift towards more climate-friendly operations, and support for carbon taxes.

The dedication of more resources to plastics, petrochemicals, and other commodities is intended to offset anticipated lower future demand for gasoline. Presently, only about 2 percent of crude oil is used as a feedstock for these products.[44] The embedded carbon is not immediately emitted to the atmosphere, which leads oil firms to argue that this use is more climate-friendly. However, a good portion of the carbon is released over longer time frames, depending on the nature of the product and how it is used, and the plastics production process itself generates significant emissions (Table 6.2). In addition, advocacy against plastics is increasing in response to its detrimental ecological effects, especially in the oceans.[45] To counter this, oil companies have been advised to embrace plastics recycling as a way to sustain market share against alternative products not based on petroleum.[46] But the reality is that recycling, substitution of alternative materials, and increasing public concern worldwide are already turning the tide against industry hopes that plastics will be a lifeline.[47]

Voluntary commitments by oil and gas firms to reduce their carbon footprints are increasing in response to increased public awareness of climate change. The Oil and Gas Climate Initiative is a group of a dozen CEOs from leading oil and gas firms, including ExxonMobil and Chevron, who have pledged to work toward a renewable energy transition in a manner consistent with the Paris Agreement.[48] It is certainly a sign of the times that oil and gas firms see the need to do this, but a look at the Initiative's specific pledges reveals more rhetoric than substance. There is a commitment by 2025 to reduce the carbon intensity of upstream production operations by 9–13 percent from 2017 levels, but not to include downstream emissions from the actual use of the product—where the vast majority of the damage is done (Table 6.2).[49]

Among all oil and gas companies, British Petroleum (BP) has made waves by announcing an intention to lower the carbon intensities of its products. BP has specifically stated that it will cut oil production by 40 percent and dramatically increase investment in renewable energy, including offshore wind, over the next decade. This is significant because it signals that the company now thinks its financial success depends on moving away from fossil fuels.[50] Other large European oil companies, such as Royal Dutch Shell, Equinor, and Total, also have recognized that their futures ultimately depend on reimagining themselves as energy transition firms.[51]

So far, though, these firms are the exception: in a sampling of over 100 energy companies, only 20 percent recognized that getting to net-zero emissions is necessary.[52] The largest American firms, ExxonMobil and Chevron, are among those taking only limited steps.[53] For example, Chevron adopts the position that investments in renewable energy are a "support" to its oil and gas business, and it also does its carbon accounting in terms of how much pollution is produced per unit of fuel energy. This is misleading because it allows the company to claim it is cleaning up its operations, even if overall energy production and carbon pollution are still increasing.[54]

The third initiative of oil and gas firms is to support a carbon tax proposal from the Climate Leadership Council (CLC).[55] The tax would be based on the carbon content of the fuel and would be collected from all fossil fuel companies. The subsequent revenues would be returned to American households. The CLC clearly offered the proposal with oil and gas interests in mind, given their initial suggestion—that has since been withdrawn—that adopting the tax should shield these firms from liability for climate damages.[56] Calls for a carbon tax are also coming from an increasingly widespread swath of corporate America.[57]

At first glance, it appears surprising that oil and gas companies would support a carbon tax.[58] One reason is that a well-established tax would provide them with more confidence about the shape of future markets, enabling better decision-making with respect to new investments (Table 6.3). If politics dictates that there must be a price for carbon pollution, oil and gas corporations prefer a straightforward tax to complex administrative regulations—especially if some of the tax money is returned directly into the US economy by way of dividends paid directly to American households. This has the potential to spur fossil fuel consumption by those receiving the payments. Along with the new tax, the

Table 6.3: Why oil and gas firms may support a modest carbon tax

- Generates good will against sharp reprisals in the event of climate chaos
- Long-term certainty for investments in exploration and infrastructure
- Acquire part of coal's remaining market share in electricity production
- Drives some progressive climate advocates away from supporting any carbon price
- Lack of good alternatives for petroleum in the near term
- Rebate for carbon embedded in plastics and other petroleum products
- Carbon capture and storage rebate supports continued natural gas production
- Ability to pass on part of the tax to customers

CLC plan would cancel existing regulations on some greenhouse gases while preventing implementation of others.

For oil and gas firms, the question is whether the increased business certainty and favorable public relations is worth the potential loss in profits, as well as the advantage given to renewable energy companies, who would pay no tax. Since the proposal is not yet embodied in a federal bill, it is easier for some oil and gas firms to provide support now. But the CLC is searching for members of Congress who would introduce their legislation, fleshing out the proposal with suggested provisions that would be favorable to oil and gas interests. These include a pricing plan that would sound the final death knell for the coal industry, allowing natural gas to capture part of its electricity market share. Other provisions include credits for carbon embedded in plastics and other commodities, and for any use of carbon capture technology. If the carbon tax is modest, these provisions might allow many oil and gas firms to maintain their traditional business models for decades.[59]

Fossil fuel energy companies may perceive another benefit from advocating carbon pricing: it helps to divide climate advocates. Some progressives who are predisposed against market-based solutions to climate change are even more likely to oppose carbon pricing when it looks like the fossil fuel companies support it.[60] We can be sure that the firms are well aware of this dynamic. In this sense, the CLC plan provides a perfect harbor—it costs oil and gas firms nothing to support it and allows them to look good by appearing to support climate change mitigation. And their expression of this support may make it less likely that the strategy will ultimately be realized.

Advocacy: Divestiture

The beginnings of investor retreat from the oil and gas industry provide climate advocates an opening to accelerate the trend. A major argument for divestment is that oil and gas stocks are overvalued because investors are committing funds under the assumption that the known reserves will actually be developed. This assumption creates a large financial *carbon bubble* estimated at $1–4 trillion, an amount comparable to losses from the recession of the late 2000s.[61]

The carbon bubble represents the dollar value of investments in assets, such as oil and gas reserves and physical infrastructure, that would never yield returns if there is a rapid flight away from fossil fuels. If this flight indeed occurs, it would amount to bursting the bubble, with potentially devastating short-term impacts. Yet a delay in the transition could be worse, as continued overinvestment in the interim period would create an even larger pool of these *stranded assets*. Also relevant are *stranded liabilities*, which represent the cost of retiring the long-lived infrastructure. Oil and gas companies are legally required to plug and abandon unused wells, but a burst bubble could leave them lacking the assets to do so. This creates a danger that the costs will be transferred to taxpayers.[62]

In addition to the rising concerns about overvaluation, oil and gas investors are also casting wary eyes on developments in the insurance industry. The large reinsurer Swiss Re is looking hard at the increased payouts necessitated by recent climate disasters, especially in North America. It is pulling back from insuring coal companies, and other insurance firms are following suit.[63] Similarly, the insurance giant The Hartford has stopped insuring developers of Canadian *tar sands*, a dirty mix of sand, water, and hydrocarbons that requires energy-intensive processing to yield useful oil.[64] Coal and tar sands each have cleaner, lower-cost competitors, so it is not surprising that insurers are pulling out of these fuels first. But the concern of oil and gas investors is that this so-far limited pullout is the leading edge of a movement that may eventually include reluctance to insure more conventional fossil fuel production as well.

Central players in the financial industry are now clearly moving to highlight the risks of climate change. In the summer of 2020, more than three dozen pension plans, financial institutions, and other asset managers urged the Federal Reserve and Securities and Exchange Commission (SEC) to explicitly integrate climate change across their mandates.[65] The Commodity Futures Trading Commission, an independent regulator, has already produced a report warning about climate change risks.[66] In response, congressional Democrats are introducing legislation that would require the Federal Reserve to conduct climate change stress tests, and companies to disclose climate-related financial risks to the SEC.[67] Large money managers such as BlackRock Inc. and the Vanguard Group—both of whom wield considerable influence through shareholder votes at oil and gas companies—clearly have an opportunity to move the ball.[68]

One way that climate advocates already act on divestiture is through investor groups that engage top emitting companies. A leading initiative is Climate Action 100+, which pressures a broad range of companies to decrease emissions, improve their governance structure for climate change risks, and strengthen their climate-related financial disclosures.[69] Climate Action 100+ includes hundreds of investors that manage tens of trillions of dollars in assets, working with firms that together represent 80 percent of corporate emissions. Engagement at these companies takes many forms, including use of shareholder resolutions to force accountability when necessary.

Climate Action 100+ also works with other groups that provide technical data on emissions and systems for climate-related disclosures. For example, the nonprofit charity CDP runs a global disclosure system of the environmental impacts created by companies, cities, and other entities.[70] Seventy percent of the companies targeted by CDP have set long-term emissions reduction goals, although most of these are not yet Paris Agreement-compliant. Inspiration for these companies could come from high-profile actions taken by technology and social media giants like Apple, Microsoft, Facebook, Google, and Amazon. All of these firms have made significant pledges to reduce emissions, and Google has also stated that it will stop building customized technology for the oil and gas industry.[71]

Green banks offer a powerful approach that complements divestiture. These financial institutions have a mandate to facilitate private investment into climate resiliency and mitigation. Either the US federal governments or the states can create such banks, which has already been done in California, Hawaii, and some Northeastern states. In the 116th Congress (2019-2020), House and Senate Democrats introduced legislation for a National Climate Bank as part of the CLEAN Future Act, an overarching energy bill.[72] Green banks around the world have mobilized billions of dollars in capital and are able to give particular attention to making sure that investments make a difference in underserved communities.[73] The efforts of green banks are bolstered by studies showing that investments in renewable energy have outperformed fossil fuels in both the US and Europe.[74]

All of this is just a beginning. Other actions can include pressuring lawmakers to require that climate change be considered when regulators, such as the Federal Reserve Board, set the rules for how financial institutions must operate.[75] For example, these banking regulators can raise the bar for how much capital is needed for financial firms to invest in fossil energy. Advocates can also join a direct-action movement that pressures financial giants to stop funding the fossil energy companies.[76] Major investment banks are already declining to finance new oil and gas drilling in Alaska. This may render new leases in the pristine Alaskan National Wildlife Refuge made by the Trump administration wholly irrelevant.[77]

There are also opportunities to join campus divestment campaigns to pressure more universities to follow the University of California's decision to shed its fossil fuel holdings.[78] State pension funds are another large source of fossil fuel investments that can and should be pressured to locate their assets elsewhere. Finally, all of us can educate and empower our communities about ways to engage individually, so that our personal financial lives are consistent with commitments to a green world. Stock indices and other investment aids that track renewable energy firms and other climate-friendly companies are available to assist in this effort.[79]

2. Fossil Fuel Politics

Fossil fuel firms have enormous leverage in US politics, which they exert by contributing to the election campaigns of candidates—overwhelmingly Republicans—that support the industry agenda. In the 2018 US midterm elections, fossil fuel companies donated at least $359 million to their favored candidates, over 13 times more than was spent by renewable energy interests.[80] Some of this spending is by trade associations such as the American Petroleum Institute and Western States Petroleum Association. These associations promote the industry through advocacy and lobbying for friendly policies, such as ending the ban on oil exports. They also fight against legislation that combats climate change.[81]

Fossil fuel interests have taken full advantage of the US Supreme Court's 2010 *Citizens United* decision, which held that corporations have free speech rights.[82] *Citizens United* led to a second decision that eliminated limits on donations to *political action committees* (PACs), which pool contributions from many members to support specific candidates or legislation. The term *super PAC* is now used to denote PACs that accept unlimited contributions. Self-styled "social welfare" groups, which do not have to disclose donors, have become large super PAC contributors. These groups allow for sham philanthropy from libertarian, pro-fossil fuel activists, such as the Koch brothers empire, memorably termed the "Kochtopus." Investigative journalist Jane Mayer, author of the invaluable *Dark Money*, writes that as a result of *Citizens United*, "…the American political system became awash in unlimited, untraceable cash."[83]

Today virtually every climate protection measure of significance is heavily lobbied against by fossil fuel companies and their allies. For example, in 2018 climate advocates in Arizona successfully introduced a ballot initiative, Proposition 127, that would have required 50 percent of the states' electricity generation to come from renewables by 2030. In response, the parent company of Arizona's monopoly electric utility, Pinnacle West, spent over $30 million to defeat the measure. This is on top of the millions that Pinnacle West previously spent to help elect fossil-fuel friendly candidates to the Arizona Corporation Commission, which regulates electric utilities. Electric utilities in Michigan and Ohio have also accepted large amounts of dark money to enrich fossil fuel interests.[84] These are among the tactics that climate advocates need to be aware of and fight against.[85]

Subsidies

Oil and gas companies seek to maintain and expand an array of subsidies that have increased their profits while undermining emissions reduction goals. The subsidies make it easier for fossil energy firms to carry out new exploration and contribute substantially to "carbon lock-in." The economic viability of the fossil fuel industry is being preserved well beyond its proper lifetime by advantages built into the US tax code over many decades. Conservative estimates of direct subsidies that the US provides to fossil fuel companies amount to $20 billion per year.[86]

Fossil fuel subsidies come in many forms. For example, government payments that assist low-income households with heating costs are a form of subsidy, as are the cut-rate deals that the US provides to lease federal lands for fossil fuel production. Many other production subsidies are embedded in the tax code in complex ways. For example, oil companies benefit from provisions that allow accelerated depreciation of new capital investment (such as drilling new wells), as well as tax laws that allow special treatment for the depletion of existing wells. Both deductions cost the US government billions per year in foregone revenue. There are other examples.[87]

Decreasing or zeroing out these subsidies is conceptually equivalent to imposing taxes on carbon emissions, since both would increase the costs of fossil fuel production. Recent analysis suggests that, worldwide, eliminating subsidies would provide about 25 percent of the energy-related emissions reductions pledged under the Paris Agreement. Removal of tax incentives for new investment could be particularly beneficial in shifting those dollars away from fossil fuels and towards new, renewable energy projects.[88] A US bill that would eliminate fossil fuel subsidies is the Clean Energy for America Act, introduced by Oregon Senator Ron Wyden in May of 2019.[89] This legislation would also restructure the tax code to incentivize renewable energy over fossil fuel use.

Promotion of Climate Change Denialism

Fossil fuel companies have not been content to solely use their wealth in influencing US politics. For many decades, these firms have deliberately misrepresented and sown doubt about climate change science. Their intention is to paralyze the political system's ability to act on climate change by presenting the science as "uncertain." These tactics are lifted directly from the playbook of tobacco companies, which misled the American public for decades about the hazards of smoking. Indeed, some of the same individuals who helped direct the tobacco misinformation campaigns have also been central in the attempts to undermine climate science.[90]

Fossil fuel companies and large US automobile manufacturers are key players in a far-Right political influence network founded by the Koch brothers, further disbursing funds to a plethora of denialist groups.[91] This is established by thousands of pages of internal company documents—spanning decades—that have been brought to light. Led by ExxonMobil, the companies orchestrated a campaign of disinformation, in part by funding a large number of denialist organizations to do their bidding.[92] The general strategy has always been to exaggerate the uncertainties in the scientific findings, thus undermining the scientific consensus. The fossil fuel companies have not offered alternative views to explain warming, but merely seek to do enough to block decisive action.[93]

The evidence for ExxonMobil's disinformation campaign is particularly thorough. Scientists who formerly worked at or were consultants for Exxon have given congressional testimony that Exxon knew global warming was occurring.[94] During the 1970s and 1980s, the company conducted legitimate experiments that were published in peer-reviewed scientific journals. But Exxon's message changed dramatically when it began paying for editorial-style advertisements in The New York Times. In these writings, which targeted policymakers and the general public, Exxon instead expressed serious doubt that anthropogenic global warming was real or solvable. These documents were intended to influence the national conversation away from taking bold climate action.[95]

The analysis of ExxonMobil's publications, internal documents, and editorial advertisements leaves no doubt that their writings fostered key aspects of climate denialism. Despite the knowledge of their own experts and the overwhelming consensus of climate scientists worldwide,[96] Exxon persisted in arguing that climate change is not happening, not serious, not human-caused, and not solvable.

This repeated, exaggerated emphasis on uncertainties in climate science data undermines the public's general understanding of science. Science never delivers absolute proof but rather provides its conclusions with an appropriate assessment of confidence. We have no choice but to make policy decisions in the face of some uncertainty, yet the fossil fuel company campaigns effectively demand an unreachable standard of absolute certainty, where any doubt at all is reason enough for inaction.

The misinformation campaign also benefits from the media perspective that all sides of a controversy should be discussed. If scientists publish work that demonstrates that the Earth is warming dangerously and others contest this, the media seemingly have an obligation to report both sides of the argument. But what the media (and public) fail to understand is that both sides have *already* been examined within the expert court of scientific peer review. The possibility that the observed warming is a natural phenomenon has been thoroughly looked at, and the overwhelming scientific consensus is that it is not.[97] Media outlets and political parties lack the expertise to challenge this. It is necessary to trust the experts on this matter, as we do in so many other walks of life. It is telling that Democrats and Republicans alike trust the findings of biomedical science and support robust funding for biomedical research. Only when the profits of a key wealthy constituency are on the line does a scientific issue become politicized.[98]

The misinformation campaign has been supremely successful in convincing Republican party elites to make climate change denial into a political litmus test—a job made easier by the root psychology of the Trump era, in which fake news and conspiracy theories were elevated to the status of "alternative facts."[99] Polls show that climate change is now the single most partisan issue in the country, an unsurprising finding considering that over half the climate change videos on YouTube promote misinformation.[100] There is little doubt that quite a few congressional Republicans know that climate change is a real threat and would like to act, but they lack the courage to speak up because of the threat to their political fortunes.[101] Consequently, as long as conservative elite opinion toes the fossil fuel industry line, there may be little hope for bipartisan agreement on major legislation.[102]

Another tactic of fossil fuel companies and their trade associations is to launch specific attacks on the credibility of prominent climate scientists. For example, Dr. James Hansen, whose voice for climate advocacy is among the strongest in the scientific community, has been subjected to a storm of politically motivated vitriol for years.[103] Dr. Hansen's 1988 testimony at a Senate hearing was a key event in initially moving the issue of global warming into public consciousness. More recently, climate scientist Michael Mann—another leader in the field—was attacked for his "hockey stick" depiction of the sharp increase in global temperature in the past few decades. Having no substantive argument against this, climate change denialists have resorted instead to baseless, *ad hominem* attacks on Dr. Mann's credibility.[104]

As mentioned above, fossil fuel companies fund a large number of organizations that promote climate change denial (Table 6.4).[105] Although these groups all systematically promote disinformation, many are incorporated as 501(c)(3) educational nonprofits.[106] The tactics of the Heartland Institute should be called out in particular. In 2017, they distributed a booklet to school teachers and children alike titled "Why scientists disagree about global warming."[107] The false title alone reveals that the booklet's agenda had nothing to do with education.

Table 6.4: Five organizations promoting denial of climate science[a]

Group	Activities
Heartland Institute	Promotes climate change misinformation in public schools, hosts a yearly denialist conference on climate change, hosts "America First" energy conference to promote fossil fuels, hired German citizen Naomi Seibt as an "anti-Greta" to promote denialism
American Legislative Exchange Council	Provides resources on climate change denial to state legislatures, promotes legislation falsely claiming climate change is beneficial, promotes a "teach both sides" approach in public schools
Heritage Foundation	Provides biased, superficial analysis on fossil fuels and renewable energy to media outlets promoting denialism, editorializes against Paris agreement, characterizes environmentalists as "extreme"
CATO Institute	Disputes global warming science and questions the basis for taking action, published fake addendum to a US national climate assessment, argues for plenty of time to solve climate change
Americans for Prosperity	Finances attacks on ballot measures to tackle climate change, works to obstruct EPA regulation of climate pollutants, attacks the RGGI cap and trade carbon pricing program

[a]Some information is from Greenpeace, Koch-funded climate denial front groups, https://www.greenpeace.org/usa/global-warming/climate-deniers/front-groups/

In the last few years, the climate denial rhetoric has begun to soften, driven by Democratic recapture of the House of Representatives coupled with recognition by Republicans that climate change denial hurts them among some key voter demographics. The new language no longer denies that climate change occurs, or even that humans might play some role, but rather argues that we do not know how much humans are responsible.[108] There is little evidence, however, that this tactical shift reflects a real change of heart. The emphasis on uncertainty remains. The policies entertained now by congressional Republicans—while representing a possible opening after many years of party-line denial—still do not recognize the need to sharply cut fossil fuel use.[109]

Advocacy: Lawsuits

Actions taken by oil and gas companies to deliberately mislead the public about climate change are now the basis for an expanding number of legal actions across the country.[110] Inspiration for many of these lawsuits comes from successful action taken against the tobacco industry. The tobacco cases established that the public interest could prevail by showing that the companies' misleading claims about cigarette smoking endangered public health. In a final master settlement, tobacco firms had to pay nearly $250 billion, fund anti-smoking campaigns, and accept restrictions on advertising and lobbying.[111]

Fossil energy companies' use of the same strategies to manufacture doubt about climate change leaves them open to legal attack due to claims of climate change-related injuries to human health and infrastructure. These lawsuits are important simply as an expression of the rule of law, all the more so because of Trump administration attacks on democratic norms. As with the tobacco cases, they can potentially lead to judicial orders compelling fossil fuel firms to pay for the damages they have caused to both human health and the environment.

Many of the cases against fossil fuel companies rest on laws that protect citizens against consumer fraud and nuisance. They are often brought by states and local municipalities and ask the companies to pay money damages to compensate them for climate-related costs. For example, the cities of San Francisco and Oakland are asking for oil and gas companies to pay a fund that would be used to build the infrastructure necessary to protect against sea level rise.[112] It is important to clearly see what is at stake in these cases. Costs of climate change will certainly skyrocket in the next few decades. The key question is: who must pay? If the fossil energy firms evade liability, then the costs will be borne by governments and ultimately financed by US taxpayers.

Another line of legal challenges is based on the notion that fossil energy companies have defrauded their investors. Here, the most high-profile actions are suits in Massachusetts and New York against ExxonMobil, arising from the company's decision to actively mislead the public about climate science (as described above). In December 2019, a New York court cleared ExxonMobil by ruling that the state had not demonstrated that it was the company's intention to mislead shareholders. However, similar suits are in progress in Massachusetts, Minnesota, and the District of Columbia. The Minnesota and DC cases are notable because they are the first to explicitly highlight the disproportionate impact of climate change on low-income communities and communities of color.[113] Exxon's victory in New York may be hollow, as it has already lost big in the court of public opinion.[114]

The most famous climate change lawsuit is *Juliana vs. United States*, advanced by the nonprofit environmental law firm Our Children's Trust.[115] A group of 21 youth plaintiffs is suing the federal government because it has known about the destructive effects of climate change for a half-century, yet has chosen to do almost nothing about it. This has endangered the lives and livelihoods for all of today's youth. In legal terms, this is known as "state-created danger." *Juliana* is also based on the *public trust doctrine*, a legal theory with ancient roots in Roman law. The basic idea is that natural resources should be held in trust for present and future generations, and the role of government is to act as a fiduciary to ensure that the trust is preserved. This stands in sharp contrast to conventional doctrine in environmental law, where governments issue permits that allow manufacturers to pollute the air and water within specified limits.[116]

The *Juliana* plaintiffs are not asking for monetary damages. Instead, they demand that the US institute a national Climate Recovery Plan that is in line with the best available science and embodies principles of social justice.[117] The case inspired a proliferation of similar lawsuits across the nation and the world.[118] The *Juliana* plaintiffs have won victories in a number of different venues, and, as of this writing, are entering into mediation with the Biden administration to reach a settlement.[119] The court's argument was that climate change should be solved by the executive and legal branches of government, not through a judicial order. Yet this does not address claims that the government's willful failure to address climate change violates Fifth Amendment principles of due process and equal protection under the laws,[120] *Juliana*, and scores of other legal actions, persist.

3. The Role of Government

In this section we will consider how government can limit the fossil fuel industry agenda, especially oil and gas. A key challenge is that much of the US oil and gas resource is privately owned. Owners of underground mineral rights can explore and produce the fuels or lease these rights to third parties in exchange for royalty payments. Fossil fuel markets are then largely influenced by private contracts—not by any overarching government management plan. So federal, state, and local governments have mainly influenced fossil fuel production indirectly—for example, by imposing environmental regulations or passing laws to tax or subsidize the industry.[121]

Nonetheless, understanding how governments can influence the fossil fuel industry agenda is essential for advocates. The executive branch can use the authority conferred by environmental laws to strictly regulate the emissions of climate pollutants. However, it can also undermine the intent of those laws—as the Trump administration demonstrated—by turning administrative agencies into servants of polluters. At the federal level, government has authority over interstate commerce and foreign trade, which affects import and export of fossil fuels and trafficking among the states. The potential use of emergency powers and the control of fossil fuel production on federal lands are other areas where the federal government can exert influence. In contrast, the construction of new fossil fuel infrastructure is largely in the hands of the states. As we will see, this is an area where climate advocates have been especially effective in limiting the reach of the fossil fuel industry.

Federal Emergency Powers

The federal government can enhance its abilities to curtail fossil fuel use and speed the renewable energy transition by declaring climate change a national emergency. Emergency orders operate as a form of executive power, rooted in Article II of the US Constitution. The executive branch, led by the president, has the responsibility to "take care that the laws are faithfully executed."[122] The use of emergency powers was advocated by several 2020 Democratic presidential candidates and would mirror the efforts of 68 US cities, towns, and counties in taking such actions, including New York City, Miami, and Austin.[123] Beyond the constitutional authority, a federal law known as the National Emergencies Act also provides the president with discretion to declare emergencies subject to congressional review.[124] In practice, Congress has almost never exercised this oversight, even though dozens of such emergencies are now in force. No limits on what qualifies as an emergency are specified in the text of this 1976 law.[125]

A consequence of the National Emergencies Act is that no fewer than 123 specific statutes presently on the books can be activated by a president upon declaration of a national emergency; more statutes would be activated if Congress also so declared.[126] Environmental law professor Dan Farber has done a preliminary assessment of how these laws might be used if climate change were to be declared a national emergency. He finds that the potential exists for the president to direct military construction funds toward building renewable energy infrastructure. Other possibilities include suspending oil leases, restricting carbon-heavy transportation, and limiting international economic transactions involving fossil fuels.[127]

How well would this approach work? In 1952, the Supreme Court invalidated President Truman's seizure of private steel mills to meet Korean War needs, ruling that his nationalization orders were unconstitutional.[128] This precedent suggests that attempts to commandeer private energy companies under the banner of a climate emergency might also fail judicial scrutiny. Top-down attempts to curtail fossil fuel use faster than alternatives are available would also be very disruptive. Drawdown of fossil fuels—even under emergency conditions—requires comprehensive planning to minimize economic and human costs of the energy transition.

Despite these limitations, presidential and congressional declarations of a climate emergency would still provide an empowering framework mandating that climate impacts be considered in all federal legislative and executive actions. Employing the vast resources of the federal government to build infrastructure has precedent in the construction of the interstate highway network and would be more likely to be implemented without a successful legal challenge. For example, federal resources might be dedicated to constructing a robust new national electricity grid that can function reliably without any fossil fuel generation. This would dramatically accelerate the renewable energy transition while spurring further action at the state and local levels.

Additionally, a climate change emergency declaration would likely make it easier for the federal government to halt new oil and gas leasing and ban fracking on federal lands and could also facilitate early termination of existing leases. Although only about one-quarter of oil and one-eighth of natural gas production is on federal land,[129] such actions would send a forceful signal to fossil fuel markets, possibly hastening divestiture. The federal government might also phase out gas- and coal-fired electric power generation from facilities that it owns. For example, the Tennessee Valley Authority is a corporate agency of the US government and owns and operates no fewer than eight coal-fired power plants. A climate-friendly US administration could accelerate the retirement of these plants while investing instead in carbon-free energy resources.[130]

Basics of Environmental Regulation

The modern era in US environmental law began in the early 1970s, with the enactment of major federal statutes such as the Clean Air Act and Clean Water Act. In many cases these laws have been successful in cleaning up pollution from industry, transportation, and other sources. For example, the initial Clean Air Act mandates dramatically improved air quality in large cities across the country.[131] In the 1980s, when it became clear that industrial emissions were generating acid rain and opening the Antarctic ozone hole, new sections were added that successfully addressed these problems. Similarly, the nation's navigable waterways are much cleaner because of Clean Water Act restrictions that limit pollutant concentrations to acceptable levels.[132]

Federal and state legislatures write the environmental laws, but administrative agencies put them into effect. The influence of agencies is so widespread that they have sometimes been dubbed the fourth branch of government. Agencies implement laws by writing and publishing regulations, which are the actual detailed rules that the regulated industries then have to follow. The Clean Air Act, for example, mandates that ground-level ozone from smog must be reduced enough so that the health of vulnerable populations is not compromised. It then falls to a federal agency to decide what the precise ozone level should be and what controls it must impose to get there.[133]

Environmental advocates should be aware of what agencies do because of a fundamental process known as *notice and comment*.[134] When federal or state agencies write, amend, or rescind rules, they must publish them as proposed actions in the daily journal of the federal government—the *Federal Register*—or in analogous state records, which begins a period of public comment. When this time concludes, the agency must respond to comments made. Sometimes, comments are so extensive or negative that the agency will reformulate its rule to take them into account. For major projects, there may be a number of commenting opportunities with more than one agency involved. Often, both state and federal approvals are necessary.[135]

Agency regulations can be challenged in court. Challengers typically argue that the agency overstepped the boundaries of the enabling law, so the rule must be invalid because it does not conform with the intention of the legislature. Since the law is written broadly, these challenges are not difficult to put together. During Republican administrations, agency regulations that weaken oversight of the fossil fuel industry have commonly been challenged by progressive environmental groups. Conversely, fossil fuel trade associations often sue against stricter environmental rules enacted by Democratic administrations. For major federal environmental laws, most key decisions are made by the District of Columbia Court of Appeals. Occasionally, an issue is important enough to be further reviewed by the Supreme Court.

Specific details of agency regulations can make a big difference in how well human health and the environment are actually protected. It is important to note that the process of writing, rescinding, and amending rules takes place without congressional involvement. Changing the law itself is a much more public undertaking compared with revising agency regulations that most US citizens are unaware of. Over 100 environmental regulations were targeted for rollbacks by the Trump administration, a major assault that increased industry profits at the cost of public health. Enforcement of regulations, another key agency function, was also cut back.[136] In contrast, congressional Republicans did not attempt to rewrite the Clean Air Act or any other environmental law, even when they controlled both the House of Representatives and the Senate in the 2017–2018 term.

Federal regulation of air pollutants—including greenhouse gases—is the province of the Environmental Protection Administration (EPA), under the authority of the Clean Air Act. The EPA made extensive efforts to regulate greenhouse gas emissions during the Obama administration. Direct regulation of greenhouse gases was made legally possible by a landmark 2007 Supreme Court ruling, which held that such gases are pollutants within the scope of the Clean Air Act.[137] This opened the door for EPA to make an *endangerment finding*, an analysis demonstrating that greenhouse gas emissions are indeed harmful to human health and the environment.[138] At that point, regulations became required.

The most famous greenhouse gas regulation was the Clean Power Plan, an Obama administration effort to reduce greenhouse gas emissions from the electric power sector. The target in the law was a modest 32 percent reduction in carbon dioxide emissions from electricity generation by 2030, compared to 2005 levels. Under the regulation, EPA required states to set efficiency standards that would reduce carbon dioxide emissions from coal-fired power plants, as well as substitute both natural gas and renewables for coal as the primary fuel for electricity generation.[139]

We will never know how well the Clean Power Plan might have worked, because it was stayed indefinitely by the Supreme Court in early 2016 following lawsuits by fossil fuel interests. Under the Trump administration, the EPA substituted a much weaker regulation—the Affordable Clean Energy (ACE) rule, which was in turn successfully challenged and invalidated by a US federal appeals court in January 2021.[140] The fate of the Clean Power Plan shows a key weakness of regulations: with no need to consult Congress, they are easily overturned by an administration hostile to their intent.

There are also real concerns that the lawsuits against the Clean Power Plan, had they gone forward, might have been at least partly successful. While the 2007 Supreme Court ruling allowed that greenhouse gases could be regulated in principle under the Clean Air Act, this was not a blank check that any particular proposed regulation would be valid. The Clean Power Plan specified building blocks for state compliance that would have required gas-fired power plants to replace coal and renewable generation to replace both of these. But this begins to remodel the entire electricity generation industry, which is arguably outside of what the Clean Air Act empowers EPA to do. Had a federal court found that this was indeed the case, it would have ruled that at least part of the Clean Power Plan was illegal.[141]

The hard reality is that the Clean Air Act was not written to address a problem with the wide scope of climate change, which requires reconfiguring entire industries. Employing the law to specifically regulate low-abundance greenhouse gases—like methane and hydrofluorocarbons—is probably the best way that it can be used to limit climate change. Advocates for reducing greenhouse gas emissions by way of regulations would do better to press Congress to pass new legislation that specifically targets carbon dioxide emissions and empowers the EPA to carry it out. For example, in early 2020 the Clean Economy Act was introduced in the Senate with the support of 32 Democrats. It would specifically authorize the EPA to use the Clean Air Act to meet a net-zero emissions target by 2050.[142]

Federal regulations are only one option in a broad arsenal of possible approaches to control carbon dioxide emissions from electric power plants, manufacturing, and other sources. As mentioned above, regulations have been effective in curtailing conventional air and water pollution, but they are not always the least expensive way to solve environmental problems. Regulations work in the same way for all manufacturers or power producers in a specific industry, so they tend to limit individual firms' abilities to innovate and find cheaper, more efficient solutions.[143] Higher costs to comply with regulations are then inevitably passed on to consumers. Other approaches to controlling carbon dioxide emissions include regional and federal carbon pricing plans, and, for the electric power sector, state or federal standards requiring the use of increasing amounts of renewable energy to power the electricity grid.[144]

Regulation of Methane Emissions

Regulating anthropogenic methane is challenging because many of its sources are not easily tracked—like industry smokestacks—but instead come from agricultural soils, livestock, waste management operations, and landfills.[145] The amount of methane leaking from oil and gas production operations, pipeline transport, and industry end uses is also difficult to estimate because there are so many places where leaks can occur. However, measurements by independent scientists suggest that EPA is substantially underreporting methane emissions, in part because its new methods for making the estimates are overlooking very large superemitting sources.[146] This is consistent with paleoclimatology data on global atmospheric methane sources measured from ice cores, which shows that the portion of methane coming from anthropogenic sources today is substantially higher than previously thought.[147]

Little was done to control methane emissions in the US until the Obama administration released a comprehensive strategy in 2014.[148] This included specific regulation of fossil fuel sources and landfills and voluntary programs for emissions from other sectors. Following this blueprint, in 2016 the EPA issued regulations that required new solid waste landfills, as well as oil and gas facilities to implement the "best system of emissions reduction" available. This term refers to specific preferred emissions reduction technologies that are both highly effective and ready to be implemented. For oil and gas operations, the rules included restrictions on flaring and venting gas at wellheads, in addition to enhanced leak detection and repair of pipelines and other infrastructure.[149] At the same time, Obama's Bureau of Land Management (BLM), an agency within the Department of the Interior, issued a rule requiring gas capture and leak repair at wellheads and facilities on federal land.[150]

The Obama regulations were a significant step forward, although they did not address the copious methane emissions from coal mines.[151] While both the EPA and BLM rules were repealed by the Trump administration, the repeal of the BLM rule was challenged and subsequently invalidated by a federal court, which ruled that the process followed to enact the repeal was "wholly inadequate."[152] The repeal of the EPA rule in August, 2020, meant that oil and gas companies were no longer required to install technology to detect and fix methane leaks.[153] However, several states have stepped up to impose their own controls and the omnibus federal budget and Coronavirus relief bill passed at the end of 2020 contains new provisions requiring companies to use advanced technology to find and fix methane leaks.[154] Reversal of Trump administration methane abatement rules is likely to be a high priority for EPA regulators in the Biden administration.

Among the other anthropogenic greenhouse gases, ozone has been regulated by the EPA since the 1970s because of its detrimental health effects and buildup during urban smog episodes. Phasing out fossil fuel use will sharply decrease or eliminate excess "bad" ozone from the lower atmosphere. Approaches for managing nitrous oxide and hydrofluorocarbon emissions will be covered in later chapters.[155]

State and Local Actions

Most onshore and near-offshore oil and gas production is regulated by individual state governments, which have the power to approve exploration and development of new resources. States also regulate the construction of fossil-fuel powered electricity generation and manufacturing plants, and any pipeline infrastructure within their borders. The power of states to regulate this production based on human health and environmental considerations means that they could prevent fossil fuel extraction entirely on both state and private land, although this has not yet been attempted.

A number of coastal states restrict oil and gas drilling in their offshore waters. Ten states now have such restrictions, which cover a majority of the Atlantic and Pacific shorelines and part of the Gulf Coast.[156] Several Eastern states have also specifically banned fracking on land, though they do permit other operations. Most notable is New York, which has chosen not to extract its sizable shale gas resources. Citizen advocacy played an important role in Governor Cuomo's decision to institute the fracking ban.[157]

State-level control helps advocates oppose new fossil fuel infrastructure, since state legislatures and regulating agencies are easier to access for lobbying efforts. The challenge, though, is that states extract royalties when private firms produce fossil fuels. Thus, dependence on this income leads them to acquiesce to the industry agenda. An extreme case is Alaska, where revenues not only boost local and state governments, but are also distributed to citizens as an annual dividend that typically ranges from $800–$2000. This provides a strong incentive for Alaska's politicians to invite fossil fuel companies to continue their activities.[158] Alaska state oil revenues of $2.4 billion in 2018 dwarf both the immediate costs of climate change on state infrastructure and of relocating communities threatened by flooding and erosion.[159]

The perceived benefits of fossil fuel income to Alaska and other fossil fuel-rich states are deeply entrenched in their political cultures and will be difficult to directly dislodge. However, one can argue that states should try to rely less on fossil fuels for revenue because energy markets are volatile and long-term trends favor renewable resources. These notions play on the idea that fossil fuel revenues are ultimately uncertain.[160] Another approach is to advocate for ways to replace this revenue with income derived from a climate-friendly policy, such as state or federal carbon pricing. One may also remind lawmakers and business leaders that their states can still earn royalties by developing solar or wind industries.

Local municipalities can help limit fossil fuel expansion by banning construction of new infrastructure. Portland, Oregon, set a precedent in 2016 when its city council banned new bulk fossil fuel terminals with capacities to store over 2 million gallons. The council also blocked expansion of existing liquified natural gas and petroleum fuel terminals. Portland did this by making changes to its zoning code, a practice that could certainly be adopted by other cities across the country.[161]

Another approach gaining momentum is the enactment of local ordinances that would prevent installing natural gas infrastructure in most new buildings.[162] A number of California municipalities—including Berkeley, San Jose, and Santa Monica—have led the way on this. The policy is also being considered in various Northwestern cities, with initiatives for greening new buildings proposed at the federal level as well.[163] Incentives such as federal and state tax credits would help promote this initiative more broadly and extend it to replacing existing infrastructure. This would follow precedents that use tax credits to incentivize home solar panels and electric vehicles.

State preemption is sometimes used to thwart local efforts to limit fossil fuel infrastructure. In this context, *preemption* is when state law displaces local ordinances.[164] For example, a state legislature may enact a law expressly stating that fracking or new fossil fuel infrastructure may not be banned by any municipality. This has already been done in Texas and Arizona.[165] Preemption battles can be anticipated when progressive cities attempt to enact bans in fossil fuel-rich states that are controlled by conservative majorities. However, preemption can sometimes be successfully challenged on the theory that significant local interests should outweigh state authority.

Transporting and Exporting Fossil Fuels

A significant amount of infrastructure is needed to move fossil fuels around. Coal is shipped on trains and barges; oil is transferred mainly through pipelines and trucks; natural gas travels through pipelines. Except for metallurgical coal—which has a sizable export market—most coal mined in the US is used within the country, and little is imported.[166] However, with the lifting of a longtime US oil export ban in 2015 and the rapid recent build-out of liquified natural gas (LNG) export terminals, robust international trade now exists in both oil and gas.

The US federal government plays a key role in regulating interstate transport of natural gas through pipelines and in constructing LNG export terminals. However, its role in regulating coal and oil transport is much more limited. This is understandable for coal, which does not move through a fixed pipeline infrastructure. But there is also no general federal approval process that must be negotiated to establish where oil pipelines should be sited. To be sure, pipeline developers may need to get federal approval in certain cases—for example, if their routes traverse federal land or burden protected environmental or cultural resources. But in general, the decision-making power for coal and oil transportation, as well as storage hubs, rests at the state and local levels.[167]

This state of affairs has been exploited with great success by climate advocates in the Pacific Northwest. There, a thin green line of resistance is all that prevents coal, oil, and gas producers in the interior of the Western US from building a massive infrastructure of fossil fuel storage hubs and export facilities—often on pristine, environmentally-sensitive land.[168] The successes in blocking these projects have been remarkable. One by one, no fewer than six proposed coal export terminal projects fell to a wall of activist resistance in the past decade.[169] Similarly, an oil train terminal in Vancouver, Washington, on the shore of the Columbia River, was defeated by opposition from citizen advocates supported by local and state lawmakers.[170] Of course, these successes build on a strong foundation: an aware citizenry that elects environmentally progressive lawmakers to state offices. But the same principles of local control apply throughout the country. Consequently, as the impacts of climate change increase, more resistance will appear.

A combination of local, state, and federal laws regulates natural gas pipeline and export infrastructure. The Federal Energy Regulatory Commission (FERC) is the US agency with authority to permit pipelines and terminals.[171] Gulf Coast states—the hub of the nation's oil and gas infrastructure—regularly support FERC approvals of an ever-denser pipeline and export terminal infrastructure. However, on the West Coast it is a different story. Project after project has been defeated by the intensive efforts of citizen advocates and wary, environmentally conscious state governments. No LNG export terminal has yet been constructed anywhere on the California, Oregon, or Washington coasts.

LNG export terminals exemplify major federal actions that must comply with a key overarching environmental law: the National Environmental Policy Act (NEPA).[172] This requires preparation of an Environmental Impact Statement (EIS), which can run thousands of pages long. The EIS analyzes how environmental resources are impacted by a proposed new project and what can be done to mitigate these impacts. In July 2020, the Trump administration issued a rule designed to weaken NEPA by expanding the number of projects that would be exempt, and by eliminating requirements that certain kinds of impacts termed "cumulative" and "indirect" would not need to be considered. This is being challenged by environmental groups, and like the BLM methane rule repeal, the challenge is likely to be successful, because the new rule flouts the clear language of the law.[173]

NEPA remains the law. For huge endeavors like LNG export terminals, the list of potential impacts is enormous, including air and water resources, biological habitats for endangered species, indigenous culture resources, and many others (Box 6.2). The public commenting processes afford climate advocates numerous opportunities to intervene. And local, state, and federal agencies all must provide permits. With determined opposition, these projects can be held up for a long time. The efforts are worth it because blocking LNG export can lead would-be foreign buyers of natural gas to reconsider whether they need the fuel at all. In this way, US climate advocates can have global impact.

Box 6.2: The Jordan Cove LNG Terminal

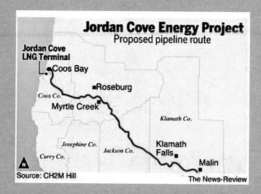

Proposed route for a new gas pipeline to connect the LNG export facility at Coos Bay, Oregon, to the pipeline network for natural gas in Oregon and nearby states.

The Jordan Cove Energy Project (JCEP) is a quintessential example of the complex web of federal and state regulation that is involved in permitting an LNG export facility. JCEP was initiated in 2007 as an LNG import facility; after that proposal was denied, the Canadian energy company Pembina—the initiator of the project—resubmitted the proposal to FERC in 2013 as an LNG export hub. This change was driven by the explosion of fracking in the US and the consequent glut of domestic natural gas that became available for export (see Box 6.1).

JCEP is slated to be located in Coos Bay on the sparsely populated Oregon coast. Because the modest local electric grid is insufficient to meet the facilities' large power requirements, a new, dedicated natural gas electricity generation plant to drive liquefaction of exportable gas is required. In turn, this would require construction of a new gas pipeline hundreds of miles long, connecting the LNG terminal with the network of existing natural gas pipelines in Oregon and neighboring states (see the Figure).

The project was denied by FERC in 2016 but was reopened a year later. Throughout its lifetime, JCEP has been steadfastly opposed by environmental activists, Native American tribes, landowners whose property would be encroached upon, tourism, fisheries and recreational interests, as well as various state agencies. In 2019, the Oregon Department of Environmental Quality (DEQ) denied a key permit because it decided that the project could not meet Clean Water Act standards. Nonetheless, FERC has issued a final Environmental Impact Statement under NEPA, and the Department of Energy has approved the project. Jordan Cove's ultimate fate remains in limbo and will be resolved in the courts—perhaps on the question of whether the federal government can preempt a ruling made by an Oregon state agency.[178]

Chapter 7

Carbon Pricing

The sharp decline of tobacco consumption in America demonstrated something crucial: if you want to discourage the use of a product that is pervasive and poses an enormous risk to human wellbeing, then use the power of government to make it more expensive. This strategy applies to kicking the fossil fuel habit as well.[1] Fossil fuels proliferate because we burn them without paying for the damage they cause to human health and the environment. The remedy is not to immediately ban the fuels—which we obviously still depend on—but to make them expensive enough for low carbon alternatives to rapidly take over.

Carbon pricing provides a powerful foundation for other climate policies to function more effectively. A great deal is known about how to design effective programs, because the approach has been employed in many countries around the world for several decades.[2] In this chapter, we will examine the two main pricing policies and discuss how they can be established or expanded in the US today. We will also explore carbon pricing politics and look at how citizen advocates can contribute to putting new pricing laws into effect.

1. Why Price Carbon?

The famous idea of the *tragedy of the commons* helps us understand why carbon pricing is essential (Figure 7.1).[3] The classic example is shared grassland: a *commons* used by ranchers to graze cattle. The grassland regenerates naturally over time as long as it is not grazed too intensely. But if the number of cattle grazing is too high, the ecology of the grassland collapses. Then no animals can graze, since the shared resource has been destroyed.

This tragedy arises because each rancher acts in his or her own interest to expand the number of cattle they own, thus earning greater profit. Even if ranchers recognize the

Tragedy of the Commons

Shared Resource

Sustainable Use

Depleted Resource

Use of the commons is below the carrying capacity; all users benefit.

Use of the commons is approaching capacity, and it remains sustainable only if sharp limits on additional cattle are imposed.

The commons is depleted, possibly irreversibly, and can no longer sustain grazing even at low levels.

Figure 7.1: Classic depiction of the tragedy of the commons. The left panel is analogous to the Earth before emergence of human civilization, with stable, natural levels of atmospheric CO_2. The center panel represents the situation today, with CO_2 levels near sustainable limits. The right panel depicts the commons after overgrazing has caused collapse, analogous to the state of Earth's climate if greenhouse gas emissions are not rapidly curtailed. Figure inspired by a similar depiction; see https://un-denial.com/2018/01/07/on-garrett-hardins-tragedy-of-the-commons/.

ecological limit of the grassland, it is still in their interest to allow their cattle to graze unrestrained. After all, even thinner, underfed cattle still earn some profit. Because every self-interested rancher makes this same calculation, the grassland collapses.

There are many examples of this principle in contemporary life. For our purposes, we can think of a clean atmosphere with a low, ecologically sustainable carbon dioxide level as the global commons. The self-interested actors are individuals, corporations, and nations who earn profits by burning fossil fuels and clearing stored carbon from the land. Our life-sustaining atmospheric commons is threatened today by escalating carbon dioxide levels that have driven the Earth out of energy balance, creating climate havoc.

There are two potential solutions to this classic problem. First, the land could be privatized so each rancher owns his or her own section. Ranchers then have the incentive to maintain grazing within the ecological limits of their piece of land. However, this is obviously not an option for our atmosphere since emissions from anywhere on Earth become mixed together through natural circulation.

The other solution is shared governance. Ranchers agree to limit the number of cattle each can graze and they vest power in some governing entity that polices the commons, ensuring compliance. Our problem today is that shared governance of the atmospheric commons has, so far, been too weak to overcome the selfish interests and ideologies of the individual players. Thus, strong resistance exists to an effective governance solution that has been identified by the great majority of experts in climate modeling and economics.

In the language of economics, the ability of individuals and groups to freely pollute the atmospheric commons with carbon dioxide is called a *negative externality*, a cost absorbed by parties who had nothing to do with creating the pollution. It represents a market failure because prices of fossil fuels and carbon-rich commodities do not include the high cost of environmental damage from carbon dioxide emissions. A corrective is therefore necessary. The most cost-effective, efficient solution is to tax the polluters, making them responsible for the damage they cause. Carbon pricing thus embodies a fundamental principle of environmental law: the polluter pays. Of course, the price must be set high enough to drive emissions down until a sustainable, global atmospheric commons is reached.

In any carbon pricing approach, the cost of polluting is tied to how much carbon dioxide is emitted. With the policy in place, burning coal will cost more than burning natural gas because coal generates more carbon emissions for the same amount of energy production. In contrast, solar and wind installations emit no carbon dioxide during their operations and are not taxed at all. Thus, when considered over the life span of a facility, renewable energy plants become much cheaper to operate as compared to fossil fuel installations. The higher the carbon tax, the better solar and wind looks to investors and governments.

Other benefits abound. Unlike other policies, carbon pricing generates revenue, which can be invested in green technologies. Revenue may also be used to advance social equity goals, reduce other taxes in ways that may stimulate the economy, or return dividends to highly impacted communities and (potentially) to all households in the jurisdiction that adopt the program.[4] As we will discuss later in this chapter, revenue allocation is subject to a great deal of political debate, since it implicates fundamental values about what government should or should not do.

Carbon pricing uses markets to slow the development of new fossil fuel infrastructure and it complements efforts to block individual projects administratively.[5] Indeed, a well-designed aggressive carbon price will reduce the need for that advocacy, as many projects will no longer go beyond the drawing board. In addition, an aggressive carbon price will spur production decreases and early shutdowns of existing fossil fuel-driven facilities. The IPCC puts it this way: "…policies reflecting a high price on emissions are necessary in models to achieve cost-effective 1.5°C pathways."[6]

Well-designed carbon pricing programs are not a panacea. Where alternatives are readily available—such as wind and solar power to generate electricity—an aggressive carbon price is indeed highly effective in bringing about emissions reductions. It will drive electricity producers toward low-carbon alternatives and stimulate governments to accelerate the deployment of carbon-free power. However, even a fairly high carbon price will not quickly reduce emissions from transportation because good carbon-free alternatives are not yet established. The industry and building sectors are expected to respond better than transportation, but not as well as electric power.[7] Clearly, the renewable energy transition will be best enabled by combining carbon pricing with other policies that target specific parts of the economy.

The power of carbon pricing in greening today's electricity grid makes it a key policy element for the first decade of the energy transition. Electrification of end uses, such as transportation, is much less desirable where the grid is still mainly powered by fossil fuels. But if carbon pricing rapidly boosts solar and wind power in the 2020s while the infrastructure for electric vehicles is simultaneously improved, we will be poised for large emissions reductions from transportation in the 2030s.[8]

Table 7.1: Carbon tax *versus* cap & trade		
Program feature	**Carbon Tax**	**Cap & Trade**
Transparency	High	Low
Administrative cost	Low	High
Certainty of emissions reductions	Moderate	High
Carbon price	Fixed	Fluctuates depending on supply and demand in auctions
Regulatory strategy	Rising tax in proportion to fuel carbon content	Number of emissions allowances decreases over time
Coverage	Comprehensive throughout the economy	Excludes small emitters
Potential to stimulate green technology innovation	High	Moderate to high
Time to implement after law is enacted	Short	Long
Potential for political gamesmanship	Low to moderate	Varies depending on how allowances are allocated

A sharply escalating carbon price can be the single most effective driver of the renewable energy transition. It can turn the rudder of the great ship of state in the green direction, away from fossil fuels. Once this turn is irrevocably made, the ship will still need expert steering from policies that restore farms and forests, enable energy-efficient manufacturing, and reimagine urban centers powered by clean fuels. But none of that can work effectively unless fossil fuel burning experiences a sharp decline. Aggressive carbon pricing is critical because it will price fossil fuels largely out of existence, thereby allowing all the other policies to do their work.

2. Approaches to Carbon Pricing

There are two ways to price carbon (Table 7.1). First, government can impose a direct *carbon tax* on producers and importers of fossil fuel. The amount of carbon dioxide generated by burning each type of fuel is well-known, and the tax is then imposed in dollars per ton of emissions. For administrative simplicity, the tax is best imposed as far upstream as possible—at the port of entry for imported fuel, the coal mine, or the initial refinery for oil and gas.[9]

The carbon tax then propagates through the economy (Figure 7.2). Imagine that you own a factory that buys natural gas to drive its processing machinery. With a carbon tax in place, you will pay more for the gas because your supplier—having already paid the tax—is going to pass the increased cost on to you. In turn, you will charge your customers more for the finished product since it cost you more to make it. Similarly, if your home heating

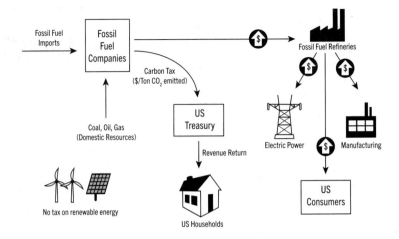

Figure 7.2: How carbon taxes work. Fossil fuel producers or importers increase the prices for their products in response to the tax (top and right side). Higher energy prices are passed on to consumers, but in fee-and-dividend plans this is compensated by return of the tax revenue to US households (center). Carbon-free energy sources, such as sunlight and wind generation facilities, pay no tax because they do not emit CO_2.

furnace runs on natural gas, in the absence of other provisions (like use of tax revenues to provide homeowner rebates), you will have to pay more for that fuel.

An important feature of any carbon pricing program is that the price, even if it starts low, should be increased each year. In part, this just ensures that the effect of the price is not diluted by inflation. But increasing prices are also appropriate because emissions farther in the future will occur in a more stressed climate system. The cost of an additional ton of carbon dioxide emissions (the *marginal cost* of emissions) is higher in the future than it is today.[10]

Ongoing price increases make it easy to see how the tax can incentivize innovation. Your factory faces escalating fuel costs into the foreseeable future, so you will naturally look at how to increase efficiency. In time, you will investigate whether you can convert to cheaper electricity generated by a greening grid, or perhaps to the use of untaxed renewable biogas instead of fossil natural gas. In fact, you will lose customers if you do not make these changes because you will have to keep raising prices to stay in business. At the same time, entrepreneurs in green technologies are attracting more investment dollars since the escalating carbon price has effectively created new buyers, such as your factory, for those innovations.

We should note that carbon taxes can't directly specify the extent to which emissions will be reduced. Certainly, the strong economic incentive will lead to less fossil fuel use, but the precise amount is impossible to know in advance. To address this, economic modeling can be used to come up with a pricing plan aggressive enough to drive the desired emissions decrease. The law should then require emissions monitoring to assess whether these decreases actually occur. If they fall short, another provision can require steeper tax increases in the following years to catch up with the desired reductions.

The second approach to carbon pricing is popularly known as *cap and trade*. Here the government sets a cap on the total amount of carbon dioxide that can be legally emitted during a given year. It then requires regulated businesses, such as large industrial facilities

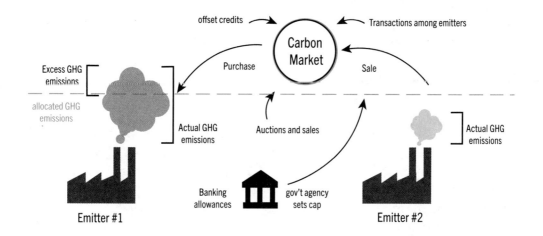

Figure 7.3: How cap and trade policies work. The horizontal dotted blue line represents the emissions limit imposed by a government agency (bottom center). Emitter #1 is unable to sufficiently reduce emissions and must purchase allowances from Emitter #2 in a carbon market. Allowances are surrendered upon emission, and the total number of allowances (the "cap") decreases over time.

and electric power plants, to purchase or otherwise acquire emissions allowances. The allowances are then returned to the government when the fossil fuel is burned (Figure 7.3).

A large manufacturer would typically be permitted to emit one metric ton of carbon dioxide upon surrendering one emissions allowance. Only a limited number of emissions allowances are created to match the desired emissions cap. Over time, the government steadily decreases the number of allowances in circulation. In contrast to carbon taxes, the required emissions reductions in cap-and-trade programs are set directly in the law. This provides greater certainty that targeted reductions will be met.

A central feature of cap and trade is *emissions allowance trading*. The government is mainly concerned with the total reduction in emissions within its jurisdiction, not about which particular covered facilities reduce emissions by greater or smaller extents. An efficient facility that is able to reduce emissions might then hold more allowances than it needs, while poorly run plants unable to curtail their carbon emissions would not hold enough to keep profitably operating. In that case, the poorly run plant can purchase allowances from the efficient facility in a carbon trading market. Allowance prices fluctuate in the short term because they are determined by supply and demand, but trend upward over longer periods because fewer allowances stay in circulation.[11] This is opposite to a carbon tax, where prices are precisely set in the law and emission levels vary in response (Table 7.1).

3. US Emissions Trading Systems

The US currently has well-established, independent cap-and-trade programs in California and the Northeast—both of which have been effective in driving modest emissions reductions in their respective regions. This section explores the nuts and bolts of these programs, illuminating their many similarities and differences (Table 7.2). New states are welcome to affiliate with either program. If carbon pricing appeals to you, working for your state to gain affiliation is a great way to move us toward a greener future.

Cap and Trade in California

California established itself as a climate leader in 2006 with the passage of a landmark law, Assembly Bill 32 (AB32).[12] AB32 created a framework requiring that greenhouse gas emissions be reduced to 1990 levels by 2020. Upping the ante, in 2016 Governor Jerry Brown signed even more ambitious legislation—Senate Bill 32 (SB32)—which committed the state to reduce emissions 40 percent below 1990 levels by 2030.[13] And SB32 is itself a stepping stone within a larger vision for 80 percent emissions reduction by 2050.

The cap-and-trade program was initiated in 2012 after six years of intensive planning by the lead agency, the California Air Resources Board (CARB). It is just one of many legislative and regulatory initiatives implemented under AB32 and continued under SB32.[14] Other laws include mandates for renewable electricity, incentives for energy efficiency, climate-friendly transportation fuels, reductions in short-lived climate pollutants (such as methane), and land use policies to sequester carbon.[15] Climate change is clearly very well integrated into governance in California, which provides a model for other states. In 2017, California met AB32's requirement to reduce emissions back to 1990 levels before 2020. From 2004 to 2017, emissions were reduced by over 12 percent.[16]

Table 7.2: California and RGGI cap & trade programs		
Program feature	**California (WCI)**	**RGGI**
Participating states and provinces[a]	California, Quebec	NY, NJ, MA, CT, VT, NH, RI, ME, DE, MD, VA
Coverage	Economywide	Electricity sector only
Emissions reduction timetable	40% below 1990 levels in 2030	30% below 2020 levels in 2030
Allocation of emission allowances	Many distributed free	Few distributed free
Offsets	Allowed but limited	Allowed but limited
Price collaring[b]	Broad range of allowed prices, high price cap	Narrow range of allowed prices, low price cap
Potential for political gamesmanship	High	Low
Open to new affiliates	Yes	Yes
[a]Pennsylvania is considering RGGI affiliation		
[b]Price collaring refers to setting both a price floor and price ceiling for allowances		

Why did California want carbon pricing when other laws covering specific economic sectors were already being enacted? The added value comes from the legal guarantee that overall emissions must be reduced to target levels. This gives confidence to commerce and manufacturing interests, who can then act knowing that the program will operate for at least the next decade and has passed legal scrutiny.[17] The revenue from selling allowances also helps fund the other programs, boosting the always-limited state budget. Some of the funds are directed to low-income communities to soften the impact of higher energy costs. Revenues have been increasing over time, with $1.4 billion in projects implemented in 2018 alone, and a cumulative investment of $5.3 billion by 2020.[18]

Let's look at the California program with a focus on the features important for any cap-and-trade initiative (Table 7.2). First, the overall scope of the law is central. SB32 is economy-wide and captures over 80 percent of all emissions from carbon dioxide, methane, nitrous oxide, and fluorocarbons. The law covers industrial sources, all large facilities that produce electricity, and distributors of fossil fuels. Small plants emitting under 25,000 tons of carbon dioxide per year are not covered, giving a break to small businesses and reducing administrative costs.

The timetable for emissions reductions is another crucial feature. The cap on allowances was reduced only modestly in the first phase of the program, but sharper reductions under SB32 began in 2020. By 2030, the mandated 40 percent cut in emissions from 1990 levels means that only about 200 million allowances will be issued in that year, compared to 334 million in 2020 (Color Plate 17).[19] These reductions are consistent with carbon budgets for a 2.0°C world, though they are unlikely to meet the more stringent 1.5°C scenarios.[20]

The way emissions allowances work is a bit complicated, but a basic understanding is necessary to gain a good grasp of the program. The regulated industries are required to turn in one emissions allowance for every ton of carbon dioxide they emit, but how do they acquire these allowances in the first place? It may surprise you to learn that CARB simply gives away many allowances for free. Some are handed out to the large electric and natural gas utilities on the condition that they give rebates to consumers to compensate for rising prices. This approach can be seen as progressive—the company still pays, and the funds go to households.

However, CARB gives other free allocations to manufacturing industries in proportion to how well they have reduced emissions historically, as well as how exposed they are to competition from outside of the state.[21] Free allowances are rationalized on the grounds that they prevent *leakage*, which is departure of the companies to states that don't price carbon. Leakage illustrates a key tension in cap-and-trade programs: it is desirable to keep regulated industries in-state, but distributing free allowances also opens the door to preferential treatment for well-connected, politically-favored firms.[22]

Another way that regulated industries acquire allowances is through auctions. In this case, the allowances must be paid for, and like an auction for fine art, they go to the highest bidders. But unlike an art auction, CARB imposes rules to constrain the range of prices. There is a *price floor* of $10 per ton, which increases by 5 percent each year. No allowances can be sold for less than that, and this ensures that they cost enough to preserve the incentive to reduce emissions. There is also a *price ceiling*, set at $65 per ton in 2021.[23] If the ceiling is reached during the bidding, CARB must offer to sell additional allowances at that price. The reasoning is that too-high prices would be passed on to consumers, inviting political blowback that could jeopardize the program. But unfortunately, adding more allowances in the auction is effectively the same as raising the emissions cap, which permits more carbon pollution.

Finally, regulated companies can also buy allowances in secondary private markets.[24] These markets permit companies that can't keep emissions low enough to buy allowances from greener firms that have them in excess, because they have been able to cut their emissions below what the cap requires. Secondary markets also allow outside speculators to buy and sell allowances with the intention of making profits. The need to monitor these secondary markets adds to the administrative cost of the program. Companies can also buy and retain allowances for future use. This is problematic because it allows them to put off making emission reductions.

The complex nature of cap-and-trade programs is a major reason they face criticism. There is one other major issue: California's cap-and-trade program permits the limited use of *carbon offsets*. These are credits that regulated companies get by investing in emissions reductions that are not inside the pricing program. The idea is that companies are allowed to emit more greenhouse gas pollution in California if they pay for reductions elsewhere. Properly done, offsets can be a good way to spur investment in difficult areas, but they have also been quite controversial (Box 7.1).

Because the California cap-and-trade law coexists with and enables many other programs, it is difficult to specifically credit it with the emissions reductions achieved.[25] However, given the overall success, it is not unreasonable to give the program a fairly good grade so far. Looking ahead, the decisive issue is how California will manage the much more ambitious plan that is now the law through 2030. The state deserves credit for putting in place an emissions reduction plan that, if continued through to 2050, is aggressive enough to do its part for a 2.0°C world. Ultimately, the program can only work if the regulated industries find ways to continually and cost-effectively cut emissions. Making this happen will require vigilance and willingness to amend the regulations as needed.[26] It will not be long before we see if the state can pull it off.

The Northeast's Regional Greenhouse Gas Initiative (RGGI)

The other success story in US carbon pricing is a regional New England-based program, sometimes affectionately known as "Reggie." RGGI is an association of eleven states with a common carbon trading market (Table 7.2). As a cap-and-trade program, RGGI shares many commonalities with the California law, including a limited use of offsets and both a price floor and price ceiling for emissions allowances.[27] We will focus here on its distinctive elements and prospects over the next decade.

RGGI only covers greenhouse gas emissions from electricity generation, so its scope is substantially narrower than California's program. This is not a major flaw since carbon pricing is presently most effective in reducing emissions from the electricity sector. Under RGGI, each state writes its own detailed regulations, including which allowances to

Box 7.1: Carbon offsets

A fundamental premise of cap-and-trade programs is to achieve emissions reductions at the lowest possible cost. This is the primary function of trading markets, which at their heart are simply a way to identify which covered facilities can reduce emissions most cheaply. But what if there are even cheaper ways to reduce emissions that are outside the scope of the program? This is where carbon offsets come in. If a covered facility can meet its obligations by identifying low-hanging fruit elsewhere—emissions reduction opportunities that cost less than lowering emissions from its own operations—then why not allow it to save money by funding the outside activity instead?

Offsets are an example of well-meaning policy that can fail because bad actors game the system. For example, if the offset money goes to fund emissions reductions programs that would have happened anyway, then there is no added benefit. Or perhaps a funded program in a faraway nation may have difficulty getting off the ground and is never confirmed by independent oversight. Accurate, transparent measurements of emissions may also be hard to come by. All these problems have occurred in offset programs under the European Union's emissions trading system at a cost of billions of dollars.[75] Closer to home, offsets in the two US cap-and-trade program have been criticized for permitting extra local pollution—of all kinds—from covered facilities, especially those located in low-income communities and communities of color. Offsets, then, clearly also implicate equity concerns.

Some argue that these potential problems with offsets, coupled with the costs of properly overseeing an offset program, should be enough to exclude them. Yet offsets do offer a mechanism for finding and funding new opportunities for emissions reductions that could lead to valuable innovation. California and RGGI permit offsets to satisfy 3–6 percent of emissions reduction obligations, and each program has identified well-defined project categories, such as forestry and agricultural emissions, and methane capture from mines and landfills.[76] The solution to the bad actor problem is oversight to ensure that offsets are not misused. One obvious answer is to limit qualifying offset programs to those that can be properly inspected and certified. Regulations can also require that any use of offsets be contingent on demonstrating that air and water pollution from covered facilities are well-controlled and do not burden historically disadvantaged communities.

auction and how many to allocate without charge. States are also free to specify how most of the revenue is spent. However, cohesion is maintained because all individual state regulations are based on an RGGI Model Rule that sets a unified framework.[28] States also participate in common auctions and secondary markets, and work within a framework that caps total regional emissions, with a schedule set for reducing the cap over time. RGGI benefits from a strong regional culture of political collaboration that has made it robust over time, capable of weathering changes in partisan control of individual states.[29]

In contrast to California, RGGI states have sharp limitations on free giveaways of emissions allowances, instead allocating a large majority through price bidding at auctions.[30] This greatly decreases the ability of electricity providers to petition for special treatment. It has also generated robust income, which most states have dedicated to programs that bolster energy efficiency, promote renewable energy, or provide households with assistance on electricity bills. Improvements in efficiency have led to modest decreases in electricity demand in RGGI states, despite robust economic growth. This has helped offset increased energy costs.[31]

The RGGI states have experienced dramatic decreases in carbon emissions since the agreement was established in 2005. As in California, the cap-and-trade program took effect in the context of other programs, so credit for the reductions is not straightforward to assign. Nonetheless, carbon dioxide emissions from the electricity sector in RGGI states declined by nearly half from 2005 to 2018, and the existing cap has been set to decrease 30 percent further by 2030 (Color Plate 18).[32]

We should note that while hydroelectric, solar, and wind generation have significantly increased in RGGI states, the greater part of the emissions reductions has come from substituting natural gas for coal. Coal use for electricity generation in RGGI states has plummeted to just a few percent of the total energy mix,[33] but going forward, a major challenge is to tackle the remaining emissions from natural gas. This could happen both by replacing natural gas with renewable power and by adding carbon capture and storage (CCS) as a way for natural gas-powered electricity plants to comply with the program.[34]

Advocacy: Expanding the State and Regional Programs

Both the California and RGGI cap-and-trade initiatives are open to including new states and territories. In fact, the California program is known as the Western Climate Initiative (WCI), since several other US states expressed interest in participating at an early stage of its development. Although those states ultimately did not choose to be included, the Canadian province of Quebec has been a WCI member since 2008, and Ontario was briefly affiliated for part of 2018.[35]

California and RGGI-associated states regularly solicit public input when developing and modifying program rules, which offers an opportunity for advocates to make their voices heard. This is likely to prove quite important—especially in California—where SB32 now demands unprecedented steep emissions reductions through the 2020s. For example, advocating for a higher allowance price floor, or for reducing the number of free allowances, would increase the incentive for emitters to shift more aggressively to renewables.

If you live in a state not yet associated with a cap-and-trade pricing program, working toward affiliation can be a powerful way to make a difference. The process of developing a new state program is complex, potentially lasting several years and involving many stakeholders. Getting educated about the law is essential, especially about issues such as which firms are to be regulated, the number of free allowances available (if any), and whether and how to allow offsets.[36] In Oregon, engaged citizen advocates helped pass a bill in the state House of Representatives to join the WCI in late spring of 2019.[37] Republican walkouts prevented Senate Democrats from passing the bill, but advocates' efforts led to a broad Executive Order from the Governor's office, boosting climate mitigation efforts across the state.[38] Advocacy work can pay off with unexpected benefits, even in the teeth of an apparent failure.

Neither WCI nor RGGI imposes a geographic restriction on which states may affiliate. RGGI even offers an outline for how new states may affiliate by developing their own emissions trading programs.[39] One role for advocates is to meet with the heads of the relevant state energy and environmental agencies since these individuals must ultimately connect with their counterparts in already participating states. In addition, since the process is lengthy and involves many stakeholders in the business, government, and non-profit sectors, advocates can engage these groups and seek common ground in developing the legislation. An excellent guide to communicating about carbon pricing is available to assist with this.[40]

4. Prospects for Carbon Taxes

Carbon taxes to address climate change have yet to be enacted in the US, but they have been established in about twenty other regions and nations. The leading light among these taxes is British Columbia's (BC) program, which was enacted in 2008 with a price of about $8 per ton of emitted carbon dioxide. Today the price is $30 per ton. BC relies almost entirely on zero-emissions hydropower to generate electricity, so its carbon tax may be seen as a test for how well the policy drives changes in sectors of the economy other than electric power generation. In these other sectors, studies show that carbon dioxide emissions dropped 6–16 percent more in BC than in other Canadian provinces, while economic growth stayed strong.[41]

The BC program provides a model that shows how a well-functioning pricing policy might be designed. A few other carbon tax laws have also been successful. The United Kingdom imposed a high price floor (which functions like a tax) on top of its participation in the European Union's cap-and-trade program, and this is credited with accelerating emissions reductions from electricity generation. And when Australia enacted a $20 per ton carbon tax in 2012, it also generated a sharp emissions drop from its electricity sector (though the tax was repealed by the next government, sending emissions back up again).[42]

Carbon taxes have clearly been more politically difficult to enact than emissions trading in the US. However, there are several reasons why we should not give up on them. Most obviously, a federally imposed carbon tax would be economy-wide and would capture emissions from red and blue states alike. In one fell swoop, carbon fuels would be taxed across the country, an appealing prospect compared to the state-by-state efforts needed to expand emissions trading. In contrast, an overarching federal cap-and-trade law is probably no longer in the cards, as it would interfere with the successful California and RGGI programs.[43]

There are several advantages of carbon taxes compared with cap and trade (Table 7.1). They are much more transparent and cost-effective to administer since there are no trading markets or offsets to worry about. Reflecting the complexity of emissions trading, it took many years for the California and RGGI efforts to begin operating once the enabling laws were passed. In contrast, the BC carbon tax only took five months to enact.[44]

No matter how small, carbon taxes work through the entire economy, and all fossil fuel producers and users feel their effects. In comparison, both the California and RGGI cap-and-trade programs exempt small emitters because it is too administratively burdensome to include them. In California, 80 percent of emissions are covered, yet a large majority of emitters fall below the 25,000-ton cap and are not regulated at all. This matters because incentivized small firms can be a source of significant green technology innovation.[45]

Carbon tax initiatives surged on Capitol Hill in the 116th session of the US Congress (2019-2020), with no fewer than ten bills introduced in the House of Representatives and Senate. These initiatives differ in many ways, but the most important distinctions have to do with how high the initial tax is, how fast it increases, and how the revenues are spent.[46] One of the bills, the *Energy Innovation and Carbon Dividend Act* (EICDA), attracted eighty-six cosponsors in the House, and has been the focus of intensive lobbying by Citizens' Climate Lobby.[47] Since it had by far the most support, we will focus on key provisions of the EICDA as an example, drawing comparisons where appropriate with some of the other proposals. The EICDA and several of the other carbon tax bills have been reintroduced in the 117th Congress.

The crucial, first thing to know about any carbon tax law is the pricing path. The EICDA's carbon price started low at just $15 per ton in 2020, but then increased sharply by at least $10 per ton in successive years. This means the EICDA carbon price would have reached $115 per ton or more by 2030, and would continue to escalate from there.[48] Among the other carbon pricing bills, only one is more aggressive, and most imposed substantially lower prices in 2030.[49]

Since carbon taxes don't fix emissions levels, how can we know whether the price paths proposed in the EICDA or other bills are high enough to meet our climate goals? One estimate comes from the High Level Commission on Carbon Pricing, an international effort of the World Bank. By analyzing how carbon prices have worked around the world so far, the Commission suggests that meeting the Paris Agreement commitments for a 2.0°C world requires a global carbon price in the range of US $50–$100 per ton by 2030.[50] IPCC projections also suggest that the EICDA price path, if implemented globally, would likely meet 2.0°C but not 1.5°C emissions targets.[51]

Another way to look at the effects of carbon taxes is through *energy-economy modeling*. Energy-economy models project how the US energy system responds to changes in the prices of different fuels. The prices of producing all fuels—from excavation, through delivery, to the final end user—are included in the model. When a tax is imposed on fossil fuels in proportion to emissions, the models can predict how much emissions are reduced in different economic sectors, how much electricity and fuel prices rise, and how much various fossil and renewable technologies contribute to the future energy mix.

Energy-economy modeling is particularly useful because it can estimate the effects of carbon taxes over short-term periods of a decade or less, avoiding the much more difficult and politically fraught approaches involved in calculating the social cost of carbon.[52] As the tax takes effect, its consequences can then be incorporated into estimating better price paths for future years. A team at the Columbia | SIPA Center on Global Energy Policy (CGEP) used these methods to estimate that a carbon price of $77–$124 per ton by 2030 would put the US on a path for net-zero carbon emissions by 2050.[53] Another CGEP team explicitly modeled the EICDA price path through 2030 (Color Plate 19).[54] Their results are striking—all coal use for electricity production is virtually eliminated by 2030. Nuclear power is largely unaffected by the policy, but solar and wind generation grow sharply. By

2030, US greenhouse gas emissions are 36–38 percent lower than 2005 levels, mostly from reductions in the electric power sector.

Other modeling studies show that more modest price paths reaching $60–$85 per ton in 2030 would also sharply reduce emissions from electricity generation. This reflects the projected rapid demise of the coal industry, which is in financially precarious straits and susceptible to even modest taxes. However, these lower tax rates are less likely to stimulate large reductions in other economic sectors and may not be sufficient to enable emissions cuts sharp enough for a 2.0°C world.[55] The Climate Leadership Council's (CLC) proposal falls into this category, since it starts at $43 per ton and then increases at about 7 percent per year.[56]

The EICDA's aggressive pricing—which reaches at least $215 per ton by 2040—is more likely to bring about emissions reductions beyond the electric power sector, although it is difficult to model this reliably. However, this is where *integrity provisions* of the law come into play. The EICDA targets successive year-by-year reductions until ultimately, emissions are reduced 90 percent below 2005 levels by 2050. If any yearly target is not reached, the tax is increased by a further amount in the following year. In this way the law embodies the iterative strategy of energy-economy modeling. More importantly, if targets after 2030 are not reached, the law mandates that new regulations must be written to make up the difference.[57] This compensates for a provision in the law that temporarily suspends existing regulations of greenhouse gases from stationary sources.

Aggressive price paths, insights from energy-economy modeling, and integrity provisions are key features to look for in any carbon pricing law. Another crucial element, found in all the recent carbon tax bills, is a *border carbon adjustment*. Under the EICDA, a tariff would be levied on imports from foreign countries that impose lower carbon taxes (or no carbon tax). In addition, rebates would be given to US companies that export fuels or carbon-intensive products to those countries. Both of these provisions protect US companies in international markets. They may even spur an international domino effect—to avoid paying the US tax, other countries may impose high carbon taxes of their own.[58] Similarly, a European Union border carbon tariff on imported goods, if integrated into its well-established cap-and-trade program, could spur US lawmakers to enact a federal carbon tax to avoid disadvantaging American exporters.[59]

How revenues are used is a crucial distinguishing feature among the eight carbon tax bills. The EICDA projects raising about $2 trillion by 2030, and except for administrative costs, will distribute all of it equally to American households. This is an example of a *revenue neutral* plan, where the money is not kept by the government. Because of this, the charge can be properly described as a *fee* rather than a tax, which can be politically advantageous given the usual Republican bias against taxes. Conservatives usually favor revenue neutrality more than progressives since it means that the government is not growing. However, studies also project that the dividend amount will exceed increases in energy costs for lower-income Americans. In fact, 96 percent of Americans in the lowest income quintile come out ahead under this policy.[60] This bestows a progressive element on the "fee and dividend" policy, giving it potential for bipartisan appeal.

Dividend return is the law in four Canadian provinces that began to tax carbon in 2020.[61] In contrast to the "yellow vest" street protests that greeted regressive new carbon taxes in France,[62] the dividend return policy produced no such reaction in Canada, even though the more conservative provinces where it was imposed had refused to enact their own policies. Few indeed are motivated to protest against dividend checks that appear regularly in their bank accounts. And if enacted in the US, the regular payments also suggest that there would be little support for reversing the policy. Despite these advantages, however, most of the other carbon tax bills allocate revenues differently, as we'll see in the following section.

Finally, it is worth noting that none of these federal bills would preempt the state cap-and-trade programs. A federal carbon tax is compatible with those state laws, although some adjustments in the cap-and-trade provisions might be called for. In particular, emissions allowance prices could decrease if an aggressive federal tax accelerates the adoption of green technology by covered firms, reducing their need to buy allowances and so lowering state revenues. To compensate, states might then raise price floors, eliminate price ceilings, and eliminate free allowances. The resulting "race to the top" could be quite beneficial for reaching climate goals.[63]

5. Carbon Pricing Politics

Carbon pricing is a far-reaching policy, and wherever enacted its effects are felt throughout the economy. Because of this broad impact new carbon pricing laws are always heavily scrutinized by many different stakeholders. In the states, negotiations to enact cap and trade must reconcile strong opinions regarding the schedule of emissions reduction, which industries to cover, what offsets to allow, and how to distribute allowances. The negotiation process is somewhat less complex for states wishing to affiliate with RGGI given its restriction to the electric power sector. But in the quest to extend either RGGI or the California program to more states, these issues are just the tip of the iceberg.

Intense debates about carbon pricing bills also reflect the distinct interests of urban and rural regions, which are affected differently by rising prices for electricity and transportation fuels, as well as by changes to the economy and employment that the laws will catalyze. Industry, agriculture, and other interests invariably seek special dispensations from their favorite lawmakers. Governance is also a key element. This may include assigning or creating new agencies to write and update the regulations, incorporate stakeholder input, and—at the state level—ensure compatibility with the common core of RGGI or WCI governance.

So far, the issues may seem like no more than what might be anticipated for any legislation with broad scope. Yet beyond these nuts and bolts, and no matter how compelling the arguments in favor, carbon pricing remains an inherently hard sell. To understand why, note that other climate policies—such as incentives for renewable energy—are able to create common ground because they clearly foster new business opportunities and deliver exciting visions of futuristic green technologies. In contrast, carbon pricing seems to offer only the unappealing prospect of a new tax.

Whether in the states or at the federal level, one political dynamic is clear: the oil and gas industry will not collaborate in its own demise. Despite nominal support for a carbon

Table 7.3: Allocation of carbon pricing revenues

Revenue use	Political appeal	Advantages and Features
Payroll tax cut or other tax credits	Conservative	Revenue neutral, stimulates economy, appeals to business interests, regressive
Dividend return to households	Bipartisan	Revenue neutral, most low-income households benefit, can be designed to appeal more to progressives
Infrastructure spending	Bipartisan	Meets critical national need, can include energy-related infrastructure, regressive
Renewable energy investments	Bipartisan	Directs revenue to climate change solutions, regressive
Research and development	Bipartisan	Directs revenue to climate change solutions, regressive
Transitional assistance	Bipartisan	Aids displaced workers from fossil fuel industries, or other disadvantaged areas of the economy
Low-income assistance	Progressive	Addresses environmental justice concerns
Social security trust fund	Progressive/ Bipartisan	Bolsters income security for retirees

tax plan promoted by the CLC, fossil fuel firms relentlessly use the dark money network to pressure influential conservative Republicans.[64] In Congress, this leads to all manner of dissembling and confusion in Republican ranks, as Senators and Representatives stake out positions to mollify voters concerned about climate, without drawing the wrath of the well-organized hard Right. In the states, several carbon pricing proposals have been defeated by these forces. As mentioned above, Senate Republican walkouts in Oregon prevented Democratic majorities from enacting a bill that would have linked the state with California's cap and trade program.[65] And in nearby Washington, a $28 million cash infusion led by oil giant BP skewed voter perceptions on a ballot initiative that would have enacted the first robust carbon tax in any state.[66]

Among all the details in a carbon tax law, the most strident disputes are usually about how to spend the revenues (Table 7.3). Like the yearly budget battles in statehouses across the country, the revenue issue evokes the persistent division over the proper role of government, an issue of particular salience given the very substantial revenues involved. In 2018, the RGGI program generated $248 million in investment income from auctions, while California's cap-and-trade program generated $1.4 billion.[67] The projected revenues from the EICDA are much larger, starting at about $75 billion in the first year and increasing to over $400 billion by 2030.[68] These numbers are impressive even by federal budget standards.

There is a very wide range of ideas for how to spend the sizable federal carbon tax revenues, most of which were included in one or more of the ten carbon pricing bills introduced in the 116th Congress.[69] The dividend return option in the EICDA is modified in one of the other bills to include an income cap, which directs more funds to lower-income households. Another revenue-neutral approach is to offset the new income with tax credits or payroll tax reductions. This is a feature of several of the new bills; proponents argue that the tax cuts will provide an economic stimulus and boost GDP. Many of the other bills invest the income in clean energy projects, research, and infrastructure. These seem like very logical approaches since they directly boost climate change solutions, but they are also regressive because they do not compensate low-income groups for the higher fuel costs passed on by fossil fuel firms. If a carbon pricing bill is ultimately enacted, it is likely to divide revenues among many of these options.

Although the most strident opposition to carbon pricing comes from the Right, a good number of those on the Left are also not fond of the approach. This derives in part from progressives' concern for communities who have historically borne the brunt of environmental pollution.[70] The concern leads to a general distrust of free market solutions such as carbon pricing, because the often weakly-regulated operation of free markets is itself seen as a source of injustice. Progressives often prefer strong, well-regulated government programs, viewing them as the most potent safeguard against discrimination.[71]

Progressives' opposition to carbon pricing has led some groups to take positions in alignment with fossil fuel interests. For example, in 2016 the Sierra Club pulled its support for Washington legislative initiative I732 because the bill did not give top priority to the needs of low-income communities and communities of color.[72] In Oregon, the Portland chapter of 350.org would not support the recent legislative initiatives to join California's cap and trade system—even though the bills contained many provisions advocated by progressives—in part because of concerns that allowance trading and offsets are unfair to vulnerable communities.[73] This opposition matters because it demonstrates that, for some progressives, justice issues are so critical that even strong, precedent-setting climate bills should be rejected if they don't contain adequate protections for vulnerable communities. Yet the communities that 350.org and Sierra Club want to protect are those that will suffer the most if robust carbon pricing legislation continues to be delayed.

Fortunately, the opposition to carbon pricing by some progressive organizations is not universally supported on the Left. Indeed, over 40 percent of Green New Deal cosigners in the US House of Representatives also supported the centrist EICDA, including some of the most progressive members of the chamber.[74] This shows that there is indeed common ground to build from. Centrist and EcoRight advocates, including CCL's grassroots volunteers, might listen more to progressives' warnings about trusting fossil energy companies' claims of support for carbon pricing. And some progressive advocates, for their part, might gain from recognizing the necessity of compromise and the importance of working with moderates and conservatives. Shifting the context of discussions to a win-win orientation is the key to breaking the political impasse. In this winning frame, the great range of options for revenue allocation provides something for every group to get on board with. And ultimately, it is not the use of the revenues but the ripple effects of the tax itself that are most crucial for driving the expansion of the green economy.

Chapter 8
Carbon-Free Power

Most electricity in the US is still generated from fossil fuels. Yet the dramatic expansion of solar and wind power over the last decade demonstrates that an historic transition has begun. While natural gas is now the leading power source, renewable energy sources have collectively begun to surpass coal for the first time. Moving forward, it is crucial to maintain the pressure to replace coal and natural gas-fired plants with carbon-free power. Transforming the electricity grid is a top priority because it will enable the carbon-free *electrification* of end uses that currently depend on fossil fuels. This is true across the board—in residential, commercial, and industrial sectors of the economy.

In this chapter we will first look at the design of the US electricity system, especially the power sources that are available to create a net carbon-free grid by 2050. Then we will consider what reforms and new policies are necessary to make this happen. This is where advocacy comes into its own, and we will identify the key pressure points in the system where it is possible to exert leverage. In the next chapter, we will consider some of the end uses in our homes, businesses, industries and transportation networks that can be electrified or otherwise transformed to work within a carbon-free framework.

1. Powering the US Electricity Grid

Getting a good grasp on the electricity system is no small task. The physical infrastructure is complex, and there are many stakeholders who want to influence how it works. Let's start with the big picture before getting into some of the details that will help with advocacy efforts.[1]

The first thing to know is that electricity is an energy carrier. It moves energy through a physical network, or grid, taking it from the primary sources—fossil fuels, nuclear, and renewables—and delivering it to the end users.[2] Today the US grid is powered by fossil

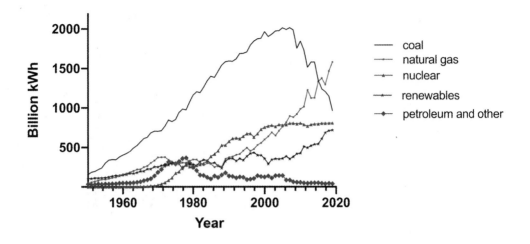

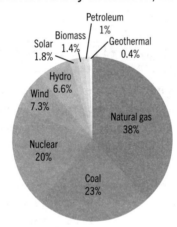

U.S. Electricity Generation, 2019

Figure 8.1a (top): Contributions of primary energy sources to the US electricity mix from 1950–2020. Plotted from data provided by the US Energy Information Administration.

Figure 8.1b (bottom): Contributions of primary energy sources to the US electricity mix in 2019. Plotted from data provided by the US Energy Information Administration.

fuels and carbon-free sources in a 60:40 ratio, with coal use dropping sharply, nuclear power holding steady, and natural gas and renewables on the upswing. The renewables are led by wind and hydropower, with solar power, biomass, and geothermal energy making smaller contributions (Figure 8.1a and 8.1b). These overall data for the country hide large variations in how electricity is generated in each state.[3]

Electricity is moved over long distances on high-voltage transmission lines, which minimize the inevitable power losses that occur in transit. When the current reaches local towns and cities, it passes through *transformers*, which decrease the voltage. Then the electric power is distributed locally to end users (Figure 8.2). Consumer demand for electricity, which grid managers call the *load*, plays an important role in

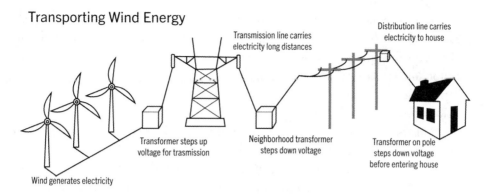

Transporting Wind Energy

Transmission line carries
electricity long distances

Distribution line carries
electricity to house

Transformer steps up
voltage for trasmission

Neighborhood transformer
steps down voltage

Transformer on pole
steps down voltage
before entering house

Wind generates electricity

Figure 8.2: Electricity as an energy carrier. Sources of electrical power—pictured here as a wind farm—are often located at considerable distances from end users. Grid infrastructure includes high-voltage transmission lines, local distribution lines, and devices for converting electrical energy, such as transformers.

policymakers' thinking about how to redesign the system for optimal efficiency with carbon-free power.

The task of converting the grid to net-zero emissions by 2050 faces significant challenges. The present, aging system is optimized for power transmission from large, centralized fossil fuel plants, but we are now moving to more distributed generation, as exemplified by the growth of home solar panels. Another issue is that the US does not have a unified grid, but rather a system of three largely independent subgrids in the Eastern US, Western US, and Texas (Figure 8.3). This limits operational flexibility and compromises efficiency because little power is transmitted across the boundaries, preventing the abundant wind and solar resources that are concentrated in specific regions from serving the needs of citizens who live elsewhere.[4] Finally, the intermittent nature of solar and wind power demands a great deal of new investment in *energy storage* technologies, which capture and save electrical power for future use. This is because the supply and demand for electric power must be kept in careful balance.

Grid managers must also think about balancing *baseload power* (which meets the minimum demand) with *peaking power* (which is needed when demand is high). The grid in any particular region is supplied by multiple primary sources, which vary in their suitability to fulfill either of these two basic needs. This balancing of supply and demand means sources don't operate at 100 percent, as some power always has to be kept in reserve in case demand spikes.

Power demand usually changes in predictable ways, diminishing in the evenings and on weekends, and peaking between 7:00 a.m. and 10:00 p.m. on weekdays. During the first months of the COVID-19 pandemic, many local power grids saw changes in demand patterns, and grid managers had to adjust by altering which generating sources were used at different times. Electricity providers often try to engage consumers in this process by incentivizing them to use more power during off hours. In the industry, involving the customer in this way is called *demand response*.[5]

Looking ahead, demand is expected to increase as the population grows and more end uses are electrified, although this should be tempered by improvements in efficiency. The challenge is that coal and natural gas generation must be phased out even as demand increases, and reliable service must also be maintained throughout the transition. The reliability issue is crucial because misleading arguments about the ability of wind and solar power to sustain the grid are a mainstay among those with vested interests in a system dependent on fossil fuel.

With this brief overview in mind, let's look at the carbon-free options for powering the electricity grid. *Generation capacity* is a central term in understanding the following descriptions of power sources.[6] This is a number in units of watts, which gives the maximum power a source could provide if it were on 100 percent of the time. Think of a new photovoltaic solar array advertised as a 10-megawatt (MW) power source for your community (10 MW = 10 million watts). This is called the nameplate capacity. However, if the array operates only 20 percent of the time—as is typical for most solar power arrays—then the amount of energy produced is also only 20 percent of the maximum (recall that energy is just power multiplied by time).[7] Industry insiders call this 20 percent fraction the *capacity factor*. On average, the 10 MW array is enough to power about 1900 homes, but it would power more if the *capacity factor* were higher.[8]

Solar Power

The most important technology in solar power is the *photovoltaic cell* (PV cell), which converts sunlight directly into electricity.[9] The cell is made of a semiconductor material, such as silicon. When exposed to sunlight, electrons in the semiconductor are freed and made to flow in a circuit, creating electricity. PV cells have been consistently improving, with new commercial panels able to convert about a fifth of the light hitting them into electricity—a big increase compared to early versions. More innovation, including printable PV panels, is in the works.[10]

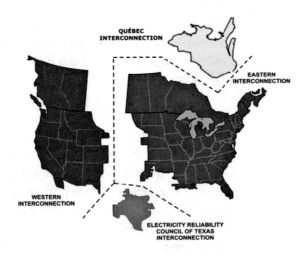

Figure 8.3: Interconnections in the North American electricity grid. The interconnections are distinct power grids, which tie together the many utilities that provide electric power. Few connections are made between these separate power grids. Credit: US Department of Energy.

Table 8.1: Carbon-free power sources for the 2040 US electricity grid

Source	Dispatchable	Expansion potential	Technical readiness	Levelized cost per MWh	Concerns
Solar PV	no	very high	high	$36[a]	raw materials, land use
Solar thermal	yes	high	low	>$100[b]	land use, water use, wildlife
Onshore wind	no	very high	high	$40[a]	land use, wildlife
Offshore wind	no	very high	high	$122[a]	wildlife, marine ecosystems
Hydroelectric	yes (seasonal)	low	high	$53[a]	greenhouse emissions, ecology, wildlife
Geothermal	yes	low	high	$37[a]	water use, air pollution, land use
Geothermal EGS	yes	high	very low	unknown	induced seismicity, pollution, social costs
Nuclear fission	yes	low	high	$82[a]	safety, social costs, raw materials, waste disposal, water use
Nuclear fission (small modular)	yes	likely high	developing	unknown	safety, social costs, raw materials, waste disposal, water use
Natural gas with CCS	yes	high	developing	$38+[c]	methane leaks, safety, social costs, pollution

[a]Average levelized costs in 2019 dollars for new generation resources entering service in 2025.

[b]Levelized cost is for utility-scale solar.

[c]Levelized cost is without CCS. Levelized cost estimates with CCS are 25-50% higher.

Electricity sources are often compared in terms of *levelized cost*. This is a measure of all the expenses involved in generating a given amount of electrical energy, including the capital cost of constructing the power plant, ongoing maintenance and operating expenses, and the cost of the fuel.[11] It is typically given in units of dollars per megawatt-hour ($/MWh) or cents per kilowatt-hour (c/kWh) (Table 8.1). While the PV panels can be costly to build, the fuel is free. This is a big reason why solar power's levelized cost has become competitive with natural gas, dropping by a factor of ten between 2011 and 2019—during this period capacity grew forty-fold. Investments in the high upfront cost of solar power are no longer considered risky because the systems are clearly paying off in the long term. The industry is poised for further rapid growth in coming decades.

Table 8.2: Power sources on today's electricity grid

Source	Capacity (GW)[a]	% Capacity	Capacity factor[b]
Natural gas	477	42.5%	57%[c]
Coal	229	20.4%	48%
Petroleum	37	3.3%	1-13%[d]
Nuclear	98	8.7%	94%
Wind	104	9.3%	35%
Hydropower	80	7.1%	39%[e]
Pumped hydro	23	2.0%	8-17%[e]
Solar PV	58[f]	5.2%	24%
Biomass	13	1.2%	60%
Concentrated solar (CSP)	2	0.2%	21%
Geothermal	2	0.2%	74%

[a]GW indicates gigawatts. One gigawatt is 1 billion (10^9) watts. Data are for 2019.

[b]US averages for 2019

[c]For efficient combined cycle generation.

[d]The range reflects different types of generators

[e]These values reflect seasonal variation

[f]Includes an estimated 23 GW of distributed solar

Perhaps the most obvious characteristic of PV solar is that it only works when the Sun is shining, which is the main reason why the capacity factor is so low. You can gain insight into the meaning of the low solar capacity factor by comparing it to the percentage of total generation that solar contributes to the grid. In 2019 PV solar contributed under 2 percent of the energy to the grid (Figure 8.1), but its generation capacity was over 5 percent of the total US capacity of 1123 gigawatts—one gigawatt, abbreviated GW, is one billion watts (Table 8.2). In order to provide enough electrical energy, solar arrays will always have to be overbuilt well beyond their nameplate capacities.

Another central aspect of the electric power grid has to do with whether a particular power source can be turned up or down at will—a feature called *dispatchability*.[12] A major concern about PV solar power is that it is not dispatchable. This connects to another issue: the very nature of electric current requires that the amount of energy generated for the grid closely matches demand. If not, blackouts occur, so grid operators constantly turn their power sources up and down to match what consumers want.[13] Fossil fuel-fired plants can easily be tuned in this way, but PV solar and wind power cannot be. Clearly, the more these renewables replace fossil fuels on the grid, the harder it becomes to make the necessary adjustments. Fortunately, this problem can be solved in a number of ways, as we will see in the following section.

Making electricity from PV solar does not generate greenhouse gases or other pollution. However, the PV cells are made with rare metals, so their manufacturing involves mining operations that can have substantial environmental impacts. Reliable supply of the metals is also a concern.[14] And while PV solar arrays don't impact freshwater resources, they do compete with other uses for land and can interfere with native vegetation and wildlife.[15] Environmentally friendly ideas for siting PV arrays include repurposing degraded industrial sites and *floatovoltaics*, which, as the name suggests, refers to floating panels on water.[16]

There is another form of solar energy known as *concentrating solar thermal power* (CSP). A CSP plant uses mirrors to focus sunlight onto a unit called a receiver. A fluid circulates continuously inside the receiver, picking up solar heat and using it to make steam, which in turn drives a generator to produce electricity (Figures 8.4a and 8.4b).[17]

Unlike PV solar, solar thermal plants are able to provide dispatchable power. This is possible because the solar heat can be stored for some time in the circulating fluid, allowing electricity to be produced and then released to the grid according to demand. If demand is low, the plant may be used to drive any industrial process requiring heat. For example, research is ongoing on how to use solar thermal heat to supply the energy needed for desalination, an attractive prospect given the threat that climate change poses to freshwater supplies.[18]

The US presently has about a half dozen large CSP plants with capacities in the 200 MW to 400 MW range, all located in the Southwestern desert. But while costs are decreasing, they are still too high to be market-competitive with established technologies.[19] And unlike PV solar, CSP can't be used in small home settings, depriving it of a major market. Nonetheless, ongoing research is targeting all the subsystems in the plant, with particular attention on the solar mirrors (the most expensive component) and the type of circulating fluid that carries the heat.[20] Environmental challenges associated with CSP growth include a high demand for cooling water, an acute issue given that the plants function best in desert environments with reliable year-round sunlight.[21]

Wind Power

In a windmill, or *wind turbine*, the movement of air spins a rotor to make electricity. Size matters for these machines because the larger the area swept by the blades, the more electricity can be produced.[22] Strikingly, the electrical energy output increases with the cube (third power) of the wind speed, so when the speed doubles, the energy output increases by a factor of eight. The largest individual onshore wind turbines now exceed 5 MW capacity and reach a quarter kilometer high at their tips.[23] The wind blows faster at higher elevations, creating an incentive to make the turbines as big as practicable in these areas.

Wind energy has experienced substantial cost reduction and robust growth over the past few decades. Indeed, wind scientists and engineers believe that with continued optimization and integration it could ultimately provide as much as half of the world's electricity needs.[24] Like solar power, wind energy is not dispatchable and has a low

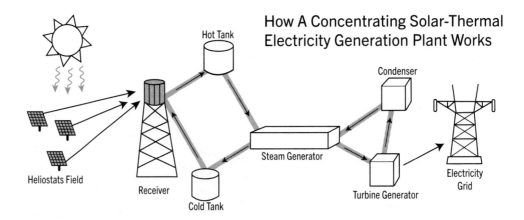

Figure 8.4a: Basic design of a solar thermal electricity generating plant. Solar energy is collected and focused on a receiver, where it heats a circulating fluid that is used to generate steam (center). In turn, the steam is used in a second loop to drive a turbine for electricity generation.

capacity factor (Tables 8.1 and 8.2). This underlines the central importance of energy storage technologies.

Some wind development happens on a small scale, involving a single unit that provides energy to a farm or other facility, but most new wind power for the electricity grid will continue to be developed at large facilities called wind farms. The largest US wind farm, the Alta Wind Energy Center in California, has 600 units spread over 3,200 acres, with a total capacity of 1.55 GW. However, much of the US wind resource is concentrated in the central part of the country. Texas, Iowa, and Oklahoma all surpass California in total wind power capacity.[25]

The major concerns about wind power are its effects on wildlife—especially bats and birds—and its land use impacts, including aesthetics. To address these issues, new wind farms can be built away from key migration routes and the turbines can be colored and otherwise designed to minimize interference.[26] New wind farms can also be set up as dual-use land that is compatible with grazing, crop agriculture, and some recreation. A wind farm with turbines reaching eighty meters high, for example, must space individual units at least a third of a mile apart.

Leasing land for wind farms can provide supplemental income for rural landowners, and their operation offers new employment opportunities for local communities. Effective governance of wind farms is gaining importance as the industry expands, with a central issue being the balance of decision-making power between local and state authorities.[27]

Wind farms can also be located offshore, where turbines either float or are anchored to the seafloor. These facilities benefit from greater windspeeds, which make it possible for individual turbines to reach capacities up to 15 MW.[28] These higher energy outputs offset the greater up-front construction costs of building in the more difficult and remote environment. European Union nations lead global offshore wind development, with a total of 22 GW installed by the end of 2019.[29] This is quite significant, amounting to about one-fifth of US onshore wind generation capacity.

Figure 8.4b: Depiction of a large solar thermal electricity generation facility. Credit: Department of Energy

In contrast, US offshore wind development has lagged amid objections from commercial fishing interests, the need to consider marine transit routes, and the impact on marine ecosystems.[30] The proposed 800 MW Vineyard Wind Project off the Massachusetts coast, billed as the first utility-scale offshore wind project in the US, has been stalled for many years as these and other objections are dealt with.[31] Nonetheless, no fewer than sixteen wind farms are planned along the East Coast, with a combined capacity of about 20 GW. These projects will take some time to come to fruition, but they have gained significant momentum and bipartisan support. European oil and gas companies have joined the wind industry players—their expertise from offshore drilling is relevant to building turbines in the same environment. East Coast utilities are also becoming involved because they are needed to help build and manage the new electric grid infrastructure necessary to accommodate the new power.[32]

A new offshore wind industry would generate employment opportunities for Americans, and not just on the East Coast. Louisiana, a hotbed of oil and gas production, experienced huge job losses in its fossil fuel industry as prices cratered with the COVID-19 lockdown in 2020. Yet those early months of the pandemic also saw the release of several studies showing that the wind energy potential in the Gulf of Mexico is enormous, exceeding 500 GW.[33] With appropriate public subsidies and private sector incentives, the economic crisis driven by the pandemic could stimulate an entirely new, climate-friendly industry to take over as oil and gas drilling declines. This is a major opportunity for American oil and gas companies to claim a stake in the industry.

Hydroelectric Power

Hydropower plants make electricity by harnessing the energy of falling water. In the US, most of this power comes from federally operated dams, where water drops through large elevations and the flow is evened out to ensure sustained production.[34] Interestingly,

Box 8.1: Climate vs. Environment?

*Fish ladder at the John Day Lock and Dam in
Eastern Oregon, which allows salmon to migrate upstream.
Credit: US Army Corps of Engineers.*

An unavoidable tension accompanies the creation of an emissions-free electricity grid because of its impact on natural resources. Solar and wind power are poised to take on dominant roles in the new system, yet they each come with large land footprints. These impacts on land and ecology are why some climate scientists at the Breakthrough Institute advocate for expanding nuclear power, which requires only a very small footprint to provide extremely high energy yields. This is part of a larger philosophy: instead of harmonizing society with nature, human activity should be "decoupled" from natural resource use and concentrated into as small an area as possible.[137]

Hydropower is perhaps the ultimate example of the tension between natural ecology and renewable energy. The Columbia River watershed is a case study in the seemingly endless battles among advocates because of the many distinct values and uses of the river—preservation of sites sacred to indigenous cultures, recreational uses, flood control, commercial transport of billions of dollars' worth of goods, water dedications to cities, agriculture and other users, and, of course, hydropower. Most emblematic and controversial are the extraordinary measures taken to preserve the vitality of the iconic salmon and steelhead fish, many of which receive heightened protection under the Endangered Species Act.[138] Dramatic measures are taken to ensure fish passage around the dams to allow access to upstream spawning and rearing habitats (see the Figure). Power generation and other river uses are also routinely adjusted to accommodate necessary water levels at fish ladders and in reservoirs behind the 14 Washington, Oregon, Idaho, and Montana dams that together comprise the core Columbia River Systems Operation.[139]

Accounts of the myriad administrative processes needed to manage the Columbia watershed could fill entire libraries. Central to this are the Environmental Impact Statements required under the National Environmental Policy Act. The preparation of these and many other documents are governed by federal laws and implemented by the Federal Energy Regulatory Commission (FERC) and numerous other agencies.[140] Advocates who wish to weigh in on these proceedings will not lack for opportunity to make their voices heard. Friends of the Columbia Gorge is a grassroots advocacy group that works tirelessly to protect and steward the many resources the Gorge offers.[141]

most dams in the US were built for flood control and navigability, not electricity. Indeed, hydropower facilities are unique among renewables because energy production is not their only priority—agriculture, industry, cities, and recreational interests all make demands for water allocations. At the same time, laws such as the Endangered Species Act require habitat protection and preservation of stream ecology (Box 8.1).[35] Hydroelectric dam managers must balance these many competing interests.

Wind power recently replaced hydropower as the largest source of renewable electricity (Figure 8.1). Hydropower output did expand dramatically between 1950 and 1990, but then leveled off as opportunities to construct new facilities diminished.[36] As you might expect, hydroelectric power is heavily concentrated in the Pacific Northwest. In fact, about half the nation's capacity is found in Oregon, Washington, and Idaho. However, California, Alaska, and some states in the Northeast and upper Midwest—especially South Dakota— also rely significantly on this form of power. In general, hydroelectric dams have long lifetimes and are expected to remain in operation for many more decades.

The often stagnant reservoirs created by damming rivers are a source of carbon dioxide and methane emissions from decaying vegetation. This means that hydroelectricity is not really carbon-free, but generates, on average, 10–15 percent of the greenhouse gas output of gas or coal plants.[37] This average, however, hides huge variations among different reservoirs, a few of which actually generate greenhouse gas pollution at levels that match or exceed those from fossil fuel generation. In these cases, runoff from surrounding areas often causes algal blooms, leading to *eutrophication*, or depletion of oxygen dissolved in the water. Eventually, this may create habitat for anaerobic methanogens, leading to the release of methane as well as carbon dioxide. Reducing the inflow of nutrients that promote microbial growth can help limit this.[38]

How Pumped Hydro Storage Works

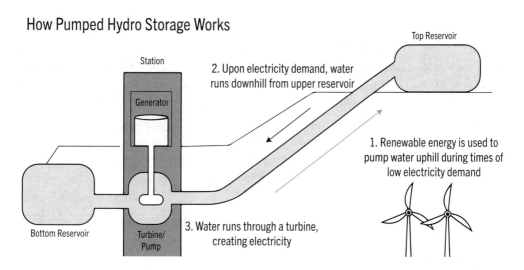

Figure 8.5: Operation of a pumped hydro facility, a form of long-term energy storage. When electricity demand is low, surplus power is used to pump water uphill into the top reservoir. Water is then released to drive electricity turbines at times of higher energy demand. Pumped hydro can supplement primary electricity production from a dam, which can vary due to seasonal differences in river flows.

The Department of Energy (DOE) has a vision to increase US hydropower capacity by 50 percent from its current level.[39] While some of this could come from adding new generation at non-powered dams, more is likely to come from *pumped storage*. In this technology, water stored in a lower reservoir is pumped to the top of the dam, where it is held and then released to generate electricity according to demand (Figure 8.5).[40] This makes *pumped hydro* a fully dispatchable resource, while the primary generation from the dam is subject to seasonal variations in river flow that can limit the ability to provide power on demand. Because it takes electrical energy to pump the water uphill, *pumped hydro* decreases the net power output of an electricity-generating dam,[41] but the gain in storage capacity makes this worthwhile.

Pumped hydro facilities can also be constructed at locations without dams; these are called closed-loop systems.[42] Because there is no on-site electricity generation to pump water to the elevated reservoir, the idea is that this pumping would be done at times of low demand that coincide with an excess of renewable power. As the grid becomes more carbon-free, concerns about using fossil fuel generation to create pumped hydro storage capacity will diminish.

Geothermal Power

Geothermal power relies on heat that emanates from Earth's molten core and from naturally occurring radioactive elements like uranium, thorium, and potassium, which are present in rocks at low abundance. This underground heat visibly breaks through to the surface at hot springs, geysers, and volcanoes, but is present everywhere.

The temperature of the underground geothermal resource determines what it can be used for. Lower-temperature resources below 150°C are widespread throughout the US and are suitable for commercial and residential space heating and some industrial applications like steam generation. Heating is often accomplished with geothermal heat pumps. In these systems—often used in smaller home, school, or commercial environments—a closed loop extends underground and circulates liquid that acquires the ground heat and transfers it to the building.[43] Like solar thermal technology, geothermal energy is useful for both heating and electricity generation. Expanding the uses of these low-temperature resources would benefit the climate by replacing oil and natural gas for heating. Geothermal heat pumps can also provide summer cooling if used at a lower depth of about thirty feet, where temperatures remain between 10°C and 15°C year-round.

The total generating capacity of geothermal power in the US is low, and it contributes less than 1 percent of electrical energy on the grid (Figure 8.1). However, this power is fully dispatchable, and the plants work with high-capacity factors (Table 8.2) because the geothermal power switch, so to speak, is always in the "on" position.

In the US, high-temperature hydrothermal resources for electricity generation are almost entirely found in California, Nevada, and a few other Western states, and are recognizable from surface characteristics such as geysers and hot springs.[44] To produce electric power, wells are drilled into a hydrothermal reservoir where hot water and steam are trapped below a layer of caprock. The water and steam are extracted and used to drive

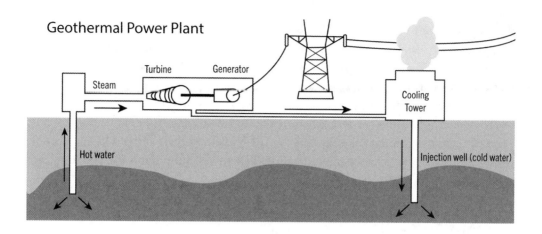

Geothermal Power Plant

Figure 8.6: Operation of an electricity generation plant that relies on geothermal power. Hot underground water is pumped to the surface to generate steam for electricity generation, and the cool water output is returned underground via a nearby injection well.

an electric generator, and the output cool water stream is injected back into the reservoir where it is naturally reheated before re-extraction (Figure 8.6). Recirculating the water through this loop prevents depletion of the underground water supply.

Further expansion of geothermal energy on the grid is possible but challenging. Unlike oil and gas exploration, finding high-temperature underground water reservoirs that lack obvious surface features is difficult.[45] This has led the DOE to champion a different approach called enhanced geothermal systems (EGS).[46] This is a variation of oil and gas fracking technology that creates large, underground manmade reservoirs. Water is then heated by circulating through these reservoirs. According to the DOE's study, the capacity to generate electricity by this approach is enormous, about five times greater than the total installed US capacity today (Table 8.2).

Unfortunately, EGS is much more challenging than fracking for oil and gas because creating the underground reservoirs is more demanding than simply fracturing rock to retrieve fossil fuels.[47] While some pilot-scale experiments have been done, EGS will not be ready anytime soon, because it requires a great deal of research and development to demonstrate feasibility, and substantial public subsidies will be needed to start operations. The fracking part of EGS also carries the same hazards inherent in oil and gas, including the possibility of induced earthquakes and contamination of groundwater. Despite these challenges, the technology is worth exploring because, if successful, it could provide a reliable, carbon-free complement to solar and wind power, ultimately decreasing the need for both energy storage and the dedication of large amounts of land.[48]

The Nuclear Option

The nuclear power industry relies on *nuclear fission*, in which a radioactive form of uranium, U-235, is bombarded by a subatomic particle called a neutron. The nucleus of the uranium atom splits into smaller pieces, generating more neutrons to create a chain reaction. This releases a great deal of energy that can be used to make steam for electricity generation. The chain reaction is kept in check with control rods made of materials that capture neutrons, and the whole reactor core is bathed in water, which moderates the reaction and carries away excess heat. These are the most important components of *pressurized light water reactors*, which make up two-thirds of the 95 nuclear reactors currently operating in the US (Figure 8.7). The remaining third of the US nuclear fleet consists of *boiling water reactors*, which differ in how the subsystems for making steam are designed.[49]

Nuclear power is carbon-free, but it is not renewable because it relies on mining finite stores of uranium ore. The US reserves are small, so about 90 percent of the uranium used in US reactors is imported from other countries, especially Australia, Canada, Kazakhstan, and Russia.[50] At the 2018 rate of consumption, the world uranium resource would run out in about 115 years—but, like fossil fuels, more is always being discovered.[51] After it is mined, uranium ore is milled into a more concentrated form and then processed into a gas called uranium hexafluoride. At this point, it is modestly enriched to a level of 3–5 percent in the fissionable U-235 isotope. The enriched uranium is then converted into small ceramic pellets that are placed into fuel rods in the reactors.[52] All this mining and processing generates substantial pollution, including low-level radioactive mine tailings that still remain at hundreds of abandoned mine sites in the Four Corners region of the Southwest.

The more potent concerns about nuclear power are reactor safety, fuel diversion into weapons manufacture, and waste disposal. The US industry safety record is quite good, and it looks even better when compared to the vast amount of injury and death caused by coal and gas.[53] The single accidental release of radiation occurred in the 1979 Three Mile Island partial meltdown, which vented a small amount of radioactive gas without causing injury. The infamous 1986 Russian Chernobyl meltdown was much worse, but it

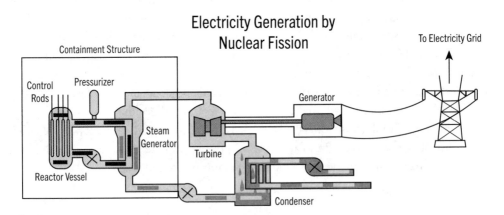

Figure 8.7: Operation of a pressurized light water reactor, the most common approach to electricity generation by nuclear fission in the US. Enriched uranium in the reactor vessel (left) is bombarded by neutrons, splitting the uranium nucleus and releasing very large amounts of heat. As in other forms of electricity generation, the heat is used to generate steam that drives a turbine for electric power generation.

occurred at a reactor built with a hazardous design not used in the US.[54] The third well-known nuclear accident, the 2011 Fukushima disaster, was caused by an earthquake and tsunami. This event caused a major loss of life, yet no deaths or cases of radiation sickness were attributed to release of radioactive material.[55] The only remaining nuclear plant on the tsunami-susceptible Pacific Coast, the California Diablo Canyon facility, is scheduled to close in 2024.[56]

The concern about the diversion of nuclear material to weapons is not about U-235, which is only slightly concentrated in reactors as compared to the high enrichment needed for atomic bombs. Commercial fission reactions, however, naturally generate a plutonium isotope that could be directly used in bombs without needing any enrichment. This isotope is created from nuclear fuel reprocessing, where usable material is recovered from spent reactor fuel rods. Partially because of this safety reason, no such reprocessing for the commercial industry is presently permitted in the US.[57] The US also does not have *breeder reactors*, which produce new fissionable material—including plutonium—and would be able to greatly extend the life of the uranium ore resource. For these reasons, the concern about weaponization is somewhat diminished, although the potential for material to be stolen and dispersed in "dirty bombs" remains.

This leads to the issue of waste disposal. Despite decades of trying to reach a consensus, Congress has been unable to agree on where to locate a national repository to safely store the material for the 10 millenniums or longer that it will take to decay, so spent fuel rods and other materials are stored on site at each of the US reactor plants.[58] The Yucca Mountain facility in Nevada has been the leading candidate for a centralized storage facility, but unsurprisingly, this choice is strongly opposed by Nevada residents.

With all these concerns, it may be hard to see why anyone supports keeping nuclear power in the US electricity mix, yet a great number of people do—including many climate scientists.[59] The main reason is simply that, however horrific some of these problems may seem, they still pale in comparison to the devastation that climate change could bring. And nuclear power does bring practical benefits. The plants have high water demands, but their land use footprints are small compared to wind and solar energy (Box 8.1), and the capacity factor is extraordinarily high: nuclear power provides 19 percent of the grid's energy, but its capacity is less than 9 percent of the US total (Table 8.2).

The US nuclear fleet is aging and expensive to maintain, with only a few new reactors slated to come into operation in the next few years. There is little support for more expensive, large power plants in the current mold, although some advocates point to the potential for dedicating nuclear-generated electricity toward making renewable hydrogen, which we will explore in the following section. This could provide a lifeline for the industry.[60] Another way forward for nuclear power is through a new wave of technology to produce small modular reactors. These new machines will still be powered by U-235 fission, but they employ much simpler and safer designs and use far less uranium fuel. Each independent module could provide 60 MW of capacity, similar to a large PV solar array.[61] For reactors planned by the leading NuScale company, design approval by the Nuclear Regulatory Commission is complete, and the first deployments are expected by the end of the 2020s.[62]

2. A Net-Carbon-Free Grid by 2040

The transition to a green grid is already underway. Coal use is sharply declining and could be completely gone within a decade. Wind and solar power are rising sharply, with few near-term technological barriers to continued growth. Hydroelectric power, geothermal energy, and nuclear fission already account for more than a quarter of the grid energy, all of it largely carbon-free (Figure 8.1). This is not a bad foundation to build on.

In the next few decades, the main battle with respect to grid power will be between natural gas and renewables. Since both solar PV and wind power have plummeted in cost over the past decade, while natural gas has not, it might seem reasonable to think that even a modest continuation of this trend would make renewables dominant. However, more solar PV and wind power implicates the need for more energy storage, which adds expenses beyond the building and operating costs of the facilities (Table 8.1). Additionally, during times of peak demand, the most effective power source has been natural gas-fired "peaking plants" which often remain idle, but can be very rapidly ramped up when needed. As we will see, some business models for electric utilities also favor retaining existing infrastructure. For all these reasons, natural gas remains a formidable competitor, and its full replacement by renewables poses a considerable challenge even if nuclear power maintains its current presence.[63]

The five carbon-free primary energy sources discussed above, most in several manifestations, likely comprise our full menu of choices (Table 8.1). Natural gas with carbon capture and storage is included as a backup, but since methane leaks can't be fully eliminated this would require costly atmospheric carbon removal beyond what is already needed for a 1.5°C world.[64] While certainly worthy of continued research, nuclear fusion and tidal power are not on the list because the engineering hurdles are too formidable for these technologies to play a meaningful role in achieving a carbon-free power grid by 2050.[65] Biomass is excluded because over half the energy it contributes comes from harvesting wood from forests.[66] This worsens warming by removing a carbon sink that can take a long time to regrow.[67] Other types of biomass such as fast-growing grasses are more climate-friendly, but are in limited supply and have other uses apart from generating electricity[68]

Planning the 2050 grid is not a simple matter of phasing out coal and gas while adding wind and solar energy. Hydropower, geothermal, and nuclear power will each also have to grow if they are to preserve their shares in the face of higher demand. The big issue, however, is the intermittent nature of wind and PV solar power. As they continue to expand on the grid, the *nondispatchability* of these sources could create increasingly severe problems in matching supply with demand.[69] Since intermittency also means *low capacity factors* (Table 8.2), the need to dedicate land to wind farms and solar arrays will skyrocket (Box 8.1). Offshore wind, fortunately, offers a way to mitigate that impact.

The current grid is like the food system we had before refrigeration, when perishable food was wasted if not eaten right away.[70] There was often plenty (high capacity), but there were few ways to preserve it (low storage). Energy storage, which comes in many varieties, is one way to solve this problem. We have already discussed pumped hydro, which provides 95 percent of the grid storage capacity today.[71] This is a well-developed technology that can and should be expanded. But it is limited to suitable locations, and attracting investors is challenging because of high upfront costs and multiyear timeframes for building each facility.

Fortunately, many other energy storage options are available.[72] Lithium-ion batteries, flow batteries, and solid-state batteries are three ways to store electricity in the form of high-energy chemicals. The chemical reaction is powered by the input current from the primary source and the energy is released when needed by running the reaction in reverse. All the batteries are rechargeable, of course, although their lifetimes vary. Smaller lithium-ion batteries power most electric vehicles (EVs) and have the early edge over their competitors. There is much interest in connecting EV batteries to the grid, a provocative idea that might ultimately provide millions of small energy storage factories on wheels.[73] Other approaches include thermal storage, which we discussed in connection with concentrated solar power. There are even small mechanical flywheels that store energy in their fast rotation and can be tapped to generate electricity by running in reverse.[74]

All these energy storage approaches work on relatively short timeframes, from minutes (flywheels) to a few hours (lithium-ion batteries), or a bit longer (flow batteries or pumped hydro). But what happens when there are decreases in sunlight or wind that last days, or weeks, or longer? This is the major reason why some energy scientists are uncomfortable with calls for a 100 percent renewables grid with high wind and solar dependence.

This concern could be addressed if a longer-term storage option were available. Long-term energy storage for the grid is currently present in the form of fossil fuel-fired power plants, where the fossil fuels can be thought of as "stored sunlight." On a carbon-free grid, the long-term storage could instead be provided by hydrogen, an energy-rich molecule that can be burned to make electricity with only water as a byproduct. Today hydrogen is made from natural gas, but the process generates large amounts of carbon dioxide emissions. Hydrogen production using carbon-free energy is a major technological goal for the renewable energy economy.[75]

Including nuclear fission on the list of 2050 power sources is sure to be controversial.[76] However, eliminating nuclear power would mean either keeping the equivalent generation capacity in natural gas or betting on a large and rapid expansion in grid storage capacity. Neither conventional hydroelectric nor geothermal power have much growth potential (Table 8.1), so eliminating both nuclear fission and natural gas means that more than 90 percent of grid power would have to come from solar and wind. And as the share of grid energy from these nondispatchable sources increases, much more energy storage will be needed to avoid blackouts when sunlight and wind are low. The new modular nuclear reactors, on the other hand, can provide dispatchable power,[77] and are therefore well-suited to fill the gap opened by the retirement of large reactors and the elimination of natural gas.

There are, of course, some game changers that would upend this analysis. Unanticipated rapid breakthroughs in concentrated solar power, enhanced geothermal, nuclear fusion, or tidal power could provide new primary energy sources. Further, cheap grid batteries that run for days or weeks might emerge, and cost-effective production of renewable hydrogen on a large scale might be realized sooner than expected. One study projects that the rapidly dropping costs for wind, PV solar, and battery storage could make a 90 percent carbon-free grid achievable by 2035.[78] If that work proves prescient, then it may be possible to replace nuclear power with renewables sooner than many analysts think.

For at least the next decade, though, the most effective path to stay within carbon budgets is to push the swift expansion of PV solar and wind power while curtailing natural gas and finishing off coal. There will be time to reappraise the need for nuclear energy once solar and wind have gained perhaps a 40-50 percent share of the grid, which is still low enough to avoid the most difficult technical problems of energy storage.[79] So let's dive into the policies and politics and see how we can help make this happen.

3. Policies for Renewable Electricity

Congress has not enacted many laws advancing renewable energy, leaving it mainly up to the states to act. Beyond funding of basic research in new energy technologies, the most important federal role has been to provide investment and production tax credits for solar, wind, and other forms of renewable energy. Investment tax credits offer tax rebates for a percentage of the investment in new infrastructure, while production tax credits provide rebates based on the amount of energy generated. These policies have certainly helped the solar and wind industries become competitive with fossil fuels but are typically implemented for just a few years and often allowed to lapse, limiting their effectiveness. For example, the large omnibus budget and Coronavirus relief bill that Congress passed at the end of 2020 extended the production tax credit for onshore wind by just one year, and the investment tax credit for solar power and some other technologies by two years. Offshore wind power fared better, however, securing a five-year extension of its 30 percent investment tax credit, which covers projects that begin construction any time before January, 2026.[80]

The leading role of the states in renewable electricity policy has led to a good deal of often successful policy experimentation in forward-looking states, but it also means the transition to a carbon-free grid is happening at vastly different rates around the country. The West Coast, New England, and mid-Atlantic states have generally adopted ambitious, binding schedules for adding carbon-free power to their electricity mixes and have achieved substantial greenhouse gas reductions. But most other regions still depend strongly on coal and natural gas.[81] The resistance to change in those parts of the country reflects their reliance on fossil fuels for jobs and revenue.

Among the most successful state laws are those setting required timelines for electric utilities to transition to carbon-free or renewable sources. These mandates are sometimes coupled with favorable investment or production tax credits, similar to those enacted at the federal level, or with guarantees of favorable prices for qualifying renewable sources. Policies have also been enacted to promote the two-way movement of electricity—not only would power move from the central source to end users, but also from electricity-generating users (with solar panels) back to the grid.[82]

The rise of wind and solar power, including home solar power, raises urgent questions about how to best design and operate a grid dominated by intermittent sources. It also highlights the importance of a stronger federal role in the transition to renewables. Bringing a unified, country-wide electricity grid into existence would certainly require more federal involvement. In addition, many fossil fuel-rich states are unlikely to promote renewable power with the necessary urgency unless compelled to do so by federal law. It seems clear that advocacy for a carbon-free electricity grid will have to include efforts at both the state and federal levels.

Renewable and Clean Energy Standards

The most effective state programs for renewable energy are *renewable portfolio standards* (RPS), laws that require electricity suppliers to acquire power from renewable sources. Clean energy mandates like these are a type of performance standard—a law or regulation that sets minimum requirements for energy efficiency or emissions levels from industrial processes, cars, appliances, etc. These standards are broadly applied to control the spread of many pollutants.

Thirty states and the District of Columbia had an RPS in place in 2020.[83] There are two features of these laws that are most important: the percentage of electricity that must be generated from renewable sources, and the number of years the state has to get there. For example, fourteen states have enacted standards that require utilities to acquire at least 50 percent of their electricity from renewables. Among these, the District of Columbia, Hawaii, and Virginia have perhaps the most ambitious programs, requiring 100 percent renewables by 2032, 2045, and 2050, respectively. However, other states' RPS are much less stringent—the standards in Missouri and Michigan require just 15 percent renewable power by 2021. And twenty states have no RPS at all, although some have set voluntary goals.

Many sources of power can be included in an RPS. While wind, solar, and geothermal power are always part of the mix, biomass, nuclear power, efficiency improvements to any source, or even certain types of coal power are sometimes included.[84] When these sources are included, the laws are sometimes designated as *clean energy standards* rather than renewable portfolio standards. Many states also combine an RPS with a clean energy or renewable energy *target*, which can be set as a nonbinding goal, executive order, or legislative mandate. For example, California's RPS requires that 60 percent of its electricity come from *renewable* sources by 2030; it also has a legislative mandate for 100 percent *clean* electricity by 2045. Here the definition of "clean" includes nuclear power, large hydrothermal generation, and natural gas with carbon capture and storage.[85]

If a state wishes to emphasize the development of a particular generating technology, it can use a "carve-out" provision in its RPS to specify that a certain percentage of power must come from that source. A decade ago, RPS carve-outs for PV solar helped drive the sharp reductions in cost for this technology.[86] Looking ahead, solar thermal or small modular nuclear plants could be candidates for carve-out provisions that would boost these approaches.

How do we know if utilities are complying with RPS requirements? This accounting is done through *renewable energy credits*, or RECs, which are meticulously tracked everywhere in the US.[87] When renewable sources like wind farms and solar arrays go into operation, they are awarded RECs in proportion to the amount of electrical energy they generate—typically one REC per megawatt-hour.[88] Then, when a utility buys renewable power to comply with the RPS, it buys the associated RECs at the same time. The renewable generator is paid for both its electricity and for its RECs, which are a separate commodity. Once the RECs have been bought by the utility, they are retired and can't be used again. And if a utility fails to acquire enough RECs, it can be required to pay penalties.

Here is the interesting thing about RECs: because they are a distinct commodity, they can be separated, or "unbundled," from the renewable electricity they are associated with, and be bought, sold, and traded on their own. If a company wishes to advertise itself as "green," for example, it merely purchases some RECs, acquiring and displaying the certificates to prove it. In many parts of the country, electricity consumers can pay an extra charge to account for the fraction of electricity on the grid that comes from fossil fuels. Having paid this, consumers can then claim they use 100 percent renewable power—never mind that some of the electricity was actually made from natural gas. The extra payment ultimately goes to the wind energy producer, not the gas-fired power plant.

A problem with RECs today is there are too many of them. Nationally, the production of renewable power has outpaced the requirements of state RPS laws, and the resulting abundance of unbundled RECs has driven down their price. In fact, prices are often so low that the extra revenue earned by the renewable generator is too small to make a difference in its decision to build more facilities.[89] The solution to this is for more states to enact aggressive RPS laws.

Many state RPS increase the fraction of required renewable energy over time, so capacity must also keep expanding. When it is time to build, an auction can be held where, for example, wind energy companies compete to provide RECs at the lowest price. The wind companies know they will get the RECs once they start producing, and that they will have guaranteed REC buyers in the utilities, who must purchase them to comply with the law. The result is a contract for the lowest-cost renewable energy that may extend for ten or twenty years, making it possible for the wind generation company to secure the investment it needs to start building. The price of the RECs is also determined in this auction.[90] The RPS may look like a top-down government mandate but it is actually built on this market mechanism.

If RPS laws work so well, couldn't Congress enact one that could apply nationwide? In early 2020 House Democrats proposed exactly that, drafting a national Clean Electricity Standard as part of a huge new bill called the Clean Future Act.[91] The bill sets a baseline carbon intensity standard equal to coal-fired power and then awards credits to all generation that is cleaner, in proportion to the amount of carbon emissions it generates. These credits are analogous to RECs in the state RPS laws. Full credit goes to completely carbon-free generation, like solar, wind, and nuclear power, while natural gas-fired power would get a small credit. The bill requires a fully carbon-free grid by 2050, so credits for any carbon-polluting generation will decrease over time and eventually go away.[92]

The Clean Future Act has received half-hearted praise from some on the Left who want to cut net emissions to zero well before 2050 yet prefer this approach to carbon pricing.[93] As a matter of substance, most economists disagree with this view, holding instead that an aggressive economy-wide price on carbon is the most efficient way to end fossil fuel use. But efficient policies are worth nothing if the political determination to enact them cannot be mustered —and as we have seen, federal carbon pricing faces formidable hurdles.[94] In the debate over how to proceed, a national Clean Electricity Standard should certainly be welcomed as a worthy alternative to the carbon tax laws described in the last chapter. It remains to be seen which approach will win the day.

Advocacy: Influencing State Electricity Policy

There are several ways for climate advocates to influence state electricity policy. Like other laws, RPS mandates are debated in legislatures, written up in bills, and enacted with a signature from the governor. If you live in a state without an RPS, or with an RPS that is weak or set to expire, it is worth thinking about engaging at this level. A weak or nonexistent RPS does not stop renewable energy facilities from being built, but it does mean decisions about what to build are not shaped by an overriding concern about climate. Rather, financial and political considerations will drive these choices. While the costs of new coal-fired power plants are now so high that they are no longer likely to be built anywhere in the US, the same is not true for natural gas-fired facilities.[95] And the influence of natural gas company lobbyists on state legislatures should not be underestimated.

When discussing the bigger picture, it is important to take into account the present circumstances where you live.[96] How much does your state depend on coal or natural gas plants for its electricity supply? Is there a sizable contribution from nuclear power? How old are the generating facilities now, and when will they likely need to be retired and replaced?[97] Are some renewable sources favored over others because of the climate and geography?

Politics also plays a considerable role. In conservative states, it may be better to consider a clean energy standard instead of a strict renewables mandate. If the new modular nuclear power plants come online and work well, perhaps their inclusion in an RPS could broaden its base of support. And if it looks like no other approach can succeed, then even natural gas might be considered, provided that new plants are built from scratch including carbon capture and storage. This increases costs, but federal tax credits are available, and some aspects of the technology have a long-term role to play in stabilizing the climate.[98]

Of course, the landscape for state RPS will change dramatically if a national clean electricity standard is enacted. If mandates in the federal law are strict or the law preempts state action, then state RPS would be unnecessary or not possible to enact. But another possibility is if federal law sets a minimum standard, allowing states to set stricter policies if they so desire. What should not be acceptable is a weak federal law that preempts states and cancels their ability to enact more aggressive timelines.

Another way to engage at the state level is by joining consumer advocacy groups that monitor the work of *public utility commissions* (PUCs), which are agencies of state governments that regulate the electricity providers—the utilities that deliver power to homes and businesses. Almost three-quarters of US electricity customers are served by *investor-owned utilities* (IOUs), which are large for-profit private companies. A large and influential trade organization, the Edison Electric Institute (EEI), represents these companies' interests, in part by lobbying state and federal lawmakers. Nineteen of the approximately sixty EEI member utilities have established carbon-neutral goals for their power mixes, most by 2040-2050.[99] Other electricity providers include the *publicly owned utilities*, which are operated by the federal or state governments, or by municipalities. A third category of providers are electric cooperatives, which are member-owned nonprofit utilities that serve local, rural areas and are responsive to the specific needs of their communities.[100]

PUCs have a mandate to regulate utilities to ensure that electric power is delivered reliably.[101] One way that they do this is through *ratemaking*—establishing an electricity price that is just and reasonable, allowing the utility to profit without causing financial hardship for consumers. Ratemaking rules allow utilities to recover both their operating expenses and the capital they must expend to provide the service.[102] The process is crucial because it greatly influences utilities' decisions about which generating sources to buy power from. For example, consumers may advocate for particular kinds of renewable power, but if ratemaking calculations show that the impact on electricity prices is too great, support for any new building projects may rapidly evaporate.[103]

About three decades ago, almost all utilities were vertically integrated, meaning that they not only delivered electric power to customers, but they also owned the power sources and the transmission and distribution lines. They were near-complete monopolies—regulated by PUCs to prevent abuses—and offered price stability and long-term certainty. Today, 33 states still follow this model.[104] If you live in one of these states, as a consumer you have no choice about where to buy your electric power. The utilities plan the addition of new power and the retirement of old sources, subject to PUC regulation and constrained by the need to acquire RECs (if they are operating in a state with an RPS mandate). Because these utilities make profits based on investments in their own infrastructures, there is a lot of unnecessary overbuilding, and it is more difficult for PUCs to convince them to retire gas and coal plants early. As a result, consumers end up paying more than necessary for their electricity.[105]

The other 17 states have undergone varying degrees of *restructuring*, also called *deregulation*, which is the process for allowing competition into the electricity markets. Utilities in these states have sold off some of their generating assets, but often maintain ownership over transmission and distribution lines, which remain monopolies. In these states, some consumers can decide where to buy their electricity, creating a competitive market that determines the price of power. This has greatly spurred the growth of renewable electricity. Deregulated states include Texas, California, Oregon, and many others in the Northeast and parts of the Midwest. Restructuring has changed the business model for investor-owned utilities because the ratemaking process is modified to accommodate the realities of the new power markets.

Let's take a look at how all this can be useful in advocacy. Suppose you live in a state with traditional vertically integrated utilities that still depend on coal, or in a deregulated state where your utility still owns coal-fired power plants. Running the coal plants is increasingly expensive compared to operating renewable energy facilities—and might even be more expensive than building a new solar array or wind farm from scratch. The utility might consider retiring its coal plants early, or it may think about running them less often.[106] In either case, it will need the approval of state regulators, who must ensure that electricity supply remains reliable during the transition. The proposed plans will be considered in a regulatory proceeding, which includes opportunities for the public to participate.

The proceedings in such cases don't need to be particularly adversarial. The utility already wants to get away from coal on the grounds of profitability, and climate advocates can simply add that this shift should happen as fast as possible. Whether the coal plant closes its doors in five years versus ten makes a big difference in carbon dioxide emissions. A tricky aspect, however, is the potential for the coal plant to become a *stranded asset*, with its value reduced or eliminated because of the early retirement.[107] The average lifetime of coal-fired power plants is about 50 years, so financial losses can often be significant.[108] Consumers may be on the hook to absorb some of the loss since reasonable utility profits are built into the ratemaking process. This could pit climate and consumer advocates against each other, as fast retirement of the plants may not be the lowest-cost option.[109]

Things can get more adversarial if coal companies try to join the process at utility regulatory commissions. This has happened in Indiana, where the coal industry is arguing against depreciation of coal-fired power plants and challenging findings about the much lower cost of natural gas plants. The industry is also raising questions about the reliability of intermittent renewable generation.[110] If nothing else this development shows how the traditional alliance between utilities and coal companies is over. Investor-owned utilities make their money by providing electricity to consumers and are perfectly happy to use whatever power sources maximize their ability to do so.

Advocates may also choose to engage in the process by which utilities develop integrated resource plans, which are required in about two-thirds of the states.[111] These documents describe the expected needs for new power generation resources and provide timetables for retiring old plants, accomplished over a timeline of ten or twenty years. Importantly, they must also justify why particular power sources are chosen and how those decisions align with state climate goals.

Modernizing the Electricity Grid

What happens in the states is crucial, but it is not the whole story. For a full grasp of the electricity system, we must also look at the national environment and at the actions of individual households, where homeowners can now generate their own power with rooftop solar panels. The glue connecting these different levels is the physical infrastructure of the electricity grid.

Many have sung the virtues of a national electricity grid. A small grid is vulnerable to disruptions because the available resources to draw on are more limited. In contrast, larger grids are better able to absorb both fluctuations in demand and the increasingly variable supply that is connected to the growth of wind and solar power. Given that wind and sunlight are unevenly distributed across the country, drawing more on those resources will also require buildout of the current transmission infrastructure.[112]

Up-front capital costs to build a better grid are significant, but the expected savings from improved efficiency are large as well. Studies suggest that a nationally integrated grid would reduce consumer electricity bills by 10 percent, and more savings would arise from fewer power outages.[113] Extreme weather impacts are among the risks faced by the grid, so modernization can improve climate change adaptation even as it speeds mitigation by enabling faster integration of solar and wind power.

As noted above, very little power presently moves across the boundaries that separate the three existing subparts of the grid. High-voltage *direct current* (DC) transmission lines, which carry power with less loss, are sorely needed to carry electricity for long distances across the present boundaries. Building out the *alternating current* (AC) line infrastructure within the Eastern and Western interconnections would also be beneficial (Figure 8.3).[114]

Advanced computer algorithms for forecasting electricity demand over periods ranging from hours to years will also play a key role in modernizing the grid. Forecasting is challenging because it depends not only on accurate next-day weather predictions but also on unpredictable human activities. Nonetheless, it is already possible to estimate next-day loads with accuracies of 1–3 percent. Accurate forecasting saves money by allowing better calibration of which sources to turn up or down during the day. In the long term, it creates more savings by enabling better decisions to be made by suppliers, financial institutions, and other supporting players in the electricity system.[115]

Americans for a Clean Energy Grid, a national advocacy organization, focuses on modernizing the electricity grid in North America.[116] Among other efforts, the group is currently sponsoring a Macro Grid Initiative to build support from policymakers and the public.[117] At the national level, the key player that these advocates need to win over is the Federal Energy Regulatory Commission (FERC), which regulates the interstate transmission and wholesale transactions of electricity.[118] In this instance, *wholesale* refers to the sale of power from generating facilities to utilities, a central feature of a deregulated system. In contrast, electricity sales from the utility to consumers in residences, business, and industry are *retail* transactions, and are overseen by PUCs as described above. Given its role in regulating wholesale prices, one way the FERC could act is by adding a charge to account for the carbon intensity of the power source. This would be consistent with its mandate to ensure just and reasonable rates since market-based pricing should properly assign a cost to the carbon pollution caused by fossil fuel generation.[119] To speed up the construction of new long-distance transmission, FERC also has authority to override states should the latter impose burdensome regulations that can slow development.[120]

FERC also oversees state programs under the Public Utility Regulatory Policies Act (PURPA). This has encouraged small-scale renewable energy generators by requiring state-supervised utilities to buy their power; the law has also encouraged states to open their markets to independent power producers and required the owners of transmissions lines to grant access to these generators.[121] Eventually this led to the creation of seven independent system operators (ISOs), which enabled the competitive electricity generation that is the hallmark of deregulation.[122] Today, two-thirds of electricity consumers in the US are served by these organizations. ISOs are charged with reliably administering the regional power systems and wholesale electricity markets, and with formulating plans to meet future needs for clean electricity generation and reliable delivery.

Creating more transmission links and encouraging the development of a smarter network will require ISOs to collaborate with each other. Climate advocates can look in on this work, since ISOs are under a FERC directive to encourage participation by stakeholders—including consumer groups. For example, a Consumer Liaison Group in New England has met regularly with a regional ISO over a period of ten years.[123] The group seeks to identify issues that are relevant to consumers and to disseminate information on ISO initiatives that inform electricity costs. Encouraging ISOs to prioritize grid modernization more aggressively is another way for climate advocates to make their voices heard.

Distributed Solar Power

The proliferation of home solar panels is one of the most tangible aspects of the renewable energy transition. The residential and commercial PV solar market, also known as *distributed solar*, had an estimated capacity of 23 GW at the end of 2019—about 40 percent of total PV solar capacity (Table 8.2).[124] Due to economies of scale, the levelized cost of distributed solar is about three – to six-fold higher than PV solar installed in large arrays.[125] Nonetheless, innovative policies have made it possible for many homeowners to overcome high installation costs and benefit financially in the long term.

The most important policy is known as *net metering*. This policy allows home and business owners to return excess power from their solar panels back to the grid. Net metering is just a way to monitor electricity usage—the meter runs backwards when the home solar panels generate excess power, and the consumer is only charged for the net amount of electricity used. This works because the credit per unit of energy returned to the grid matches the price that the utility charges. For many homeowners, credits are generated during daylight hours and banked against nighttime use, when power is drawn from the grid as usual. By expanding demand for solar energy, state net metering laws have helped drive cost-saving innovations in PV manufacturing and caused a huge growth in employment in the solar industry.[126]

When solar panels are installed in a home or business, the option usually exists to install a *solar battery* as well.[127] In this case, when more power is generated than used, the excess amount charges the battery before any is sent to the grid (Figure 8.8). This battery power can then be used later, instead of drawing from the grid. The battery is also useful as a reserve power source in the event of an outage, though it does not provide additional

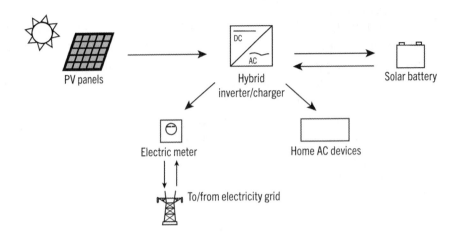

Figure 8.8: Design for a home solar photovoltaic system. The inverter (top center) converts the DC power from the PV panels to the AC power used by home appliances and the electricity grid. The solar battery is an optional component—its storage capacity offers the possibility of lower reliance on the grid. The electric meter integrates the amount of current that flows into and out of the home to the local grid.

savings on top of net metering. However, some utilities are beginning to charge higher rates during peak demand hours,[128] and some states do not have net metering but rather give home solar owners only partial credit for the energy they return to the grid.[129] In these cases, the homeowner can have net financial gains despite the additional cost of the solar battery.

In a remarkable move, California now requires all new houses and multi-family residences that are three stories or lower to be built with solar panels.[130] The increase in upfront housing costs is significant, but on average that cost is expected to be recovered in energy savings over a 15-20 year period. Incorporating solar panels into the cost of home-building also avoids some additional expenses, such as permitting and added financing, that are incurred when panels are installed later.[131] Another benefit of distributed solar is less need for electricity infrastructure, since homes and business are no longer so reliant on the larger grid. This also means that less energy is wasted due to losses during transmission.[132]

The new California regulations also allow home and business owners to participate in *community solar* projects as an alternative to rooftop PV panels.[133] In this approach, an array large enough to serve a local community is constructed, and the power generated goes into the local grid instead of being used by each individual building. In some states, homes and businesses that participate in community solar benefit through a program called *virtual net metering*, where credits for excess returned power are allocated among participants according to their respective stakes in the project. Another benefit of community solar is the potential for forming a *microgrid*—a local electricity network that is connected to the larger grid, but can also disconnect from it if conditions warrant.[134] Microgrids are clearly beneficial if the larger grid shuts down for any reason, and they are also useful to grid operators at times when high demand leads to transmission bottlenecks.

All renewable energies have benefited greatly from state RPS programs, but these laws have not emphasized distributed generation or energy storage. Carveouts in these areas would fill this gap and are well worth advocating for.[135] At the federal level, an investment tax credit for distributed solar systems that includes energy storage is available.[136] As mentioned above, the omnibus federal budget bill passed at the end of 2020 included extension of this solar tax credit for a period of two years. An additional tax credit for energy efficiency upgrades for residences was also included.

Electricity has always been something that very few people think about and almost everyone takes for granted. Yet choices about the grid that are looming in the next few years and decades will determine the speed of the energy transition perhaps more than any other factor. In the next chapter, we will look at how far electricity can penetrate homes, commerce, and industry—and how an increasingly carbon-free grid can drive the emergence of a society finally freed from its dependence on fossil fuels.

Chapter 9
Carbon-Free Lifestyles

Americans are profligate fossil energy users. In the US, carbon dioxide emissions per capita are over twice those in Europe and China and more than triple the world average.[1] This is perhaps not surprising considering our sprawling geography, abundant fossil fuel resources, and consumerist culture. In this chapter, we will look closely at where emissions are coming from and how they might be mitigated (Figure 9.1).[2] As we will see, each economic sector is unique and faces its own challenges in decarbonizing.

There is a lot of detail to examine in each area, but we will tie things together by returning to a vision for what a carbon-free America could look like by the later part of this century. As we saw in the last chapter, carbon-free power can come from sources like solar and wind, yet electricity cannot do everything we need energy for. The other key piece is to use this electricity to make renewable forms of hydrogen and carbon—the elements we mostly get from fossil fuel hydrocarbons today.

We will begin our discussion with the industrial sector. Two general concepts for decarbonizing this sector are electrification and efficiency of energy use, which will be illustrated with examples from high-emitting industries. We will then look at how emissions are generated in transportation, buildings, and cities, all of which are linked by urban development. Many cities have developed climate action plans, which expand public transportation options while decreasing reliance on automobiles. The plans often incorporate social equity, emphasizing that decarbonization and adaptation should not come at the expense of more vulnerable populations. There are many ways for urban planning to lead the way in a just transition that meets climate benchmarks.

1. Industry

Industry emissions can be divided into three categories. Over half come from on-site burning of oil and gas, generating heat to drive conversions of raw materials into products.

Total US Greenhouse Gas Emissions by Economic Sector in 2018

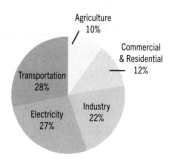

Overview of Greenhouse Gas Emissions in 2018

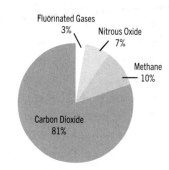

Figure 9.1: Breakdown of 2018 US greenhouse gas emissions by economic sector (left) and by identity of the pollutant. Credit: Environmental Protection Agency.

Next, some emissions come directly from the processes themselves rather than the energy needed to drive them. These are called *process emissions* and are specific to each industry. Finally, emissions can also be generated by the downstream use and disposal of manufactured products. This is especially important for fluorinated greenhouse gases such as hydrofluorocarbons.[3] The total energy consumed by any industry also includes electricity, which is accounted for separately (Figure 9.1). The relative amounts of on-site fossil fuel burning versus electricity use are also specific to each industry.

Electrification is crucial to reducing industry emissions and will become increasingly effective as the grid is converted to renewable sources. There is good potential for electrifying most manufacturing processes, but this will sometimes require surmounting some difficult technical obstacles. The great diversity of manufacturing operations precludes a one-size-fits-all response, requiring instead an industry-by-industry approach that is more costly and time-consuming. More complexity arises because energy supply is often integrated into manufacturing processes. For example, burning fossil fuels on-site generates a great deal of heat beyond what is often required to drive the process, and this heat can be used elsewhere in the operations. Electricity from the grid, however, produces little excess heat. That can be an advantage in other settings but weighs against easy electrification in some industries.[4]

One way to meet these challenges is by increasing energy efficiency.[5] A key way to do so is with combined heat and power plants, sometimes known as *cogeneration* facilities. The basic idea is to generate electricity in close proximity to the industry site and then use the leftover heat from the nearby electric generator to drive the manufacturing process (Figure 9.2). Another approach is to first provide fuel to an industrial furnace and then use the waste heat from that furnace to generate electricity for other parts of the facility. When provided separately, electricity and heat typically have efficiencies of 45–55 percent, so about half the energy is wasted. In contrast, well-designed cogeneration facilities have efficiencies of 65–85 percent.[6]

Cogeneration is already implemented throughout the US in both residential and commercial buildings and in industry settings. However, neither PV solar nor wind are suitable power sources; they do not make electricity by boiling water to create hot steam, so there is no way to couple the power generation with a way to effectively transfer heat. Instead, combined heat and power systems mainly use natural gas, which compromises the improved efficiency by adding more carbon emissions. However, there are many ways to solve this problem, including generating power using natural gas from landfills, carefully selected biomass, or, eventually, renewable hydrogen. If fossil natural gas is still needed in the short term, its emissions could be substantially decreased by capturing carbon dioxide from the emissions stream.

Another way to improve energy efficiency is to increase the use of smart algorithms and sensors for optimizing energy use throughout a building or industrial process. The use of less carbon-intensive materials is another approach with broad application. Finally, an important concept in the field of energy efficiency is the *circular economy*, an idea that extends the familiar practice of recycling. It invokes a cascade of options, beginning with ways of prolonging the useful lifetime of a product and concluding by redistributing the product to other users, refurbishing it, or finally breaking it down to recyclable materials that provide input for new rounds of manufacturing (Figure 9.3).[7]

Some argue that electrification will be slow to take hold because in the US today, electricity is nearly four times more expensive than natural gas per unit of energy provided.[8] This is a misleading argument, however. Because of their much higher efficiencies, the overall costs of heat pumps, electric furnaces, and electric water heaters in homes and businesses are already lower than the natural gas–powered alternatives, despite the still-higher price of electricity.[9] The electricity advantage will widen further as the grid shares of wind and solar power grow and their costs continue to fall. The difficulty in electrifying some industries arises from the complex nature of the processes, not the higher cost per unit of energy.

Steel, Cement, and Petrochemicals

Let's take a brief look at the US industries that generate the most carbon pollution from their process emissions.[10] The top three are iron and steel production, cement manufacturing, and chemicals produced from hydrocarbon feedstocks, or petrochemicals (Table 9.1).[11] Other industries that contribute significant emissions include those that manufacture ammonia—which is mainly used for fertilizers—and those that produce lime, which has many uses. The synthesis of two widely used nitrogen-containing chemicals, nitric acid and adipic acid, generates significant emissions of nitrous oxide. Many other industrial processes contribute to greenhouse warming in smaller amounts.[12]

Steelmaking generates both methane and carbon dioxide from the processing of iron ore into useful iron and also from the reaction between iron and coke (a coal product).[13] The coke reaction removes oxygen atoms that are combined with the iron in the crude ore in a chemical process known as *reduction*. It also adds carbon to the iron, thereby strengthening it and creating steel. In the steelmaking process a large part of the emissions come from the hot blast furnaces where these reactions occur.

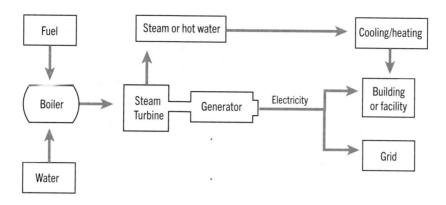

Figure 9.2: Design of a combined heat and power (cogeneration) system. Steam from the boiler is used to generate electricity, and residual heat is channeled to provide additional temperature control in the building. Credit: US Department of Energy.

Globally, energy use and emissions per ton of produced steel have substantially decreased in the past few decades, but net emissions remain high because of increased demand. The capacity for further efficiency improvements is limited, so the only ways to reduce emissions are by replacing steel with other materials in some settings or by implementing new, cleaner technologies to improve the steelmaking process.[14] In the short term, carbon capture and storage could be used to recover carbon dioxide from the smokestack gas and safely bury it underground.[15] Ultimately, the better approach is to replace the coke with a mixture of renewable natural gas and hydrogen. The hydrogen would remove the oxygens from the iron ore while the natural gas would provide the carbon.[16]

A great deal of steel in the US is made from recycled materials. This is often accomplished in mini mills that directly use electricity as the energy source. Since the input scrap metal undergoes reduction when first made, this process is decarbonized in tandem with the electricity power grid.[17]

Cement is manufactured by heating a mixture of limestone and clay at very high temperatures in a kiln, then grinding the product into a powder. This is then mixed with sand, gravel, and water to produce concrete. The key component is the limestone, or calcium carbonate ($CaCO_3$). When heated, it decomposes into lime (CaO), leaving carbon dioxide (CO_2) to be released into the atmosphere.[18] This is a very clear example of an industry-specific process emission—the carbon trapped in the solid limestone has now been converted into carbon dioxide by the heating process. More carbon dioxide is emitted from burning fossil fuels to heat the limestone kiln. In the US, cement-related emissions are slightly below those from steelmaking and don't seem particularly noteworthy (Table 9.1). However, worldwide cement manufacture is carried out on an enormous scale, accounting for 8 percent of global carbon dioxide emissions.[19]

One way to reduce emissions from cement manufacturing is to replace some of the limestone with different materials. Although currently limited to niche applications, a variety of different minerals and chemicals are being tested to replace limestone, and some may have the potential to be developed at a larger scale. There are even lime replacements that absorb carbon dioxide while they harden, thus offering the potential to sequester carbon rather than emit it. Many new companies are working to develop these technologies.[20] As with steelmaking, it is also possible to use either renewable biomass or natural gas with carbon capture and storage to reduce emissions from the heating process.[21]

Petrochemicals are industrial chemical products that are made from hydrocarbon feedstocks. Much of the carbon dioxide emitted in petrochemicals manufacturing comes from fossil fuel combustion. This generates the heat needed to drive reactions, such as the cracking of petroleum hydrocarbons to produce useful smaller molecules. Some emissions reductions in the petrochemicals industry have been achieved by using better catalysts, which drive the reactions without requiring as much external heat. However, what is really needed is a renewable carbon feedstock to replace the fossil hydrocarbons. Again, this could be provided by biomass, renewable natural gas captured from landfills or other sources, or synthetic natural gas made from renewable hydrogen.[22]

Policies to Reduce Industry Emissions

From a general policy standpoint, reducing industry emissions is challenging because most of the important commodities—like steel and chemicals—are globally traded. Policies that make emissions more expensive therefore face the possibility of the companies relocating overseas. The potential loss of income and employment makes it hard for legislators to support such policies, especially since so many blue-collar industry jobs in the US have already been lost through automation. Additionally, as we have already seen with fossil fuel production, rapid change is made difficult by the large amount of capital investment in the existing infrastructure.[23]

One general approach to the challenging industry emissions problem is to implement a *clean products standard*, as recently proposed by analysts at Rhodium group, an

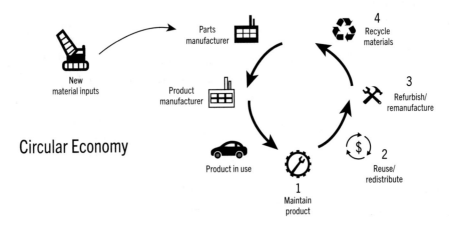

Figure 9.3: Depiction of how a circular economy can optimize resource use. New material inputs (upper left) are minimized as every opportunity is taken to prolong product life, remanufacture the product, and recycle its components. Credit: J. Rissman et al., "Technologies and Policies to Decarbonize Global Industry: Review and Assessment of Mitigation Drivers through 2070," Applied Energy 266, 114848 (2020).

Table 9.1: Major industry greenhouse gas emissions, 2018[a]		
Process	**MMT CO$_2$(eq)**	**Climate Pollutant**
All US emissions	5249	all
All industrial sector emissions[b]	1155	all
Fossil fuel combustion in industry	833	CO$_2$
Products containing HFCs	168	HFCs
Iron and steelmaking	42.6	CO$_2$
Cement manufacturing	40.3	CO$_2$
Petrochemicals	29.4	CO$_2$
Nitric and adipic acid production	19.6	N$_2$O
Ammonia production	13.5	CO$_2$
Lime manufacture	13.2	CO$_2$

[a]Data are from the EPA, Inventory of U.S. Greenhouse Gas Emissions and Sinks, 2018

[b]This excludes electricity use in the industrial sector, which in 2018 generated 487 MMT of CO$_2$.

independent research provider. The standard would require industries in high-emitting industrial sectors to decarbonize by establishing maximum levels of allowable emissions per unit of the goods produced. Unlike many existing regulations, this policy would not require specific technologies, but would rather allow companies to find technological approaches that best fit their individual operations. Like cap-and-trade carbon pricing schemes, the policy also envisions a trading market in which better-performing companies earn credits that can be sold to firms that are less able to reduce their emissions. To help domestic companies against foreign competitors, the emission standard could also be applied to imports of key materials such as steel.[24]

Another important step is to greatly increase the investment in basic and applied research and development. While the US ranks highly among nations in clean energy investment,[25] it is lagging in many areas of crucial importance, especially advanced nuclear technology, renewable hydrogen production, and carbon capture and storage.[26] Among these, nuclear power and renewable hydrogen are ways to deliver carbon-free heat, while carbon capture and storage is important for reducing process emissions where no good replacement technology is yet available. The carbon dioxide generated from limestone in cement manufacturing is a good example of this.

While universities and national laboratories are the best venues for basic research dollars, it is also crucial to find ways for government and industry to jointly translate this fundamental science into commercial success (Box 9.1). Government can also play a role by setting high procurement standards. For example, it can require that agencies seek renewable energy alternatives for their operations that involve transportation and procurement.[27] Industry regulations that require the phasing in of low-carbon heat sources are another option at both the state and federal levels. These mandates could function similarly to renewable portfolio standards, which are working well to decarbonize the electricity sector. Tax credits, like those that have boosted PV solar and wind power for electricity generation, are another option.[28]

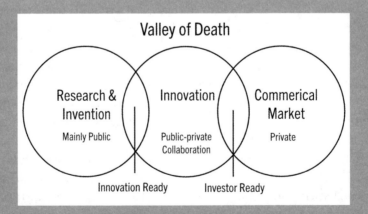

Refrigeration: Hydrofluorocarbons (HFCs)

HFCs, used mainly for refrigeration and air-conditioning, are extremely potent man-made global warming agents that were developed as substitutes for ozone-depleting chlorofluorocarbons (CFCs). Unlike the industry process emissions discussed above, HFCs are mainly emitted when the products are used, either from leakage during operations or from improper disposal. Aside from fossil fuel production, HFCs are the single largest source of greenhouse gas emissions from the US industrial sector (Table 9.1).

In the quest to control global warming HFC elimination is low-hanging fruit. There are many climate-friendly alternatives available, each tailored to one or more specific uses of the compounds.[29] US companies lead in the development of these alternative refrigerants, so HFC phaseout has been supported by manufacturing interests. Recognizing this, Congress bundled an aggressive HFC phaseout law with the second pandemic relief package and federal budget renewal—legislation that passed in the final days of 2020.[30] The international community has also approved a modification to the Montreal Protocol called the Kigali Amendment, which bans HFCs.[31] The Biden administration submitted this treaty to the Senate for ratification soon after taking office. Several northeastern states are also enacting their own bans on HFC use.[32]

Waste Management: Methane

How we deal with waste has important implications for the climate. Some waste is incinerated—emitting carbon dioxide—but most is deposited in landfills, where it decomposes to produce a great deal of methane. A small amount of additional methane and nitrous oxide emissions comes from wastewater treatment and composting (Table 9.2).[33] In 2018 about 20 percent of US methane emissions came from these processes, amounting to 2 percent of US climate pollution (Figure 9.1). The remaining 80 percent of US methane emissions came from agriculture and fossil fuel operations.[34]

The gas emanating from landfills is a roughly 50:50 mixture of methane and carbon dioxide and is commonly called *biogas*. Obama-era regulations require that larger landfills install biogas collection systems to control emissions, although these rules were recently weakened to give the landfills more time to comply.[35] Sewage and industrial wastewater treatment plants also generate biogas.[36] One use for biogas is to make electricity—in 2018, about 0.3 percent of US electricity generation came from this source.[37]

Alternatively, the methane in biogas can be purified away from the carbon dioxide and used in place of fossil natural gas for any application. This can be considered a renewable resource, though there is only enough to substitute for a small fraction of the fossil natural gas used in the US.[38] Nonetheless, the US has over 3000 landfills and the process for generating pipeline-quality gas can be economical for larger facilities. This could be an alternative to electrification for limited applications in buildings or industry. The small-scale, combined heat and power systems described above may be particularly well-suited to use this resource. Other sources of renewable methane include biomass and manure from livestock.[39]

Table 9.2: Greenhouse gas emissions from waste management[a]

Process	MMT CO$_2$(eq)	Pollutant
Landfills	110.6	CH$_4$
Wastewater treatment	14.2	CH$_4$
Wastewater treatment	5.0	N$_2$O
Waste incineration	11.1	CO$_2$
Composting	2.5	CH$_4$
Composting	2.2	N$_2$O

[a]Data are from the EPA, Inventory of U.S. Greenhouse Gas Emissions and Sinks, 2018

2. Renewable Hydrogen and Carbon

Although much effort in the energy transition must go into expanding the scope of established technologies—especially solar and wind power— we have already seen three examples of the need to accelerate the development of newer industries that are not yet profitable on a commercial scale. Two of these involve the large-scale, renewable synthesis of hydrogen gas—as a medium for long-term storage of electrical energy on a grid increasingly powered by intermittent renewables, and also as a way to deliver carbon-free heat to industrial processes that presently depend on fossil fuels. The third example is the need for new carbon feedstocks to replace fossil hydrocarbons in the synthesis of industrial chemicals and other commodities. There are also uses for renewable hydrogen and carbon in other sectors of the economy—transportation, residential and commercial buildings, and agriculture.

Hydrogen (H_2) is an energy-rich gas that—like fossil fuels—can be burned (combined with oxygen) to release energy. But since hydrogen contains no carbon, burning it only produces water as a byproduct. Hydrogen is similar to electricity because it is also an energy carrier. In fact, it was not long ago that the idea of a full *hydrogen economy* was earnestly debated. In this vision, hydrogen gas would be synthesized at central facilities, carried via pipeline to end uses in homes and industries, and used as a replacement for gasoline and diesel fuels to provide onboard power for transportation. As it turned out, renewably generated electricity won out as the leading energy carrier because it is more cost-effective and requires less complex technology. But while extremely powerful, electricity does not reach everywhere. A number of companies have been developing renewable hydrogen, and most energy experts think that it will ultimately play an important complementary role.[40]

Hydrogen already permeates today's economy. It is combined with nitrogen from air to make fertilizers, added to some kinds of crude oil (like the viscous Canadian tar sands product) to make those deposits into useful transportation fuels, and used to synthesize methanol—a key industrial chemical that is both a fuel in its own right and an ingredient for many household commodities.[41] In the US today, 95 percent of hydrogen is made from natural gas by a reaction called *natural gas reforming*.[42] Methane in natural gas (CH_4) is mixed with steam (H_2O) under pressure to generate the H_2 molecule, while the carbon atom combines with the oxygen to produce—you guessed it—carbon dioxide (CO_2).

The cleanest way to make hydrogen renewably is by using renewable energy to split water into hydrogen and oxygen, a process called *electrolysis*.[43] This process is simply the reverse of hydrogen combustion; in essence, the energy from sunlight or wind has been stored in the energy-rich hydrogen molecule (Figure 9.4). On the electricity grid, the hydrogen would then be a fully dispatchable, carbon-free power source, ready to burn in gas turbines to generate power upon demand.[44] Versions of renewable hydrogen technology are available today, but they are still quite expensive—about five or six times more costly than natural gas reforming.[45] However, adding carbon capture and storage technology to a natural gas reforming plant only increases the cost of hydrogen production by about 50 percent while simultaneously cutting emissions. This clearly has potential as an interim solution.[46] Dedicating nuclear power for hydrogen production is another option that is receiving attention.[47]

Hydrogen, however, is not enough—ultimately, we also need a renewable version of *carbon*. Our dependence on fossil fuel hydrocarbons to make household and industrial commodities is bad enough, but much worse is our reliance on petroleum for transportation. The ultimate solution to all of this is called the *circular carbon economy*.[48] This refers to making renewable hydrogen by electrolysis, then reacting this energy-rich molecule with carbon dioxide captured directly from the atmosphere, to produce *synthetic methane* (Figure 9.4).[49] In turn, the methane can be heated by burning renewable hydrogen, then combined with oxygen to produce *syngas*, a mixture of carbon monoxide (CO) and hydrogen.[50] Finally, the syngas can be reacted to make methanol, ammonia, and a whole range of complex hydrocarbons that are useful as both fuels and feedstocks for making other industrial chemicals and commodities.[51]

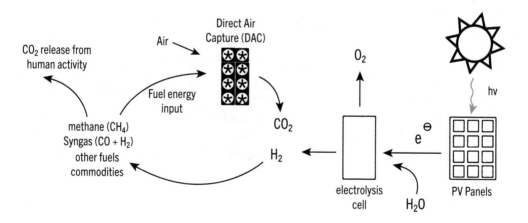

Figure 9.4: The basis for a circular carbon economy. Renewable electricity from sunlight is used to produce a current of electrons (e) that drives an electrolysis cell. The cell uses the energy from the electric current to split water into oxygen (O_2) and molecular hydrogen (H_2). H_2 can be burned directly as a carbon-free fuel, or (as shown) it can be combined with CO_2 captured from the atmosphere (DAC) to produce methane and syngas, from which many commodities can be made. The CO_2 released upon burning methane and other synthetic fuels is recaptured, and the atmospheric CO_2 level remains stable.

A major benefit along the way would be the ability to burn the synthetic methane in existing natural gas plants.[52] This idea has been studied in detail and used to model a new pathway to achieve a carbon-neutral electricity grid in California. The plan would preserve natural gas plants and even build some small new ones in the next decade or two, with the idea that those plants can then burn synthetic methane when it is available. This approach is much cheaper than a 100 percent renewables grid because a huge amount of expensive additional energy storage is needed to take the grid from, say, 80 percent to 100 percent renewables. Either synthetic methane or renewable hydrogen could be burned in these new plants because it is easy to design electricity-generating turbines that work with either gas.[53]

This all sounds terrific, but while ultimately achievable, will require many decades to realize. To be effective in advocacy, we need both this vision and the clear, objective assessment of our current reality. This brings us to the topic of current, alternative transportation fuels, the problematic role of biomass in generating those fuels, and the great promise of electric vehicles.

3. Transportation Fuels

Transportation is now the single largest source of greenhouse gases, having recently surpassed electric power generation.[54] Policymakers concerned with reducing transportation emissions often discuss their objectives in terms of a "three-legged stool": replacing fossil hydrocarbons with alternative fuels, increasing the efficiency of fuel use, and decreasing the vehicle miles traveled (VMT).[55] The last two legs of the stool are particularly important now, while we still depend so much on petroleum. There has been some progress in improving fuel efficiency, yet transportation emissions continue to increase because VMT keeps going up.[56]

Let's begin with the first leg of the stool. So far, the alternatives to fossil fuels for transportation have mainly been derived from biomass, particularly corn and soybeans, although these crops compete with food production on agricultural land. In contrast, dedicated *energy crops*, such as tall grasses and fast-growing trees, do not raise this concern. These and other biomass resources also find many uses beyond the production of liquid biofuels (Table 9.3).[57] The use of any biomass should be accompanied by careful analysis and monitoring to ensure that land carbon stores are preserved. We also should not overlook the environmental and social values implicated in large-scale land-use decisions.

Crop Biofuels and Their Discontents

Today, transportation fuels are heavily dominated by two crude oil products: gasoline and diesel. Gasoline powers cars and light-duty trucks, while diesel is used by aircraft, trains, ships, buses, and heavy trucking. About one-fourth of the diesel fuel in the US is used separately in industry.[58] The difference between gasoline and diesel is that the hydrocarbons in gasoline are smaller and have fewer carbon atoms. This affects engine design—diesel engines have no spark plugs because the larger hydrocarbons don't vaporize easily. In contrast, the smell of gasoline comes from molecules that have evaporated and ignite after getting a spark from the starter. Gasoline and diesel have completely separate markets, and the biofuels to supplement or replace each of these fuels come from different sources.

The number one biofuel in the US is ethanol made from corn. No fewer than 16 billion gallons are produced annually, an amount that requires dedicating 40 percent of the midwestern corn crop to this end.[59] The ethanol is made by yeast fermentation (the same process used to brew beer) and once recovered it is blended with gasoline at a level of 10–15 percent. This blending is required everywhere in the country by federal laws from 2005 and 2007 that enacted a *renewable fuel standard* (RFS). These laws created a lucrative guaranteed market for the corn ethanol industry.[60]

Ethanol blending was initiated because it improves the ability of gasoline to burn efficiently in internal combustion engines. Ethanol replaced a prior additive known as MTBE, which similarly enhanced combustion but had toxic effects.[61] However, the corn ethanol program comes with social costs that especially impact vulnerable and low-income Americans. Compared to fossil fuels, corn ethanol production generates more ozone, particulate matter, and sulfur oxides, all air pollutants that harm human health.[62] The program has also led to higher food prices, especially for flour and rice, since dedicating so much agricultural land to corn crowds out wheat and rice production.[63]

Corn ethanol delivers little value to compensate for the trouble it causes. According to the RFS, the 10 percent ethanol portion of the fuel mix with gasoline only has to generate a 20 percent reduction in emissions compared to pure gasoline. This means that the overall emissions-reduction mandate is only 2 percent—but even that much is not achieved, because Congress included an exemption for a large number of highly polluting ethanol plants built before the law went into effect. Unsurprisingly, a US government report confirmed that overall RFS emissions-reductions targets would not be met.[64]

Ethanol is also made from agricultural crop residues, dedicated energy crops (like switchgrass grown on nonarable land), or municipal solid waste. These sources yield

Table 9.3: Biomass resources	
Sources	**Uses**
Food crops (corn, soybean, canola)	Feedstock for chemicals and commodities synthesis
Energy crops (switchgrass, poplar)	Ethanol production (blend with gasoline)
Municipal solid waste	Biodiesel production (blend with petroleum diesel or B100)
Agricultural residues	Electricity generation
Forest thinning	Biochar production
Production forests	Pipeline gas
	Bioenergy with carbon capture and storage (BECCS)

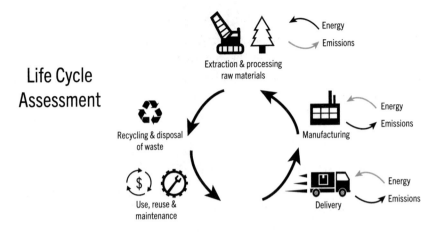

Life Cycle Assessment

Extraction & processing raw materials

Energy
Emissions

Manufacturing

Energy
Emissions

Delivery

Energy
Emissions

Use, reuse & maintenance

Recycling & disposal of waste

Figure 9.5: Depiction of how lifecycle analysis is used to determine carbon intensities (CI, see toxt). The greenhouse gas impacts of biomass-based production or other production chains are considered, including raw material extraction, manufacturing processes, transportation to the site of use, burning or other use of the product, and elimination of waste. Carbon emissions associated with the use of external energy to drive these processes are included in the calculation.

cellulosic ethanol, which is much more climate-friendly, in part because little or no fertilizer is needed to make it.[65] Its production also does not compete with land needed to grow food. Unfortunately, cellulosic plant tissue is hard and tough, and the large-scale extraction of fermentable material is difficult. In 2019, only about 400 million gallons of cellulosic ethanol were produced or imported to the US, even though the RFS law required twenty times that amount.[66] Perhaps Congress did want a better, more climate-friendly ethanol product, yet today 39 of every 40 gallons of ethanol are still produced from corn.

The less heralded cousin of ethanol is *biodiesel*, a blend of hydrocarbons derived mainly from soybeans, with contributions from corn, palm, and canola oils. Biodiesel is mixed with the diesel fraction of crude oil, most commonly as a 20 percent blend known as B20. However, biodiesel can also be used in pure form (B100) as a direct drop-in substitute for petroleum diesel.[67] In contrast to ethanol, biodiesel has a high energy content that is comparable to fossil fuel. In 2019, the RFS law required production or import of 2.1 billion gallons of biodiesel.[68] This is much lower than the ethanol volume requirement because the yield of biodiesel per acre of soybean cropland is about five times lower than the yield of ethanol per acre of corn.[69]

Making biodiesel from oil-rich crops is more climate-friendly than producing ethanol from corn. In part, this is because the oil is directly extracted from the soybeans, in contrast with the more energy-intensive ethanol fermentation process. Another factor is that soybeans, as leguminous plants, do not require nitrogen fertilizers. Compared to diesel fuel from fossil petroleum, burning biodiesel also produces much less particulate matter. This includes a 60–80 percent reduction in black carbon emissions, which have a very strong global warming potential.[70]

Crop biodiesel clearly offers significant benefits and is certainly a much more climate-friendly use of land than corn ethanol. However, allowing imported biodiesel to satisfy RFS requirements turned out to be a huge mistake because it created an incentive for developing countries to sell biodiesel oil in the US market. In Indonesia, tropical rain-forest was cut down to make way for palm oil plantations, bringing about an enormous loss of stored carbon into the atmosphere.[71] Soybean cultivation around the world has also led to widespread deforestation and displacement of indigenous peoples.[72] High US duties on imported biodiesel oils have sharply reduced imports in the past few years, but the damage has already been done.[73]

Carbon Intensities and Climate Impacts

Biofuel additives in transportation fuels clearly present a mixed picture. Cellulosic ethanol and domestic crop biodiesel offer climate benefits, while corn ethanol and imported biodiesel are fairly obvious mistakes. The reaction to all this has not always been rational. Some just look at the bad examples and argue that we ought to give up on energy from biomass altogether. At the other extreme, many industry players would like to classify anything from biomass as "renewable" so it can qualify for tax credits or other benefits.

The use of biomass certainly does not involve mining fossil carbon and ejecting it into the atmosphere. Biomass also regrows on its own as part of the natural surface carbon cycle. If this were the full picture, then the industry argument would have some merit. However, regrowth of biomass can take a long time, creating a carbon deficit in the mean-time. Fossil fuels are also expended to produce, refine, transport, and use the biomass product, and the emissions from all that activity cut into the benefits. And we have already seen some of the wide-ranging environmental and social costs of taking material from the land. Such detriments include water pollution, biodiversity impacts, increases in prices for food or fiber, and the triggering of deleterious changes in land use elsewhere in the US or the world.[74]

The take-home message is that including biomass in the carbon-free energy economy is going to require some pretty involved analysis to understand the costs and benefits of biomass use for any particular application. There is no universal answer, and every pro-posed use will have to carefully consider precisely which biomass will provide the material input (Table 9.3) and what the climate, social, and environmental costs of producing and using the resulting fuel or product will be.

Fortunately, the climate costs of biomass use can be estimated reasonably well with a number called the *carbon intensity* (CI). This indicates how much carbon dioxide is emitted to generate a given amount of energy.[75] CIs are expressed in grams (g) of carbon dioxide equivalent ($CO_2(e)$) produced per million joules (MJ) of energy generated from burning the fuel.[76] One joule is the amount of energy it takes to lift one kilogram (2.2 pounds) by 10 centimeters. Fuels with lower CI values are more climate-friendly, and if you could lift that kilogram weight with a fully solar-powered and renewably manufac-tured device, the CI for the process would be zero.

CIs are determined by *life-cycle analysis* (LCA),[77] which is a general method for deter-mining the environmental impact of any product (Figure 9.5). The idea is to track the energy requirements and climate impacts of a product through its complete lifetime, from raw material extraction, through manufacturing, distribution, use, and recycling or disposal. This generates a CI number that ranks products in terms of their relative carbon impacts. For example, CI values are higher for biofuels produced in states that generate more of their electricity from fossil fuels. Having to transport a biofuel long distances to where it is used also increases the CI. How and where an alternative fuel is made makes all the difference.

LCA is at the core of a path-breaking California law known as the *low carbon fuel standard* (LCFS), which ranks alternative fuels according to their CIs.[78] The LCSF works by comparing the life-cycle greenhouse gas emissions of an alternative fuel to either gasoline or petroleum diesel, depending on which is substituted (Table 9.4).[79] All fuels are produced in several locations by manufacturing processes that vary in their efficiencies, which creates a range of CI values for each fuel type.

Under the LCFS, California fuel refiners and importers must demonstrate that the overall CI value of their fuels is decreasing over time to match the program goals. To comply, they must add increasing amounts of alternative fuels into the mix. When they do this, they earn credits under the program, with one credit representing the reduction of one metric ton of carbon dioxide emissions. If fuel refiners and importers cannot reduce emissions with alternative fuel, they are then allowed to purchase credits in markets set up by the government agency that runs the program. The markets also offer a way for alternative fuel producers to earn income by selling their excess credits.[80]

Presently, the LCFS requires 20 percent reduction in CI by 2030, as compared with a 2010 benchmark. Since the CI values of gasoline and petroleum diesel are each about 100, this means that the overall CI of transportation fuel in California must be reduced to 80 by 2030 (Table 9.4). As of 2018, the actual achieved reduction in CI was about 5 percent, close to the target specified in the law for that year.[81]

The LCFS program continues to expand as new alternative fuel production pathways develop. The law has encouraged producers everywhere to create more climate-friendly fuels to sell in the large California market. It has established a science-based, numerical CI standard that applies the same way to all fuels. To its credit, the LCFS program does not pick winners and losers but rewards innovation of all kinds in direct proportion to the climate friendliness of the new technology.

The emissions reductions timetable in the LCFS is modest but reflects the fact that we do not yet have good alternatives for petroleum gasoline and diesel that can be produced at a large scale. The situation will change, especially for gasoline—when electric vehicles more aggressively penetrate the market. When the LCFS law is renewed in 2030, it will be crucial to impose a sharper decrease in CIs and perhaps to implement different timetables for alternative fuels that replace gasoline and those that replace diesel, respectively.

How do the crop-based biofuels measure up with respect to their CIs? Midwestern corn ethanol has CI values in the range of 70–80 g CO_2(e)/MJ, a 20–30 percent reduction compared to gasoline (Table 9.4). In contrast, one California corn ethanol pathway, where heat for ethanol fermentation is provided by natural gas from landfills, has a CI of just 53 g CO_2(e)/MJ.[82] This improved CI generates more credits for the California producer. It also illustrates the harmful carbon impacts of transporting midwestern ethanol to the West Coast and of producing ethanol on a midwestern electricity grid that relies more heavily on coal and natural gas. Cellulosic ethanol is much more climate friendly than ethanol from corn, with CI values in the 20–40 range.[83]

What about biodiesel? CI values for midwestern biodiesel extracted from soybean and canola are in the range of 50–60 g CO_2(e)/MJ, significantly better than corn ethanol. These sources account for most US biodiesel, but other, small-scale pathways generate fuel with much lower emissions. For example, biodiesel made from used cooking oil is very climate-friendly, with CI values as low as 15 g CO_2(e)/MJ (Table 9.4)—an 85 percent cut compared to fossil petroleum.[84] The supply of waste oils and similar fuels is limited, but a number of crude oil refiners are nonetheless retooling to accommodate them.[85]

Table 9.4: Carbon intensities of alternative fuels	
Fuel	CI (g CO_2(e)/ MJ)
Gasoline	99.4[a]
Petroleum diesel	100.5[a]
Midwest corn ethanol	70-80
Cellulosic ethanol	20-40
Midwest crop biodiesel	50-60
Biodiesel from used cooking oil	15-30

[a]The values for gasoline and petroleum diesel are 2010 benchmarks. See the text.

The Growing Reach of the LCFS

To the credit of its designers, the LCFS law allows new low-carbon technologies to enter the program and earn credits. This makes it a valuable framework for moving toward a future economy built around renewable carbon and hydrogen (Figure 9.4). For example, credits are awarded for compressed natural gas (CNG) that is produced from biogas. This provides incentives for landfill operators, such as the ability to earn profits in the California market if they install methane-collection technology. A wide range of vehicles, with transit buses being perhaps the most widely known, can run on CNG.[86] Gas held in an onboard pressurized tank is expanded and fed to the engine, which otherwise operates in the same way that it would with gasoline.[87] Credits are also given for another product made from biogas called liquified natural gas (LNG), a related fuel that is more suitable for vehicles that require longer driving ranges.[88]

The CI values for CNG and LNG vary widely, with most in the range of 30–60 g CO_2(e)/MJ. Much of the variation comes from differences in the processes needed to remove the carbon dioxide and other impurities in the raw gas.[89] Because the LCA calculations give credit for avoided methane emissions, some efficient pathways have low or even negative CI values. One example is renewable gas produced in California from the decomposition of green waste, with processing and delivery to existing natural gas pipelines done directly on-site. This pathway has a CI value of approximately zero.[90] The operators of such plants earn large credits under the LCFS, which they then sell at a profit in the trading markets. The volumes produced are small, but the incentive to improve and expand the facility is apparent.

A potential technology not yet included in the LCFS is the heating of biomass to directly produce syngas, which could then be used to generate methane and other products (Figure 9.4).[91] This circumvents the renewable hydrogen step, but the issue is that the biomass is not fully carbon-free—the syngas production would have a CI value that depends on its source and the fossil fuel costs of making it. Some deep decarbonization pathway models rely heavily on biomass, and much of it would have to get mobilized through the syngas reaction.[92] Up to a billion tons per year of biomass may be available in the US, which would be enough to substitute a significant part of our natural gas use or up to 30 percent of petroleum use.[93] But it is unclear how much could really be taken without the huge environmental and social impacts not considered in the CI calculations.[94] So far, the biomass industry is doing little to develop this approach, and there has been almost no public conversation about it.

Many other incentives for green technology are included in the LCFS. For example, the program allows refineries that produce transportation fuels to earn credit for incorporating hydrogen made with lower carbon emissions into their production streams. Refineries can also earn credits by using other green technologies in fuel production, such as renewable electricity, electric vehicles, renewable sources of heat or biogas—the list goes on. Credits can go to providers of hydrogen fueling infrastructure, and any production pathway that includes carbon capture and sequestration in safe geologic reservoirs—or carbon capture and utilization for enhanced oil recovery—also qualifies. It is even possible to earn credits by capturing carbon dioxide directly from the air.[95]

Biofuels: Policy and Advocacy

So far, the LCFS is the best policy we have that is aimed at the first leg of the three-legged stool of transportation emissions. Among all other US states, only Oregon has passed a similar law,[96] although Washington has come close to following suit and continues to consider enactment. In Canada, British Columbia has had an LCFS in place for about as long as California has.[97]

The sound design of the LCFS has modestly reduced transportation emissions in California—no small feat given the challenge of developing large-scale green alternatives. Extending the program to other states should be a high priority for climate advocates. The LCFS could be improved by cutting the exemptions presently given for specific uses and fuel types, which include jet fuel, school buses, and some military applications.[98] The CI of the fuel mix could also be decreased faster. It will take time to develop more widely used, climate-friendly petroleum alternatives, which puts practical limits on how fast the clean transportation transition can occur. But the LCFS also gives credits for electric vehicles, which offer an enormous opportunity to further green the sector.

Continued development of technologies that use biogas resources more fully and commercialize renewable hydrogen is well worth advocating for and can command bipartisan support. However, the controversy that has developed around crop-based fuels is a challenge for alternative fuel advocates since the social and climate-related harms generated by US corn ethanol and imported biodiesel have indeed been severe. But as the LCFS shows, there are many green ways to produce alternative fuels. The remedy is not to ban biofuels altogether, but to design policies that sharply distinguish between climate-friendly and climate-unfriendly options, and which fully consider the societal impacts of these choices.

How should this be done? In addition to extending the LCFS to other states, we should also reform the federal RFS law, under which specific volume mandates fully expire in 2022. The advent of electric cars means that gasoline demand will drop sharply in the next few decades—taking the demand for ethanol along with it. Corn ethanol should thus be phased out in the next iteration of the RFS law and replaced—to the greatest extent possible—by more climate-friendly cellulosic sources. At the same time, there may be room for domestic crop biodiesel production to increase. This should only happen if the new acreage replaces corn and does not require new dedication of prime agricultural land. Lower biodiesel crop yields do indicate that corn ethanol farmers will earn less profit if they make the switch, but this policy would still preserve some income and thus be more politically feasible to enact.[99]

Biodiesel is a potent threat to petroleum diesel because it can function as a drop-in replacement for end uses that are challenging to electrify, like aviation and shipping. This means that better biodiesel policy could yield substantial climate benefits. For example, the new RFS could stimulate production by requiring higher biodiesel production volumes. Tax incentives could be offered to optimize diesel engines for biodiesel. Governments might also invest more in the potential of photosynthetic algae to make biodiesel. These algae are single-celled microorganisms that can be grown in ponds or reactors and then harvested to recover the biodiesel from their cell membranes. They grow well on wastewater, so algae-biodiesel production factories could also help curtail water pollution.[100]

Some of these policies to promote biodiesel could find bipartisan support, but breaking the hold of corn ethanol is surely among the most challenging ventures in climate advocacy. The political fortunes of farm-state lawmakers from both parties are deeply wedded to the lucrative program. Lawmakers are also kept in line by aggressive lobbying from ethanol trade organizations.[101] The newest industry initiative was a successful push for EPA to allow year-round sales of E15 (a 15 percent ethanol blend with gasoline). This initiative was immediately challenged with a lawsuit from fossil energy companies, who stand to earn less profit from every gallon of gas sold.[102] Another push is for flex-fuel vehicles, which can run on up to 85 percent ethanol. It seems that anything goes to create larger markets for ethanol. Yet, like that of gasoline it is blended with, the demise of big ethanol is inevitable. The reason is electric cars.

4. Electric Vehicles

Electric vehicles (EVs) are the future of transportation. An unmistakable sign of this came in July 2020, when Tesla surpassed Toyota as the world's most valuable carmaker.[103] Certainly in the mind of the investor, electric cars are here to stay. In 2018 and 2019, about 2 percent of new car sales in the US were plug-in vehicles, a threefold increase compared to 2015.[104] This is an important beginning toward meeting climate goals for the transportation sector, which require 90 percent of new US car and light-truck sales to be EVs by 2050.[105]

The promise of EVs depends on decarbonizing the US electricity grid. Electric cars are most prominent on the East and West coasts, where states are heavily invested in growing renewable power, but much less common in many interior parts of the country. There is less climate benefit to EVs in West Virginia or Wyoming, for example, since electricity in those states is still primarily made by burning coal.[106] However, EVs use energy much more efficiently than internal combustion vehicles, so even charging EVs on a heavily fossil fuel-based grid still offers some climate benefit.

Electric Vehicle Technology

Electric cars and light trucks come in three flavors. *Hybrid electric vehicles* (HEVs) were the first to hit the US market about 20 years ago. They feature an electric battery and drive train in parallel with a gasoline engine. When brakes are applied, some of the energy that goes into stopping the car is redirected to recharge the onboard battery, a process called regenerative braking. From a climate perspective, the continued use of gasoline is problematic, but the high gas mileage of HEVs certainly helps, and these early models helped establish consumer acceptance. Following the HEVs were the *plug-in hybrids* (PHEVs), which, as the name implies, can also recharge the on-board battery when plugged into the electricity grid. While HEVs rarely operate in battery-only mode, depending on the model PHEVs can travel some dozens of miles before the gasoline engine kicks in.[107]

The future of electric cars is surely in the *battery electric vehicles* (BEVs), which dispense with gasoline entirely. The number of BEV models has exploded recently, although Tesla still leads the fleet in both luxury and technology. The signature feature is the onboard battery pack, which now permits a driving range comparable to gasoline vehicles in top models—a crucial accomplishment for reaching the mainstream market. BEVs differ according to how the battery can be recharged. Level 1 (110-volt power) and level 2 (240-volt power) chargers can be used at home with any BEV, while level 3 fast chargers are available only at dedicated charging stations. The need for dedicated stations and the length of time it takes (30–60 minutes) to achieve a substantial (60–80 percent) charge still limit BEV growth.[108] A more robust infrastructure with faster charging times will enable more rapid expansion.

Automakers getting into EVs today are mainly competing to produce the best batteries because they are the most important component in the performance of the vehicle. The technical work is challenging because road performance and capacity for fast recharging are both central issues. Consumers demand fast charging times—which generates a great deal of heat—so the battery designs also have to incorporate cooling elements. More power is always desirable, but that achievement involves tradeoffs with battery weight.[109] Integrating the battery with all of the vehicle subsystems that demand power is another crucial element of design.[110] Automakers that best solve these challenges will be the leaders in what is shaping up to be a highly lucrative industry.

The acceleration in EV growth is not coming free of environmental impacts. Today's EVs mainly employ lithium ion batteries, but lithium mining is driving water scarcity and generating toxic runoff in several places around the world.[111] Lithium is also relatively scarce, which drives an urgent need to develop better recycling methods.[112] Similar concerns surround cobalt, which is incorporated into the positive battery electrode.[113] The potential for much longer battery lifetimes—so-called "million mile" batteries—amplifies these concerns. This breakthrough, likely to increase consumer demand, would lead to less recycling in the short term.[114]

Cars and small trucks dominate energy consumption in transportation (Figure 9.6), but they are not the whole picture. Electrification of buses is straightforward and is being put in place at large scale in China, which has already converted about a sixth of its bus fleet. But in most US cities, the transition either is barely underway or has not begun at all. Despite the many benefits—much quieter operation, no conventional pollutants, lower operating costs—resistance to the new technology is still strong. And while stop-and-start city bus driving works well with electric motors, the need to put dedicated and costly charging infrastructure in place in crowded urban environments has been challenging. Nonetheless, California has mandated that by 2029 all new buses purchased by its transit agencies must be electric, providing a model for other states.[115] Some cities elsewhere in the country are also imposing electric bus mandates.[116]

Railroads are straightforward to electrify as well. In electric railroads, the power is provided by overhead high-voltage transmission lines that run along the rail routes. Many

U.S. Transportation Energy Use

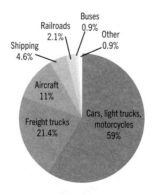

Figure 9.6: Breakdown of energy use in the US transportation sector by type of vehicle for the year 2019. Credit: Energy Information Administration, Annual Energy Outlook 2020.

European and Asian countries have moved in this direction—a seemingly straightforward choice given that electric locomotive engines are both cheaper and easier to maintain than diesel engines. While this is not yet a reality in the US, an advocacy movement has organized to help make it happen (Box 9.2).[117]

Other parts of the transportation system are more challenging. Only very small aircrafts have been electrified because the power required for commercial airliners is still well beyond the capacity of batteries to supply.[118] This is also the case for shipping, especially the large container ships that dominate energy consumption in this area.[119] In both cases, liquid diesel fuels will be challenging to displace, since their energy density is far in excess of what batteries offer. Finally, heavy trucking is a transportation mode traditionally thought to also require energy-dense liquid fuels. However, in a significant breakthrough, large-truck electrification is coming within reach of battery technology. While long-haul trucking presents a larger challenge, many trucking routes are under 100 miles and well-suited for early electrification efforts.[120] Again leading the way, California's Air Resources Board has adopted a new rule mandating the elimination of diesel trucks by 2045.[121]

EV battery technology may develop to the point where it can handle all transportation modes, but if not, hydrogen-powered *fuel cells* could ultimately provide a renewable transportation option that does not depend on batteries. Fuel cells work by combining onboard compressed hydrogen gas with air to generate electricity, which is the same reaction used to burn hydrogen in other energy-requiring applications. A few hydrogen *fuel cell electric vehicle* (FCEV) prototypes are available today. Like other EVs, these prototypes also carry a battery for auxiliary power.[122]

Although the technology to build hydrogen fuel cells is already available, it relies on expensive catalysts and is not yet cost-competitive. As we've discussed, producing renewable hydrogen fuel at large commercial scale still needs many years of development. Once produced, the hydrogen might be transported through the existing natural gas pipeline infrastructure, although this would likely require modification and is still uncertain.[123] Nonetheless, the California LCFS awards credits for renewable hydrogen, both for use in FCEVs and to produce other transportation fuels.[124]

Policy and Advocacy

There is no shortage of estimates for how fast electric cars and light trucks will penetrate the US auto market. High-end projections are that about 50 percent of new car sales will be EVs by 2035, although some other estimates are much lower.[125] Cost reductions for the battery packs are projected to be about 5 percent per year through the 2020s, helping to drive this transition.[126] However, even optimistic projections still leave half of new vehicles in 2035 running on petroleum or biofuels with varying CIs. The transition will be slowed by long vehicle lifetimes (averaging 15 years in the US), while emissions will also persist from transportation modes that are difficult to electrify. In light of all this, what policies should climate advocates work for to best speed the transition?

Box 9.2: Solutionary Rail

Commerce depends on moving goods across long distances, yet most of the available transportation options useful for this purpose are difficult to electrify. But unlike air freight, long-haul trucking, and large container ships, railroads can very easily be made to run on electricity. This is because the power can be provided by overhead lines that are strung along the track, with batteries needed only for auxiliary or backup function. Indeed, most countries have already built railroad infrastructure that is substantially electrified, including China, Russia, and many nations in the European Union. In the United States, however, only 1 percent of railroads are electrified.[175]

Solutionary Rail is a grassroots advocacy group with the sole ambition of electrifying the US railroad network.[176] One challenge faced by these advocates is that—unlike in many other countries—most US railroads are privately owned. This creates a high barrier for securing the substantial necessary funding, estimated at $1.25 billion for a 500-mile track. To overcome this, Solutionary Rail proposes the use of public-private partnerships to enable cost sharing between railroad companies and federal and state governments (See Box 9.1). This approach has been successful for other railroad infrastructure projects as well.[177] Private railroads are interested because electric rail is much more energy efficient than diesel locomotives, providing them with substantial cost savings. Public investment is justified because of the large benefits to human health and the environment that derive from ending the use of fossil diesel.

Emissions reductions from the railroad operation are only part of the benefit that railroad electrification provides. Much of the long-haul freight trucking could be replaced, eliminating those considerable emissions as well (Figure 9.6). But perhaps the most exciting aspect of Solutionary Rail is that it offers an opportunity to build out the electricity grid along the rail lines. This provides dedicated routes for high-voltage transmission lines to move renewable power from sun- and wind-rich regions to the rest of the country. This is a remarkable vision, tremendous in its scope, and equal to the challenge posed by the renewable energy transition. Executing the vision will require not just money but renewed federal climate and energy leadership that will broker among the interests of the many stakeholders.

One line of discussion with local, state, and federal lawmakers should emphasize the opportunities for governments to spur electrification with their own purchasing decisions. High-mileage electric cars and light trucks in particular become relatively cheaper than gasoline versions, since the lower cost of electricity factors in more strongly when the cars are driven longer distances. On average, the operating cost of electric vehicles is only about half that of gasoline-powered cars.[127] Advocates working at the state and local levels can point to the executive order issued by President Biden in the first week of his administration, which calls for an all-electric fleet of federal vehicles. This will take time to implement but will surely accelerate expansion of the EV market.[128]

As already mentioned, the infrastructure for electric vehicle charging needs to be built out nationwide, ideally to the point where speed and efficiency for fueling matches gasoline cars. EV growth will substantially increase electricity demand, which means more profits for investor-owned utilities. This provides these companies with an incentive for constructing charging stations. States will also be involved in this process by providing supplemental funding and by regulating decision-making through the work of the public utility commissions. As we have seen, citizen advocates can participate in this process along with their other efforts to green the electricity grid.[129] As EVs proliferate, the capacity of utilities to include demand response in their planning will increase, since each EV battery (when connected) can upload power to the grid as well as draw from it. Utility cost savings can be substantial, offering yet another incentive.[130]

Citizen advocates can also petition their state and federal representatives to provide direct EV subsidies and tax credits, which are already offered in many states.[131] A federal tax credit of $7,500 has substantially spurred EV sales, but the credit phases out for any manufacturer that sells more than 200,000 vehicles. While offering incentives for new automakers to enter the market, this policy disadvantages companies that achieved early success. Yet these are also the firms that are often leading the innovations that could drive faster industry-wide EV adoption. We are still in the very early stages of the EV transformation, and there is little rationale for limiting the scope of the credits at this time.

Other government policies that may help include giving government rebates to business owners for installing charging infrastructure, reserving EV parking spaces, and providing access to special travel lanes during rush hours.[132] These are good advocacy opportunities because decision-making power often rests with local authorities. California leads a group of 11 states that have instituted a zero-emission vehicle (ZEV) program, which requires automakers to sell a certain fraction of PHEV, BEV, and hydrogen FCEVs within the state.[133] Under the program, automakers are assigned ZEV credits, which may be banked and traded. An analysis of the program's outcomes suggests that it has been successful and offers numerous ideas for improving performance.[134]

Advocates wishing to promote EVs should go into the work with a clear view of the challenges ahead. For example, substantial electrification of the light and medium truck fleet is an important goal for the next decade, but this means large increases in electricity demand and recharging infrastructure, together with high up-front costs for the vehicles. Large fleets such as Amazon and Federal Express that are concerned with their public image are well-positioned, but most trucks are owned by small operations with limited resources. Standardization of the electricity charging infrastructure, to enable any truck to recharge at any facility, is also crucial for the technology to rapidly penetrate.[135]

Given these challenges, it is also crucial to tighten the federal corporate average fuel economy (CAFE) standards, which set gas mileage requirements. Historically, mileage standards have improved over time, reaching a peak under the Obama administration—with a standard requiring an average fuel efficiency of 54.5 miles per gallon for cars and light trucks by 2025.[136] Unfortunately, these regulations were rolled back under the Trump administration, which in turn set weaker standards allowing for more carbon pollution under the so-called SAFE (safe, affordable, fuel-efficient) vehicle rule.[137] Reversing this rule and reinstituting aggressive mileage standards for all vehicle categories—cars, light and medium duty trucks, and heavy, long-haul trucking—is a priority in the Biden administration.[138]

California's early air pollution laws, dating from before the modern Clean Air Act, entitle it to a waiver and enable it to set higher standards than the federal governments. States may choose to follow California—and 13 other states plus the District of Columbia have done so, representing 40 percent of the US auto market. Although the Trump administration attempted to revoke California's special status under the law, this action is also being reversed by the Biden administration.[139]

Finally, it is worth being aware of a new regional compact called the Transportation Climate Initiative, which involves up to 12 New England and mid-Atlantic states, plus the District of Columbia. This follows the successful Regional Greenhouse Gas Initiative, which priced carbon emissions by a cap-and-trade approach.[140] States would cap emissions from cars and other vehicles, and transportation fuel providers would have to buy allowances at auctions for any emissions beyond the cap. Revenues would then flow back to states to be used for climate-friendly programs such as EV infrastructure.[141] The initiative is just getting off the ground and, if successful, will offer a model for reducing transportation emissions in other parts of the country.[142]

5. Cities

Over 80 percent of Americans live in urban or suburban areas, highlighting the importance of cities in the national effort to limit climate change.[143] Cities connect to the residential and commercial sectors of the economy, which together contributed 12 percent of total US emissions in 2018 (Figure 9.1). Urban policies also influence transportation emissions, since zoning laws and public transit options help determine how much private automobiles are used. City climate plans affect lifestyle and behavioral choices and approach the larger questions of how to incorporate justice and equity into the design of climate solutions. Finally, cities are a crucial counterweight to the shortcomings of national or global climate plans: hundreds of cities and counties are among the many groups that declared their support for the Paris Agreement in the wake of the Trump administration's decision to withdraw.[144]

Energy-efficient homes and businesses feature prominently in city climate plans. In these settings, fossil fuels are mainly burned for space heating, hot water, and cooking, all uses that can easily be electrified using heat pumps, induction stovetops, electric furnaces, and electric water heaters. As mentioned above, these technologies are already cheaper than alternatives powered by natural gas.[145] Thus, decarbonizing the electricity grid along with switching end uses away from oil and gas offers a clear long-term solution.[146]

Because grid decarbonization will take decades, there is plenty of room now for policies to improve energy efficiency. Heating, cooling, and hot-water infrastructures in buildings have long lifetimes, and good planning is important to avoid locking in systems with high emissions. Initiatives in a growing number of cities are beginning to ban natural gas in new buildings, but there are still concerns about higher up-front costs for electric alternatives, even though savings are realized over time because operating costs are lower. Changing electricity and natural gas usage will likely also impact utility bills, a matter of some concern in low-income communities.[147]

Energy efficiency rules, such as building energy performance and appliance standards, can be set by cities, states, and the federal government.[148] Standards for heating, cooling, weatherization, and other essential components are usually put in place with strict building codes that apply both to new construction and to retrofits of existing buildings. Certain architectural forms, such as reflective roofs, passive solar heating, and compact construction with low surface-to-volume ratios, also promote energy efficiency.[149] To help cities decide on detailed building code provisions, the World Resources Institute leads a public-private partnership called the Building Efficiency Accelerator to speed the development of building energy-efficiency policy and practice.[150] As advocates, we can engage by connecting our cities with this initiative and then participating in the public process for incorporating building standards into city climate plans. We can also connect with the Global Alliance for Buildings and Construction (Global ABC), a worldwide platform where governments, private firms and other stakeholders can realize a common vision for zero-emission buildings.[151]

A significant part of the energy consumption and carbon dioxide emissions of a building arises from materials and construction. To address this, advocates should encourage city planners to plan at least some new building construction with wood instead of concrete and steel.[152] This would save substantial emissions associated with the cement-manufacturing and steelmaking industries. A new technology for building with wood is called *mass timber*.[153] It relies on gluing together pieces of cross-laminated timber to create a product with high strength and remarkable earthquake and fire resistance. New structures can be assembled from prefabricated materials, which decrease construction times and minimize waste.[154]

To date, the tallest structure built with this technology is an 18-story, mixed-use high-rise in Norway.[155] Mass timber has been incorporated into the International Building Code, which is accepted in most US jurisdictions. Oregon and Washington have led the way by preemptively adopting the guidelines, perhaps unsurprisingly given their timber resources.[156] These states recognize that mass timber provides a market for forest thinnings, which must be removed anyway to suppress forest fires. A strong emerging market for wood buildings could thus go hand in hand with environmentally responsible practices, permitting timber harvesting while maintaining long-term forest carbon stores.[157]

Minimum energy- and water-use performance standards for appliances are set for all states by the Department of Energy. These cover everything from furnaces and air conditioners to light bulbs.[158] These standards are set at little cost to the government, and they pay for themselves in both financial savings and the health and environmental benefits of reduced pollution. The standards are especially useful when made highly visible to consumers—for example, by the well-known Energy Star label.[159] The federal government also provides funds to state energy programs, which are used in a variety of ways to promote energy efficiency in buildings, appliances, and other areas.

Urban Climate Plans

City climate plans include enhancing the energy efficiency of buildings, but more holistically they incorporate *urban form*, which refers to the configuration of buildings, roads, open spaces, and transportation nodes. Together, transportation within city boundaries and transportation between the city center and outlying areas make up a substantial portion of vehicle miles traveled (VMT) in the US. This is the third leg of the three-legged stool for reducing transportation emissions. Of the nation's largest 100 cities, 45 have established climate plans with well-defined targets for greenhouse gas reductions, although not all of these are consistent with limiting warming for a 2.0°C world.[160]

What policies are available to reduce urban transportation emissions? Some of the most widely discussed ideas include a healthy mass-transit system that links outlying suburbs with the city center; mixed-use zoning (so that citizens of any neighborhood have access to amenities without driving long distances); high-density living areas near major transit centers; dedicated biking and walking paths; and an easy-to-navigate grid of short city blocks, which minimize travel distance between destinations. Ideally, these features should be complemented with other policies that reduce congestion. This could include fees to enter crowded city areas during rush hours, as well as limited parking coupled with steep parking fees to discourage private vehicles as much as possible.[161] Of course, every city is unique. A map of transportation emissions across all major US metropolitan areas helps in comparing the challenges faced by different communities.[162]

Urban growth boundaries are a good way to encourage high-density development within the city center. Boundaries help avoid suburban sprawl while simultaneously preserving surrounding open space for recreation, aesthetics, and climate benefits. A leading example is Oregon, where state law mandates that each city and metropolitan area must create a growth boundary around its perimeter. In the Portland area, 24 cities with a combined population of over 1.5 million residents are under the jurisdiction of a governing authority known as Metro, which provides region-wide planning and coordination to manage urban growth and preserve the surrounding forests and farmlands.[163] This approach is worth considering in other metropolitan areas, where, at the very least, it may reduce further sprawl. In a comprehensive review of climate action plans adopted by 29 US cities, Portland scored highest.[164]

As previously discussed, low-income communities, communities of color, and other frontline groups have been more affected by climate change than the general population while also contributing less to the problem.[165] Since these communities make up sizable fractions of many city populations, urban climate plans should incorporate actions that directly address the concerns of these groups. A number of US city climate plans already integrate social justice as a guiding principle, including those of Boston, Portland, Minneapolis, New York, and Washington, D.C.[166] It is particularly important that the plans address *redlining*, the racist practice of segregating and restricting investments in minority neighborhoods. Decades of redlining have resulted in the concentration of urban heat islands in Black communities and other communities of color.[167]

A key concern in urban climate plans is to avoid *green gentrification*, which refers to increases in real estate values—especially in city centers—that lead low-income and marginalized communities to relocate to surrounding areas that may receive less attention. The best way to combat this is by creating a community process in which all stakeholders have a seat at the table. Both the direct impacts of climate change and the often unforeseen consequences of mitigating and adapting to it should be clearly explicated. Although it does not fully address the issue, one approach to creating a vibrant, connected urban environment is to create affordable and reliable mass-transit links between outlying neighborhoods and city centers.[168]

Advocates for building equity into city planning have many examples to draw from. Austin, Texas, developed a community solar program that included home and apartment renters as stakeholders and extended access to low-income consumers. Typically, solar panels are a green amenity that benefits homeowners who have the means to invest in the technology. By extending its program in this way, Austin demonstrated its commitment to inclusiveness while also building climate awareness and further cutting emissions.[169]

The C40 Cities Climate Leadership Group, a global association of 96 large cities, has developed resources for cities to meet the goals of the Paris Agreement. This includes strategies for creating inclusiveness and protections for groups that have been differentially affected by the climate impacts of fossil fuels.[170] The resources include detailed guides for community engagement, planning, and assessment of the impacts of policies. Case studies of large cities, including New York and Los Angeles, offer more models of successful engagement that advocates everywhere can draw inspiration from.

This chapter has outlined just a few of the many opportunities for climate advocates to engage in community work and with all levels of government. However, our focus here on the built infrastructure of the country should be complemented by a closer look at the rural landscape and the many ways in which carbon can be better locked into forests and farmlands. The final chapter will offer an opportunity to review the carbon cycle and carbon budgets and to see how this understanding can be integrated into these land management practices.

Chapter 10

Carbon Removal & Solar Geoengineering

In this final chapter, we will shift our attention to methods of actively removing carbon from the atmosphere. As previously discussed, removing carbon is necessary because emissions reduction alone is no longer enough to keep warming below 1.5°C.[1] Two approaches are currently available. First, forests can be expanded, and natural, sustainable practices can make existing forests, farms, and grasslands into better carbon sinks. Some of these practices are well established and able to significantly slow temperature increases over the next few decades. Second, there is potential for entirely new industries to extract carbon dioxide from the air and bury it safely underground. Here, substantial development is still needed before the technologies are workable and cost-effective on a larger scale. The natural and industrial undertakings are necessary and urgent, both deserving of more attention than they presently receive.[2] As in many other areas, a big part of our work as advocates is to increase the political will to engage the problem.

Both natural land management and technology-based carbon removal enjoy bipartisan support. For example, in March 2019, large majorities in Congress passed the National Resources Management Act—a comprehensive bill that protects many forests and wilderness areas while also offering better access for hunting and fishing.[3] The Great American Outdoors Act, which funds the National Park system along with a host of conservation and recreation programs, was recently passed with similar support.[4] In both cases, diverse interests came together because the bills' language emphasized conservation and sustainable use—without dwelling on climate change. For climate advocates, the key was to find common ground with interest groups that had their own reasons to protect the land.

A broad constituency also supports new technologies for carbon removal, because these carbon capture methods that can profitably reduce emissions from existing industries are also adaptable for large-scale carbon drawdown.

1. The Carbon Removal Challenge

Carbon removal takes us back to the principles of how human activity affects the carbon cycle.[5] Let's recall that deforestation and changes to soil structure from industrial agriculture remove stable carbon sinks, so more carbon dioxide stays in the atmosphere. Part of our task is to undo this damage by reversing the process, planting new forests, and adopting practices that sequester more carbon on the land.

A common stance among climate advocates is that natural land management alone can restore a healthy carbon balance. This is understandable since it appears to be much more Earth-friendly than building out new industrial infrastructure. But the problem is that land use change accounts for just 30 percent of the cumulative increase in carbon dioxide levels.[6] And because the natural capacity of the land to absorb new carbon is limited, it is unable to *also* take up the additional carbon dioxide from fossil fuel burning and cement manufacture. So far, only the oceans—Earth's other large carbon sink—have been able to absorb atmospheric carbon dioxide in proportion to the increasing emissions.[7] However, many climate scientists think that, in time, the oceans will also become unable to maintain a high rate of carbon uptake. For these reasons, both natural land management and industrial-scale carbon removal are necessary.

Collectively, the methods for atmospheric carbon removal are called *negative emissions technologies*. They are the only way to reduce carbon dioxide concentrations to safe levels and will likely also be needed to compensate for residual greenhouse gas emissions that can't be eliminated by any other approach. We must recognize at the outset, however, that investing in negative emissions technologies does not justify reducing other efforts. Clearly, the higher we allow carbon dioxide levels to go, the greater the chance of passing one or more irrevocable and potentially catastrophic climate tipping points.[8] So the best approach is all of the above, with emissions reduction and atmospheric carbon removal each receiving attention. The need for rapid action on carbon removal is further underscored by the fact that some of the new technologies will take decades to develop.[9]

Let's look at some of the numbers. Climate models tell us that, to achieve a 1.5°C world, we must deploy enough negative emissions technology to reduce the carbon dioxide concentration below 2016 levels (403 ppm) by 2100.[10] In this best-case scenario, carbon dioxide levels would plateau at about 450 ppm in two or three decades before declining. Whether this aspiration is feasible depends on two things: the potential for negative emissions technologies to remove the carbon, and the capacity of underground reservoirs to safely sequester it indefinitely.

Fortunately, it appears possible to satisfy both requirements. Climate models for 1.5°C worlds project that, by 2050, 5–15 billion gigatons of carbon dioxide ($GtCO_2$) will have to be removed from the atmosphere each year.[11] Today, we have built the capacity to capture

Methods for Carbon Removal

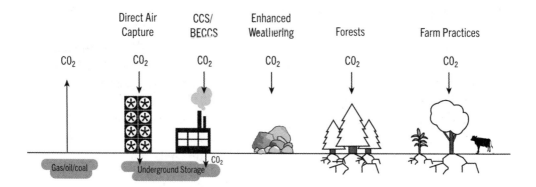

Figure 10.1: Methods for atmospheric carbon removal discussed in this chapter, all of which reverse the increase in CO_2 levels driven by fossil fuel burning (far left). Several of the industrial approaches couple the carbon drawdown with underground sequestration.

only about 1 percent of that,[12] but the buildup to necessary levels is feasible. The best estimates of plausible annual carbon removal by 2050 in the US alone are 0.5–2.85 $GtCO_2$, a significant part of the global requirement.[13]

Carbon dioxide is best sequestered in suitable underground geological formations, which are abundant. Estimated US storage capacities range from 1500–2000 $GtCO_2$ to nearly 15,000 $GtCO_2$.[14] Although safety concerns have yet to be fully addressed, there is no shortage of space to store the carbon.

Climate-sustaining atmospheric carbon removal is an enormous project and will be one of humanity's great undertakings. We can start immediately, by expanding effective and already available approaches in agriculture and forestry. In fact, almost all progress in the next decade will likely come from natural land management—especially tree restoration. Large-scale carbon removal needs more development, including an infrastructure buildout driven by more application of carbon capture and storage (CCS) in industries like cement manufacturing and steelmaking. To realize our climate goals, the new technologies may eventually need to bear a larger share of the burden (Figure 10.1),[15] and that will depend on how aggressively greenhouse gas emissions are reduced over the next few decades.

2. Natural Land Management

When land is conserved and managed with the intention of reversing climate change, the spotlight is typically on how well the practices take up and sequester carbon from the atmosphere. This focus is distinct from the traditional values that have motivated land stewardship in America, but there is a sizable body of research showing that the two perspectives are entirely consistent. Rich soil ecology, clean fresh water, lush forest canopies, productive farmland, and healthy, vibrant biodiversity go hand in hand with effective carbon sequestration.[16]

Three general ideas help us navigate options for conserving carbon on the land.[17] First, there is a key distinction between preserving existing carbon stores and increasing those stores by removing carbon dioxide from the atmosphere. For example, actions to prevent further deforestation and other land degradation (preservation) are complemented by planting more trees (carbon removal). Both are important in the overall program to manage the land in a climate-sustaining way. The distinction becomes blurred in regard to forest management, which can both preserve existing carbon stores and improve sequestration.

Another important distinction is between sequestration options that require new land use, and those that invoke better management of existing land. If new forests come at the expense of agricultural land, then higher productivity might be needed elsewhere to sustain food production. In contrast, restoration of degraded agricultural soils enhances carbon storage without requiring any new land dedications.

A third distinction is between actions that have rapid versus delayed impacts. Agreements to permanently protect existing high-carbon land stores—which include forests, peatlands, wetlands, and mangroves—have immediate effect. However, other actions such as *afforestation*, which is planting new trees on lands that have not recently been forested, may take years or decades to reach their maximum carbon storage potentials.

Agriculture and forestry experts have tried many carbon removal methods, and we are fortunate that some very thorough analyses have identified those likely to work best in the US.[18] With this guidance, we will focus on approaches in forestry, agriculture, and livestock management that have the greatest potential to sequester carbon. Carbon is also stored in peatlands and tidal wetlands, and the importance of maintaining these stores while preserving the natural ecologies should not be overlooked. However, the capacity for new carbon sequestration in these regions is relatively small (Table 10.1).[19]

Afforestation and Forest Restoration

Reforesting nonagricultural lands and careful forest management are the most promising paths for atmospheric carbon removal today. By 2025, these two approaches together have the potential to sequester 0.57 $GtCO_2$ per year in the US (Table 10.1, Figure 10.2).[20] Importantly, these gains can be realized at a relatively low cost and do not impact land used for agriculture. Urban reforestation has lower potential for sequestration since available land is more limited.[21] Nonetheless, tree planting in cities delivers outsized climate benefits because it moderates urban heat islands and thus lowers demand for air conditioning and refrigeration. Urban treescape can also absorb harmful air pollutants such as ozone and particulate matter.

For any afforestation project, the total amount of carbon sequestered increases only until the trees reach maximum size, while the surrounding forest ecosystem also matures. As this limit approaches, the amount of new carbon stored per year decreases, eventually reaching zero at a saturation point. The precise timeline for this depends on the local

Table 10.1: Carbon drawdown in forests, farms and grasslands[a]

Approach	Estimated mitigation potential by 2025 (GtCO$_2$/year)
Reforestation	0.307 (0.09 - 0.777)
Natural forest management	0.267 (0.232 - 0.302)
Avoided forest conversion	0.038 (0.022 - 0.053)
Urban reforestation	0.023 (0.019 – 0.030)
Other forestry approaches	0.030
Avoided grassland conversion	0.107 (0.055 – 0.188)
Cover crops	0.103 (0.053 – 0.154)
Biochar	0.095 (0.064 – 0.135)
Alley cropping	0.082 (0.035 – 0.166)
Cropland nutrient management	0.052 (0.017 – 0.121)
Livestock manure management	0.024 (0.018 – 0.030)
Other agriculture approaches	0.042
Wetlands, peatlands, and seagrass	0.034
Total	**1.204 (0.855 – 1.644)**

[a]Data from J.E. Fargione et al., Natural climate solutions for the United States, Science Advances 4, eaat1869, Nov 14, 2018, Table S1.

geography and climate, the specific mix of tree species, and the ability of forest managers to maintain healthy ecology. The amount of stored carbon in US forests has increased in the last few decades, with about one-third of the nation's land area forested today.[22] But continuing to pursue afforestation in the US and Canada could saturate forest carbon stores by the late 21st century,[23] although other estimates suggest that carbon might continue to accumulate for centuries.[24] Regardless, eventual carbon saturation in forests is a major reason why natural land management alone is likely not enough to solve the carbon removal problem.

Climate change impacts forests through increased wildfire risk, droughts, insect outbreaks, and altered seasonal streamflows.[25] Higher atmospheric carbon dioxide levels partly compensate for these impacts by promoting faster growth of vegetation, an effect known as *carbon dioxide fertilization*. The climate-related factors play out differently depending on local forest conditions.[26] Carbon dioxide fertilization is used by climate change deniers to argue that global warming is beneficial. However, these deniers cite the existence of this effect without acknowledging that the long-term harmful effects of climate change on forests outweigh the benefits from faster growth.[27]

Two newer principles are emerging for maximizing forest carbon sequestration while maintaining healthy ecology. One key insight is that greater forest *complexity*

Figure 10.2: Reforestation opportunities in the lower 48 states. Black dots indicate historically forested areas that are presently unforested. Credit: Fargione at al., Science, 2018, Figure S1.

leads to higher *net primary productivity*, a measure of how much carbon is absorbed from the air by photosynthesis.[28] Complexity measures the diversity present across the tree canopy, including variations in height, leaf color, and leaf size. This controls how much input light is efficiently absorbed across the full spectrum of sunlight. More diversity means more efficient use of light to optimize carbon absorption from the atmosphere. A diverse population of plant species also makes carbon sequestration more efficient since tree diversity contributes to canopy variation across the treetops.[29]

The other principle is to maintain native vegetation. Although this may occasionally sacrifice some carbon uptake efficiency, native ecosystems are often more resilient to external threats such as invasive species and forest fires, both already worsened by climate change. They are also more likely to support a broader range of biodiversity, and to gain public acceptance. It is worth noting that large scale afforestation with non-native trees might require nitrogen fertilizers. In this case, nitrous oxide emissions could negate some of the carbon sequestration benefits.[30]

Some scientists caution against placing too much reliance on forests as a climate solution. A central issue is the typically low forest albedo. If dark forests replace lighter-colored agricultural lands or grasslands, part of the carbon sequestration benefit is negated because more sunlight energy is absorbed and converted to heat.[31] Studies suggest that the balance strongly favors carbon uptake in the tropics but is unfavorable in the conifer-rich northern boreal forests. In temperate US forests, the situation is mixed. Carbon uptake is relatively favorable in the Eastern US and on the West Coast, but disfavored in the Rocky Mountains and dry Southwest, where slow-growing conifers dominate. Some trees also emit gases known as *isoprenes*, which react with other components of the atmosphere to produce the greenhouse gas ozone. Isoprene effects are complex because they also help form cooling aerosols. Forest managers involved in afforestation consider all of these factors when setting policies.[32]

How should the US maximize its forest carbon stores while maintaining healthy ecology and biodiversity? This is a challenging question for several reasons. Carbon cycling through land is among the most complex aspects of the climate system, and our best efforts thus far still leave large uncertainties about the consequences of whatever policies we choose. Another issue is that forest jurisdiction is distributed across federal, state, and local governments. The federal government owns only about one-third of US forests, while state and local governments own about 10 percent. This leaves nearly 60 percent of US forest land in the hands of private owners—both families and corporations.[33] Policies to promote carbon sequestration in forests clearly need to be developed with all of these stakeholders in mind.

At the federal level, climate advocates can work to oppose a recent EPA initiative to classify wood-based power as carbon neutral.[34] Life-cycle analysis shows that this proposed policy is not climate friendly, since forests can take many decades to regrow, creating a carbon deficit during the interim.[35] The new rule would allow the cutting of natural forests as well as the unsustainable harvesting of tree farms to manufacture wood pellets for electricity generation. Forest carbon sinks would be decimated to provide electricity that is more cheaply and easily generated by carbon-free sources. But private industry have found ways to make profits on the venture. The Enviva corporation leads the US effort to harvest southern forests in order to meet European demand for wood pellets. In North Carolina, the Dogwood Alliance is among those fighting Enviva's expansion and exposing its damaging practices.[36]

Another way advocates can take action at the federal level is by petitioning Congress to amend the most important laws for forest management—the National Forest Management Act (NFMA), and the Federal Land Policy and Management Act (FLPMA).[37] Under both laws, land and resource management plans for specific forests are developed by the Forest Service and other agencies. The process to incorporate the impacts of climate change into federal forest management began under President Obama but stalled when he left office.[38] The Obama-era process was initiated by putting forward new agency regulations under the authority of the NFMA and FLPMA laws, but, as we have seen, regulations are easily overturned and even more easily ignored by subsequent administrations.[39] Revising NFMA and FLPMA to specifically require federal forest managers to account for carbon storage over time would provide a stable framework to encourage these practices.[40]

Closer to home, climate advocates should petition their state governments to promote carbon sequestration in a similar way. California has taken a leading role with its Forest Practices Act, which recognizes the important role of forests in the state's carbon balance and mandates sustainable practices to sequester specific amounts of carbon each year.[41] The law also provides the authority to regulate privately held forests—landowners who wish to harvest their timber must comply. Other states should follow suit. Western parts of Oregon and Washington, for example, still retain substantial temperate rainforests that hold large amounts of carbon.

It is crucial to provide incentives for more carbon storage in private forests, including those owned by timber companies. Given the strong demand for fiber, policies that too rigidly exclude even sustainable harvesting are likely to promote *leakage*. This means the forest under strict regulation is preserved, but forests elsewhere will be decimated to provide the product.[42] A particular concern is that such deforestation could occur in the developing world, where it would impact vulnerable indigenous populations. It is better to include timber companies as stakeholders who share a common interest with climate advocates in preserving healthy, carbon-dense US forests that yield a sustainable product stream over time.

Public subsidies to private family landowners are essential because while the costs of planting new forest land are up-front, years or decades may pass before there is any return on investment. It has been estimated that a modest federal investment of $4–5 billion per year over 20 years would be enough for these private lands to reach their carbon storage potential. Direct payments to landowners or tax credits are two ways to provide the subsidies.[43] Another important way to sustainably manage private forest land is through *conservation easements*. These are legal agreements by which a government agency or a land trust pays the landowner in exchange for a permanent restriction on forest uses that are not climate friendly.[44]

Products that come from US forests include paper and packaging, lumber and construction materials, and high-value wood for furniture and flooring.[45] Fiber from forests can be found in a surprising variety of everyday commodities.[46] Since we depend so much on forest products, it is important to ensure sustainable harvesting methods; conducting life-cycle analyses to determine the climate footprint is one way of doing so.[47] Products that sequester carbon for long periods of time, such as wood-frame houses and furniture, are more likely to be climate-friendly. But as previously discussed, there are many intangible, unquantifiable aspects of sustainability that this analysis does not include. To account for these, the Forest Stewardship Council provides a certification for forests that are sustainably managed according to ten specific principles. Among these include the preservation of forest workers' and indigenous peoples' rights, environmental benefits, biodiversity, and long-term monitoring.[48] Forest Stewardship Council certification and life-cycle analysis are complementary tools to assure that forests are sustainably maintained.[49]

Forests are also harvested to provide wood for home heating. About 2 percent of American homes are heated by wood burning, and the fraction can exceed 30 percent in some rural areas.[50] This doesn't need to be detrimental as long as the wood is harvested with priority placed on maintaining sustainability of the local resources. Energy-efficient, well-designed wood-burning stoves can minimize the amount of wood needed and largely eliminate indoor pollution. Guidelines for environmentally friendly practice are available.[51]

State and local initiatives offer the best opportunities for climate advocates to get involved in forest conservation. Of course, we can also plant some trees ourselves. The Nature Conservancy organizes tree restoration and related environmental work projects all over the country,[52] and other local groups eager for new volunteers can be found nearly everywhere.

Figure 10.3a: An example of alley cropping (left). Credit: "Alley Cropping research plots. Guelph, Canada. 2009" by Naresh Tevathasan licensed under CC BY 2.0 Desaturated and cropped from original

Figure 10.3b: An example of silvopasture (center). Credit: "Mr. Steven's farm" by National Agroforestry Center licensed under CC BY 2.0 Desaturated and cropped from original

Figure 10.3c: An example of a riparian forest buffer (right). Credit: licensed under CC BY 2.0 by National Agroforestry Center Desaturated and cropped from original

Agriculture and Grasslands

A number of approaches are available for sequestering carbon on US agricultural and grasslands, which taken together offer a potential for climate change mitigation not far below forests (Table 10.1).[53] Grasslands are important because the perennial grasses (which regrow naturally each year) hold a great deal of carbon in their underground root systems. Given present industrial agricultural practices (Box 10.1), conversion to farmland presents the risk that much of this carbon could be lost from the soil, and then emitted as carbon dioxide. Therefore, avoiding grassland conversion and identifying less productive areas where croplands can be converted to grasses is a more effective way to store carbon. As with forests, conservation easements can play an important role. Demand for these easements is high among ranchers, who form alliances with environmental groups to protect their land from urban expansion.[54]

Planting *cover crops* is an effective way to keep carbon in the soil. These crops are planted after harvesting and prior to planting the main market crops. They protect against erosion and improve retention of water in the soil. Cover crops in the legume family also fix nitrogen in the soil, reducing the need for fertilizers.[55] Strategies have been developed for mixing different cover crops, combining cover crops with the main crop, and planning the best times for planting and removal.[56]

Improvements in cropland nutrient management are also important (Table 10.1; Box 10.1). Replacing commercial fertilizers with nutrient-rich organic materials—including crop residue, manure, and composts—provides a way to reduce climate impact while adding new carbon to the soil. Organic growers also add nutrients via supplements such as alfalfa, bone meal, fish and seaweed scraps, and rock minerals. Because organic agriculture requires more labor input and many changes to existing industrial practice, it's difficult to rapidly implement these methods without large losses of productivity. A hybrid approach in which organic methods are phased in gradually is more feasible.[57]

Organic fertilizers can be combined with *conservation tillage*, which reduces disruption to soil structure and helps retain more carbon in the upper soil levels.[58] This works by

Box 10.1: Fertilizers and industrial agriculture

Modern US agriculture took shape post-World War II, especially under President Nixon's Agriculture Secretary, Earl Butz. Secretary Butz consolidated developing trends to create a centrally managed, heavily subsidized food system, making foods into industrial commodities and rewarding large agribusinesses that could most effectively implement economies of scale. These policies built on the Green Revolution, which introduced new plant breeding techniques and other innovations for greatly expanding yields per acre. Military technology was also applied to create new pesticides, herbicides, and approaches to mechanization.[120] The consequences of all this are well-known: sharp declines in family farms, and creation of the endless rows of monocultures that dominate the agricultural landscape, especially corn, wheat, cotton, rice, and soybeans.

Negative effects on the climate from this system have been profound. When animals and plants were husbanded together, farms could take advantage of natural environmental cycles, in which animal waste was food for plants, and decaying plants were food for animals.[121] The separation of plants and animals at an industrial scale instead creates heavily polluted waste streams from both operations, including huge increases in both methane and nitrous oxide emissions.

The massive application of fertilizers creates both nitrous oxide and carbon dioxide emissions. Most fertilizer is really nothing more than an artificial, energy-rich form of nitrogen called ammonia, which has the chemical formula NH_3. After application, some of the NH_3 breaks down directly into nitrous oxide in the soils. The carbon dioxide emissions instead come from fertilizer manufacturing, carried out at huge scale today via the Haber process (see the Figure). Invented in 1909, this process made the Green Revolution possible and helped drive the enormous human population increases of the twentieth century. It relies on combining nitrogen from the air (N_2) with energy-rich hydrogen (H_2) to form NH_3: $N_2 + 3H_2 \Rightarrow 2NH_3$.[122] The problem is that today hydrogen is synthesized mainly by combining the methane in natural gas (CH_4) with water, producing H_2 and CO_2.[123] Carbon-free hydrogen production, when it becomes economical at industrial scale, will no doubt find significant application here.

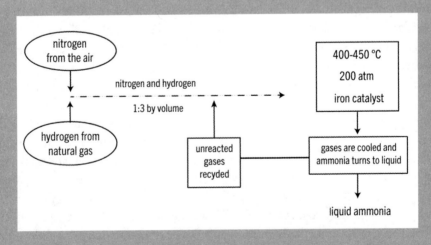

Depiction of the Haber process for the production of ammonia used for fertilizers.

slowing the breakdown of carbon-containing soil nutrients by microbes, which produces both carbon dioxide and methane. Recent studies, however, suggest that these reduced tilling or no-till practices may not provide as much benefit as previously thought, because they can result in a compensating loss of carbon in the deeper soil.[59]

Agroforestry enhances agricultural lands by adding trees to the landscape. It includes *alley cropping*, which refers to growing a variety of crops, shrubs, or trees in close proximity to one another. This brings substantial benefits—yields are higher, and healthier resulting soils store more carbon. Alley cropping stands out as a relatively inexpensive practice with significant potential to sequester carbon in the soils (Table 10.1). Among other beneficial agroforestry methods, *silvopasture* refers to including trees on land used for grazing. *Riparian forest buffers* are lines of trees planted along rivers and streams that abut farmland, which filter pollutants from farm runoff while also offering wildlife habitat (Figure 10.3a, Figure 10.3b, and Figure 10.3c).[60]

Biochar is another way to sequester carbon on agricultural lands, although it does not fit comfortably within the natural land management paradigm. Biochar is a carbon-rich, charcoal-like product that is manufactured by heating excess biomass in an environment without oxygen. When applied on farmland, the biochar carbon has a soil lifetime of over 100 years, while the natural carbon cycle would return most of this carbon to the atmosphere more rapidly. While yet to be implemented at a large scale, the approach has some significant potential. Biochar does require harvesting biomass, raising land use concerns. Life-cycle analysis of biochar also has to account for the significant energy input needed for manufacturing and transport, which adds to costs and detracts from the climate benefits.[61]

Sustainable Agriculture: Policy and Advocacy

Both the federal government and individual states should support initiatives that will sequester carbon on agricultural land. A recent analysis by the World Resources Institute (WRI) recommends that new federal funding for climate-friendly agriculture should be divided roughly in thirds: financial assistance to farmers, on-site technical aid, and monitoring and research. This prescription comes from assessing the barriers to new practices, like high up-front costs to farmers, lack of enough on-the-ground technical expertise, and the need to overcome prevailing social and cultural norms (see Box 10.2).[62]

Several existing programs at the federal level also provide ways to deliver funds. The US Department of Agriculture (USDA) administers an Environmental Quality Incentives Program, which offers financial and technical assistance to growers who wish to implement environmentally friendly practices.[63] The USDA also helps farmers and ranchers build plans for conservation and climate-friendly practice through its Conservation Stewardship Program;[64] yet another program funds conservation easements for agriculture and grasslands. All this money is funneled through the enormous federal Farm Bill—last renewed in 2018—which authorizes food and nutrition assistance, agriculture and dairy programs, conservation, forestry, horticulture, and bioenergy.[65]

Some states have also adopted healthy soils programs. For example, in 2019 Nebraska created a task force to develop a comprehensive plan for rehabilitating depleted soils, including better retention of soil carbon and consequently slower rates of carbon dioxide and methane release. Illinois expanded a conservation law regarding soils and water so that conservation districts can conduct activities to improve soil health. Vermont and New Mexico have recently passed similar legislation, and bills are pending elsewhere.[66] These efforts provide models that can help advocates petition for similar legislation in their own states.

Three-quarters of US nitrous oxide emissions come from agriculture.[67] In addition to minimizing commercial fertilizer use (Box 10.1), these emissions can also be lowered by reducing food waste. Discarded food scraps contribute one-quarter of all emissions from food production.[68] This is an enormous factor, since food production is itself responsible for no less than 26 percent of global greenhouse gas emissions. The sources include the full food supply chain, livestock and fisheries, crop production for humans and animals, and land use effects.[69]

Climate advocates who want to work in agriculture should check out the National Sustainable Agriculture Coalition (NSAC).[70] This is an alliance of over 100 grassroots groups dispersed across the country, who share resources and work to support small and midsize family farms, protect natural resources through more sustainable practices, and embrace equity and diversity in food production. NSAC provides a distinctive viewpoint with respect to large corporate agriculture, and has a specific mission to research, develop, and advocate for federal policies that benefit small farms. A well-established office in Washington, D.C. advances the shared goals of the grassroots network through lobbying and educating legislators and agency administrators.

Livestock Management

Livestock—especially cattle—are a large source of methane emissions.[71] The methane comes from two sources: fermentation of ingested food by microbes that live in the gut, and the processing of manure. Most of the methane from internal fermentation is expelled into the atmosphere by belching. As you might expect, this is extremely difficult to control. There have been attempts to reduce these emissions by varying livestock diet and adding chemicals to livestock feed to slow the growth of the microbes, but so far they have not been very effective.[72]

Much better progress has been made in capturing methane from manure treatment through the use of *anaerobic biodigesters*. These are covered lagoons or tanks of various designs where manure is collected and allowed to generate *biogas*, a roughly 50:50 mixture of methane and carbon dioxide. Sometimes other types of waste are also included in the tanks. After incubation, liquids and solids are recovered for use as compost or other useful materials. The biogas is recovered separately and can be burned on-site to generate electricity for powering farm operations. Alternatively, methane can be purified from the biogas

Box 10.2: The Growing Climate Solutions Act

The California and RGGI cap and trade carbon pricing laws include carbon offsets, which offer a way for regulated companies to cut emissions by funding activities outside the scope of the programs. As we discussed in Chapter 7, this offers a way for some emitters to meet their obligations at a lower cost than would be possible by greening their own operations. Recently, there has also been strong growth in voluntary carbon markets, as an increasing number of companies and other actors see the importance of offsetting their emissions, even though they are not required by law to do so.[124]

Small farmers and private forest owners could benefit from these carbon markets, acquiring funds to put new carbon sequestering practices into operation. However, accessing these markets often comes with a high barrier. It is difficult for growers to identify reliable information about how the programs work, and to gain access to the technical expertise needed to determine how their operations can qualify.

To address this, a bipartisan bill known as the Growing Solutions Climate Act was introduced in both houses of Congress in the spring and summer of 2020.[125] The bill directs the Department of Agriculture to establish a technical assistance network and certification program that would provide transparency and legitimacy, connecting carbon market funders with farmers and family foresters. The information flow would be beneficial both ways because professionals with expertise in the operation of carbon markets often have difficulty identifying growers able to put the offset funds to proper use. The program would include USDA-certified third-party verifiers who would confirm the practices are put in place.

By late spring of 2021, the bill was gaining momentum through congressional hearings,[126] and had gained broad support among a wide range of not-for-profit conservation groups, food companies, and farmer advocacy organizations.[127] It fills an important gap and offers a good example of how government is able to identify solutions to on-the-ground barriers to implementing climate-friendly policies. If the bill is passed, the increasing numbers of growers and family foresters who become engaged could also become an important source of practical innovation, driving new and more effective practices for carbon sequestration.

mixture and processed to enter the natural gas pipeline system or be used as compressed natural gas for vehicle fuel.[73]

US livestock production today is primarily done in industrial-sized *concentrated animal feeding operations* (CAFOs), which house hundreds to millions of dairy cows, hogs, chickens, and other animals.[74] The livestock are kept under appalling, inhumane conditions, and the operations generate a great deal of air and water contamination in addition to greenhouse gas pollution. Clearly this is an area where climate and animal advocates can support a common cause. Advocating for the welfare of farm animals and making the case for vegetarian diets are two ways to tackle the problem. However, it may be more effective to put direct pressure on corporations. For example, advocacy campaigns combining protests with behind-the-scenes negotiations have been successful in driving the now-mainstream practice of sourcing eggs from cage-free poultry.[75]

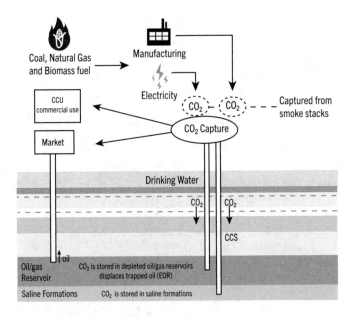

Figure 10.4: Carbon capture technology. Carbon dioxide in the emissions streams from both fossil fuel and biomass-fired electric power and manufacturing plants is captured and injected into depleted oil and gas reservoirs or underground saline formations. Alternately, the captured CO_2 may find uses in existing markets, which is referred to as carbon capture and utilization (CCU).

3. Carbon Capture Technologies

Let's now look at how industrial approaches to carbon removal can contribute to reaching climate goals. The most promising new technology is *direct air capture and storage* (DACS), which, as the name suggests, is a method for directly removing carbon dioxide from the atmosphere. DACS development can be incentivized by expanding CCS from industry smokestacks, which has already been thoroughly tested at a smaller scale. Another approach worth exploring is *enhanced weathering*, which works by speeding up the natural process by which atmospheric carbon dioxide is taken up into surface minerals. Finally, *bioenergy with carbon capture and storage* (BECCS) has many advocates but relies on extensive biomass harvesting. This may limit its ability to contribute significantly at the scale required to stabilize the climate.

Carbon Capture and Storage in Industry (CCS)

Because it is a key enabling technology, we will look first at the capture, utilization, and storage of carbon dioxide from industrial emissions. The concentration of carbon dioxide in smokestack gases is much higher than it is in the atmosphere, making it easier to capture efficiently. Regardless of how carbon dioxide is initially captured, it is fed into the same infrastructure for pipeline transport and geologic burial. CCS is the catalyst to spur the buildout that can later be used for DACS (Figure 10.4).

Many discussions of CCS emphasize its application for coal- and gas-fired electricity generation plants. However, employing CCS in electric power generation prolongs the period of fossil fuel reliance and limits the penetration of carbon-free power into electricity markets. In contrast, CCS is critical in heavy manufacturing, including cement plants, fertilizer production, and iron and steelmaking.[76] In these industries, there is currently no good alternative to carbon capture for mitigating emissions.[77] And using CCS in the manufacturing of hydrogen from natural gas could be a good low-carbon way to augment renewable hydrogen synthesis from water electrolysis. Among other applications, this would help increase the capacity of hydrogen to provide long-term energy storage on the electricity grid.[78]

CCS works by reacting the carbon dioxide in smokestack emissions with a class of organic chemicals known as *amines*. Carbon dioxide heading up the stack enters a tall, vertical tower filled with an absorbent packing, while a liquid solution of amines trickles down from the top of the tower. The packing offers a large surface area for the chemical reaction, which attaches the carbon dioxide to the amine. In a second column that runs at a higher temperature, the carbon dioxide is stripped off the amine, which is then recycled to react again. The net effect is to scrub the carbon dioxide out of the smokestack gas. At the end of the process, the carbon dioxide is recovered as a pure, concentrated liquid ready for transport and burial.[79]

The least costly and most efficient way to transport the concentrated carbon dioxide over long distances is by pipeline, where enough pressure can be applied to keep it in liquid form. For storage, the best option is injection into porous saline formations at least a half-mile deep into the Earth. These are filled with briny saltwater and occupy large spaces deep underground. The reservoirs are capped by impermeable rock, such as a layer of shale, which keeps the carbon dioxide trapped (Figure 10.4). Other trapping mechanisms include migration of the liquified carbon into porous rocks, dissolution in underground water, and reaction with minerals. These processes work together to help prevent the carbon dioxide from leaking back out.[80]

Despite these mechanisms, there remains some potential for the buried carbon dioxide to escape. This is important for policymakers and advocates to be aware of because fears of leakage can undermine public support for CCS. For example, some carbon could leak back out at the injection site, where the caprock barrier must be pierced. Old, abandoned oil and gas wells might also leak carbon dioxide, and geologic faults might potentially allow for escape. So far, however, there have been no known leakages at any burial site. State and federal regulations require continuous site monitoring and prior evaluation of the potential for seismic activity that might disrupt containment.[81] Specific attention is given to protecting any nearby underground drinking water reservoirs during siting, construction, and operation of injection sites.[82]

Carbon Capture and Utilization (CCU)

Carbon dioxide *capture* is already in place at thirteen US power and manufacturing facilities.[83] However, capital costs to build (or retrofit) an industrial plant with CCS are high, and operating costs increase because the capture process requires significant additional energy. Burying the captured carbon adds further costs. This means that subsidies or other policies are needed to incentivize the full CCS process. But in the US, serious consideration of these policies has only begun in the past few years. In their absence, some companies nonetheless have found it profitable to capture the carbon dioxide as a marketable product in its own right. Thus, the process of *carbon capture and utilization* (CCU) was born.

The biggest buyers of captured carbon are fossil industry firms that drill for oil and gas. Deep injection of carbon dioxide pressurizes oil and gas reservoirs, allowing further extraction. This process is known as *enhanced oil recovery* (EOR). In fact, most of the carbon dioxide captured so far in the US has been injected into fossil fuel reservoirs (Figure 10.4). Only relatively small amounts have been injected into saline formations, though they represent most US storage capacity and will have to be relied upon when large-scale burial becomes a reality.

EOR is not as unfriendly to the climate as it appears. The injected carbon dioxide is securely trapped, so it offsets some of the emissions from burning the extracted oil or natural gas. It is certainly an improvement over the use of carbon dioxide taken directly from natural underground reservoirs, which has been the main driver of EOR so far. Policies to require the use of captured carbon in EOR would thus have some climate benefit, while also spurring the development of pipeline infrastructure for carbon dioxide transport.[84] EOR markets will eventually wither as transportation is electrified and biomass substitutes for petroleum become more widely available.[85]

Ultimately, CCU is limited by the size of carbon dioxide markets. In addition to EOR, other markets include the carbonated beverage industry and the production of commodities from photosynthetic algae, which require carbon dioxide to grow efficiently.[86] One issue in considering CCU is the ultimate fate of the carbon. For example, captured carbon dioxide that is sold to the beverage industry returns to the atmosphere quickly. However, carbon incorporated in products with long lifespans such as cement (which replaces carbon from limestone) could have climate benefits since it would be taken out of circulation for long periods.[87]

Research is also being conducted on the use of carbon dioxide as a feedstock to make fuels and synthetic chemicals. While this may yield some useful and perhaps profitable new products, large-scale use of carbon dioxide in these applications is limited by the need to supply energy.[88] Research into CCU applications is valuable and could spur further developments in carbon capture technology. But ultimately, large-scale atmospheric carbon removal will have to be driven by CCS, not CCU.[89]

Direct Air Capture and Storage (DACS)

DACS technology operates by the same principles as CCS from industrial smokestacks. The most important difference is that the carbon dioxide concentration in the air is over 100 times lower than in the emissions of gas-fired power plants. Concentrating the dilute carbon dioxide is energy-intensive and costly, making the process much less efficient.

Small-scale applications of direct carbon capture from air already exist, such as its use on submarines to prevent toxic buildup of carbon dioxide.[90] However, the cost of DACS—estimated at \$94–\$600 per ton—is presently too high for large-scale deployment.[91] This is where innovation deriving from industry CCS comes in. For example, there are many alternatives to the use of absorbent packing in the smokestack emissions towers, as well as to other detailed aspects of carbon dioxide trapping, amine recycling, and purification. These other approaches aren't cost-competitive now, but incentives provided by expanded markets could lead to new discoveries that would make them viable.[92]

It is a promising sign that a few companies are already positioning themselves as pioneers in DACS development. Among these, the Swiss company Climeworks has partnered with other firms to capture atmospheric carbon dioxide as part of a larger process to synthesize jet fuel from renewable hydrogen.[93] The plan is to combine the captured carbon dioxide with high-energy, renewable hydrogen gas made by electrolysis with solar or wind power. This creates a beneficial cycle that may eventually eliminate carbon dioxide emissions from aircraft—one of the most difficult applications to electrify.[94]

An advantage of DACS is that the facilities can be located near favorable underground storage sites, which minimizes pipeline infrastructure needs. But the energy requirements of the process are so high that powering it with fossil fuels would negate a significant part of the climate benefit. Aggressive action to decarbonize the electricity grid in the next decade, coupled with cost-reducing innovations emerging from industry CCS, would create a viable pathway for deploying DACS at a large scale by mid-century. It is estimated that by 2050, DACS could bury 0.19–1.4 $GtCO_2$ per year in the US. Ultimately, the potential of the technology appears nearly unlimited.[95] In the long run, DACS may be the best option we have to safely stabilize the climate.

Accelerated Weathering

Weathering is a natural part of the carbon cycle, by which carbon dioxide is removed from the atmosphere over a period of tens to hundreds of thousands of years. The basic idea of accelerated weathering is simply to speed up this natural process. In any weathering reaction, atmospheric carbon dioxide combines with minerals on the Earth's surface and is directly incorporated into the rock (Figure 10.1).[96] So, the great advantage of an industrial-scale enhanced weathering industry would be that the carbon dioxide does not need to be concentrated, transported, or buried anywhere.

Weathering speeds up when the mineral surface area exposed to the air is increased. The big challenge of enhanced weathering is crushing and spreading out enough newly mined rock to make the reaction happen at a large enough scale. Neither the logistics nor the energy costs of this are yet well understood. However, after carbon dioxide is absorbed, the resulting modified minerals could be useful in building construction or other projects requiring large amounts of rock aggregate. Another application involves spreading out pulverized silicate rocks on agricultural land, where plants accelerate the weathering reaction, and the rock dust counters soil acidification (replacing the limestone presently used

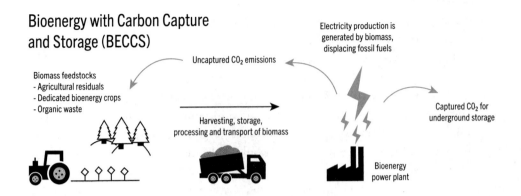

Bioenergy with Carbon Capture and Storage (BECCS)

Biomass feedstocks
- Agricultural residuals
- Dedicated bioenergy crops
- Organic waste

Uncaptured CO_2 emissions

Harvesting, storage, processing and transport of biomass

Electricity production is generated by biomass, displacing fossil fuels

Captured CO_2 for underground storage

Bioenergy power plant

Figure 10.5: Depiction of the basic elements of BECCS. Harvested biomass is burned in electric power plants or used as a manufacturing feedstock (for example, in bioethanol production). Carbon dioxide emissions from these operations are captured and sequestered underground.

for this purpose).[97] The potential existence of markets for the weathered materials might help spur the technology. In terms of policy, the best approach now is robust funding for both basic research and for understanding the environmental and social impacts of such a large-scale project.[98]

Bioenergy with Carbon Capture and Storage (BECCS)

In BECCS, biomass is first harvested and then either burned in a power plant to supply electricity or converted into biofuel. Carbon dioxide emissions from these processes are captured and sequestered underground. The biomass then regrows naturally, removing carbon from the atmosphere by photosynthesis. The process is repeated to draw down carbon dioxide levels as low as is desired (Figure 10.5).[99] BECCS can look great on paper, but there are substantial economic and environmental concerns that will likely limit its application.

The biomass sources for BECCS are residues and wastes from farmlands and forests, such as crop remains and thinnings removed for wildfire suppression. Dedicated energy crops and material from landfills and other waste streams could also provide biomass sources. Use of this biomass for BECCS would compete with its other potential applications, such as production of cellulosic ethanol, biochar, or other commodities.[100]

Like other technologies that rely on biomass, the major concern with BECCS is that the harvesting and processing operations will have harmful climate and land use effects that negate the potential benefits. As we have seen, life-cycle analysis for cellulosic ethanol production yields greater climate benefits compared to gasoline, so BECCS at ethanol fermentation plants might appear beneficial. However, other considerations weigh against this. Like DACS, BECCS at a scale useful for climate mitigation depends on buildout of the CCS infrastructure. But significant BECCS operations would also likely require extensive land use conversion, with potential loss of carbon stores through deforestation. And unless heavily subsidized, the electricity generated from biomass would not be cost-competitive with carbon-free alternatives, such as solar and wind power.[101]

Predicting the relative merits of these still-nascent technologies is a hazardous undertaking. Further exploration and pilot-scale testing of DACS, enhanced weathering, and BECCS may yield unexpected insights that would change our present thinking about the relative merits of these approaches. Undoubtedly, the urgency of carbon removal justifies investment of research dollars in all these approaches. Despite the arguments made above, an advantage of BECCS is that it relies on known technologies and, while likely requiring subsidies, is inexpensive enough to be implemented right away.[102] In contrast, we do not know how long it will be before the costs of DACS or enhanced weathering decrease enough to make them commercially feasible.

Carbon Removal: Policy and Advocacy

Despite the crucial role that CCS must play, incentives from the US federal government and states have been meager thus far. Regulations for carbon dioxide transport and sequestration are in place, though mainly in support of EOR. Funds have been provided for regional public-private partnerships to enable CCS commercialization.[103] The Department of Energy (DOE) has funded an initiative to develop a few pilot-scale CCS projects as proof that the technology can work. One BECCS project at an Illinois ethanol fermentation plant has reached fruition and is able to capture and bury over a million tons of carbon dioxide per year.[104] However, we have seen that capture and safe underground storage at a scale of billions of tons per year is required.

One initiative with significant potential is a recently expanded federal tax credit available to firms that capture carbon dioxide from industrial sources or electricity power plants.[105] Under the new law, in 2026 the tax credit will reach $35 per ton of captured carbon dioxide that is sold and used (CCU), and $50 per ton for carbon dioxide that is sequestered (CCS). These credits may be large enough to promote significant expansion of industry CCS, though they are too low to incentivize DACS directly. However, Congress has at least recognized the value of DACS by adding a provision to a December 2019 military funding bill, directing the Department of Defense to specifically pursue research in this area.[106]

It would be helpful for both the states and Congress to expand the pipeline network for carbon dioxide transport. Some Western states rich in fossil fuels have promoted this infrastructure to expand EOR, but much more development is needed. The federal government does not currently oversee the placement of carbon dioxide pipelines, so developers face a patchwork of state laws and are thus unable to invoke federal eminent domain authority.[107] This makes it easier for private landowners to block pipeline construction. Given the importance of large-scale carbon drawdown, a strong federal role in building out this infrastructure is certainly justified.[108]

A steep national price on carbon could drive CCS expansion very effectively. The aggressive federal EICDA bill—which aims for at least $115 per ton of carbon dioxide in 2030—includes a rebate of the fee for companies that employ CCS. Modeling studies project that, while overall natural gas use would stay about the same, by 2030 about 40 percent of the carbon dioxide emitted from gas-fired power plants would be economical enough to capture.[109] This would enormously spur innovations in CCS technology. This modeling offers an important benchmark to estimate how high carbon prices should be to further develop CCS, although we do not yet know how much development will be impacted by the tax.

Another promising approach is to incorporate CCS into existing state programs to green the electricity and transportation sectors. In several states, initiatives are underway to include CCS in the renewable portfolio standard for electric power generation, as well as the low carbon fuel standard for transportation fuels.[110] CCS could also be mandated by federal regulation through the EPA. In fact, this was part of the Obama administration's Clean Power Plan, which was scuttled by the Trump administration. If the enactment of new rules emerges as the approach of choice for the Biden administration, then new federal legislation should be considered to provide specific authority for the regulations.[111]

Some climate advocates are uneasy about supporting the goal to promote carbon capture from smokestacks, as this can further the interests of fossil energy firms. One way to reconcile this is to restrict CCS advocacy to essential industries, such as cement, fertilizer, or steel production, where emissions cannot presently be reduced in any other way. CCS in electricity power plants is indeed problematic since it permits methane leaks and other pollution from natural gas operations to persist, while also taking market share from solar and wind power. And while EOR with captured carbon is an improvement over taking the carbon dioxide from underground reservoirs, the climate would clearly be better off with no EOR at all. Advocacy could thus reasonably focus on restricting the federal CCS tax credit and modifying the CCS provisions in carbon tax bills to exclude electric power generation.

Another argument against CCS is based on moral hazard. If atmospheric carbon dioxide is going to be buried anyway, then the force of advocacy for clean renewables and a green economy could be diluted. The renewable energy transition would not stop, but it might happen more slowly if the idea of an Earth-saving technology takes hold. Investing resources in CCS might also mean that fewer resources are available for other approaches.[112] But the other side of the argument is simple and compelling: captured and safely sequestered carbon dioxide does not contribute to warming. And as we have seen, the most reliable science says that large-scale atmospheric carbon removal is both urgent and necessary for reaching climate goals.

CCS has the potential to divide climate advocates. As in many areas, the issue is how much to compromise in order to push an initiative over the finish line. For example, it is easy to take the high ground to eliminate CCS credits for natural gas-fired power plants, but fossil fuel interests would be in strong opposition, and then possibly torpedo a bill with potential to move the technology forward. There is obviously no easy answer here. Perhaps the only sound advice is that even if committed climate advocates take opposing sides, when the fight is over the most important thing is to come together again for the next battle.

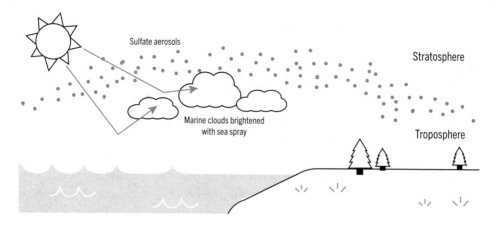

Figure 10.6: Depiction of solar geoengineering processes: sulfur aerosol injection and marine cloud brightening.

4. Solar Geoengineering

Solar geoengineering, also called *solar radiation management* (SRM), is one of the new kids on the climate block. The central idea is to alter Earth's climate system to increase the albedo. Presently about 30 percent of the sunlight reaching Earth is reflected back to space, so increasing this percentage would cause the average surface temperature to drop.[113] In the section above on forests, we noted that to be most effective, afforestation should not lower albedo. Similarly, urban development should incorporate more reflective roofing. These ideas are not particularly controversial. However, trouble brews in proposals for large-scale brightening of the atmosphere and oceans due to unforeseeable consequences if things go wrong.

There is a key distinction between solar engineering and CCS initiatives like DACS. By removing carbon dioxide from the atmosphere, DACS addresses the root cause of global warming—even if it does not count as *mitigation*, which prevents emissions in the first place. In contrast, making the atmosphere or oceans more reflective is an *adaptation*, like building a high sea wall. It does not address the basic cause.[114] If enough additional sunlight is reflected, temperatures might start going down even as greenhouse pollutant emissions increase. But this is hardly a good solution, as the continued emissions would further disrupt natural biogeochemical cycles and damage sensitive ecologies on both land and ocean. Ocean acidification, for example, would continue to worsen under such an approach.

The most discussed solar engineering option is *sulfur aerosol injection*—the manmade version of natural volcanic eruptions (Figure 10.6).[115] The aerosols would be released from aircraft flying at heights of 10-12 miles in the lower stratosphere. At present, research on this approach is almost all done by computer modeling, investigating questions like precisely which chemical form of sulfur to use—as well as where, how much and how often to release it—and how the climate system is likely to respond. It is also still curiosity-driven and not organized by any kind of mission.[116] Still, some promising, early results have been reported, suggesting that temperatures could be stabilized, and the melting of large ice sheets slowed or halted altogether.[117]

Another SRM approach is *marine cloud brightening*, which involves spraying sea salt particles into clouds over the ocean to create more cloud droplets (Figure 10.6). This has received less attention from climate researchers, although it is generally viewed as having some promise. More speculative approaches include breaking up high cirrus clouds, which absorb and trap a large amount of outgoing Earthlight. There are even proposals to continuously churn the oceans to create reflective white froth, and to build enormous orbiting mirrors that could be deployed on space stations in order to reflect sunlight.[118]

Presently, all SRM methods are poorly understood, with enormous uncertainties regarding their potential effects on the climate system. Their application could well result in different regional effects, such as altered precipitation patterns that would impact some parts of the world much more than others. Because of this, it is clearly important that strong global governance be an integral part of any SRM program. This would require enforceable agreements—a major paradigm shift in international environmental law.[119] In this context, it is important to note that all SRM approaches require continuous application over time. If global SRM were in place but then stopped abruptly, very rapid warming would follow.

SRM strategies to provide temporary relief from the worst impacts of surface heating may one day be plausible, but clearly must only be considered within the context of sustained, larger plans to eliminate the causes of global warming. They would also have to incorporate a well-planned, gradual phaseout. As advocates, perhaps our best response to the SRM question is to point out that we would be far better off to avoid such desperate measures. Let us focus together on making sure we never get there.

endnotes

Preface Notes

1. OxfordLanguages, Word of the Year 2019, https://languages.oup.com/word-of-the-year/2019/.

2. These goals would be reached by enacting the five-hundred-page policy roadmap produced by the US House of Representatives' Select Committee on the Climate Crisis—the first comprehensive plan to combat global warming generated by a congressional body. See House Select Committee on the Climate Crisis, "Solving the Climate Crisis: The Congressional Action Plan for a Clean Energy Economy and a Healthy, Resilient and Just America." Released June 30, 2020, https://climatecrisis.house.gov/report. For a modeling study projecting that this plan will reach its stated targets, see Megan Mahajan, Robbie Orvis, and Sonia Aggarwal, "Modeling the Climate Crisis Action Plan", Energy Innovation, June 2020, https://energyinnovation.org/wp-content/uploads/2020/07/Modeling-the-Climate-Crisis-Action-Plan_FOR-RELEASE.pdf.

3. James Lovelock, *Gaia: A New Look on Life on Earth*, (Oxford: Oxford University Press, 1979).

4. R.E. Zeebe et al., "Anthropogenic Carbon Release Rate Unprecedented During the Past 66 Million Years," *Nature Geoscience*, no. 9 (2016): 325, https://www.nature.com/articles/ngeo2681.

5. January 27, 2021 was "Climate Day" at the Biden White House. Following high-profile executive actions to rejoin the Paris Climate Agreement and revoke the permit for the Keystone oil pipeline, President Biden's team released a massive executive order placing the climate crisis at the center of foreign policy, national security, and economic and jobs growth, with emphasis on environmental justice and revitalization of communities dependent on fossil fuel extraction. See The White House, "Executive Order on Tackling the Climate Crisis at Home and Abroad", January 27, 2021, https://www.whitehouse.gov/briefing-room/presidential-actions/2021/01/27/executive-order-on-tackling-the-climate-crisis-at-home-and-abroad/.

6. Majority votes suffice to pass legislation in the House of Representatives, and Senate Democrats can use a process called budget reconciliation to bypass the filibuster and pass bills with a bare majority. However, this is limited to legislation with significant direct impact on the budget. Otherwise, 60 votes are required to move a bill to the Senate floor for discussion and voting—unless the filibuster is repealed, which is unlikely. Dylan Matthews, "Budget Reconciliation, Explained", *Vox*, November 23, 2016, https://www.vox.com/policy-and-politics/2016/11/23/13709518/budget-reconciliation-explained.

7. These examples all culminated in favorable Supreme Court decisions, but local efforts can also drive actions by Congress and the executive branch. See David Cole, *Engines of Liberty: How Citizen Movements Succeed*, (New York: Basic Books, 2016).

8. Melissa Denchak, "What is the Keystone XL Pipeline?" National Resources Defense Council, January 20, 2021, https://www.nrdc.org/stories/what-keystone-pipeline.

9. The data come from surveys conducted by Yale and George Mason universities, which have identified six segments of the American public that respond in distinctive ways to climate change. These are called "Global Warming's Six Americas." See Chapter 5 and Anthony Leiserowitz, "Building Public and Political Will for Climate Change Action" in *A Better Planet: Big Ideas for a Sustainable Future* (Daniel C. Esty, ed.), (New Haven: Yale University Press, 2019) 155-162.

10. David Roberts, "At Last, A Climate Policy Platform That Can Unite the Left," *Vox*, last modified July 9, 2020, https://www.vox.com/energy-and-environment/21252892/climate-change-democrats-joe-biden-renewable-energy-unions-environmental-justice. The platform discussed in this article is embodied in the House Climate Crisis roadmap for the carbon-free energy transition; see https://climatecrisis.house.gov/report.

11. For insight into Republican thinking on climate change, see Zoya Teirstein, "Feeling the Heat. How the Green New Deal Lit a Fire Under the GOP," Grist, October 14, 2020, https://grist.org/politics/republican-party-climate-change/.

12. This was the Waxman-Markey carbon pricing bill, which would have enacted a nationwide cap-and-trade program to impose a price on greenhouse gas emissions. For an account of the politics of its demise, see Ryan Lizza, "As the World Burns," *The New Yorker*, October 3, 2010, https://www.newyorker.com/magazine/2010/10/11/as-the-world-burns.

13. For denialist tactics, see Michael Mann, *The New Climate War: The Fight to Take Back Our Planet*, (New York: Hachette Book Group, 2021). For rebuttals to nearly 200 denialist arguments, see the Skeptical Science website, https://skepticalscience.com/argument.php.

14. A prominent advocate of the skeptical position is Bjorn Lomborg; see *False Alarm: How Climate Change Panic Costs us Trillions, Hurts the Poor, and Fails to Fix the Planet*, (New York: Basic Books, 2020). Journalist David Wallace-Wells has produced the most prominent recent example of gloom and doom; see *The Uninhabitable Earth: Life After Warming*, (New York: Tim Duggan Books, 2019).

15. Jason Mark, "Yes, Actually, Personal Responsibility Is Essential to Solving the Climate Crisis," *Sierra*, November 26, 2019, https://www.sierraclub.org/sierra/yes-actually-individual-responsibility-essential-solving-climate-crisis.

Chapter 1 Notes

1. For perspectives on Native American Nature writing, see Lee Schweninger, "Writing Nature: Silko and Native Americans as Nature Writers," *Multi-Ethnic Literature of the United States*, Vol. 18 (1993):, 47-60, https://doi.org/10.2307/467933 For an anthology of American nature writing since Thoreau, see Bill McKibben, ed., *American Earth* (New York: Literary Classics of the United States, 2008).

2. Julia Rosen, "Climate Change Fears Propel Scientists Out of the Lab and Into the World," *Los Angeles Times*, December 17, 2019, https://www.latimes.com/environment/story/2019-12-17/scientists-become-advocates-on-climate-change.

3. This invokes Al Gore's classic *An Inconvenient Truth*. Al Gore, *An Inconvenient Truth* (New York: Rodale, 2006). For books describing attacks on the credibility of climate science and scientists, see Michael Mann, *The Hockey Stick and the Climate Wars. Dispatches From the Front Lines.* (New York: Columbia University Press, 2012) and Naomi Oreskes and Eric Conway, *Merchants of Doubt* (New York: Bloomsbury Press, 2010).

4. William F. Ruddiman, *Earth's Climate, Past and Future*, 3rd ed. (New York: W.H. Freeman and Company, 2014), 8-10.

5. See the description of the Gaia hypothesis in the Preface for elaboration on this point.

6. A graphic showing the layers of the atmosphere is available from NASA; see https://www.nasa.gov/image-feature/goddard/earths-atmospheric-layers. NASA also put together a terrific resource for kids (and adults), available at https://spaceplace.nasa.gov.

7. See the first section of Chapter 7 for a description of the classic "Tragedy of the Commons" dilemma.

8. A good resource for learning more about the makeup of the atmosphere is https://scied.ucar.edu/chemical-composition-atmosphere-diagram. The National Center for Atmospheric Research (NCAR) in Boulder, Colorado, which sponsors this website, is a hotbed of scientific research on climate, and also has a strong program in climate education.

9. For an explanation of why some atmospheric gases have greenhouse effects and others do not, see David Archer, *Global Warming: Understanding the Forecast*, 2nd ed. (Hoboken: John Wiley & Sons, Inc., 2012), 30-32.

10. The response of water vapor to increasing amounts of the other greenhouse gases is an example of a positive feedback loop. See Chapters 2 and 3.

11. National Geographic has more information and a terrific short video on ocean circulation, see https://www.nationalgeographic.org/media/ocean-currents-and-climate/.

12. However, the buildup of ocean carbon dioxide does cause *ocean acidification*, with serious consequences for marine life. See Chapter 4.

13. The National Snow and Ice Data Center (NSIDC), at the University of Colorado, Boulder, is an excellent resource on the cryosphere. See https://nsidc.org/cryosphere/allaboutcryosphere.html.

14. Archer, *Global Warming*, 19-28, describes the layer model of the atmosphere, starting with Earth as a bare rock.

15. See https://www.wunderground.com/maps/satellite/regional-infrared for examples of IR images.

16. Earth's energy balance will be discussed in more detail in the first section of Chapter 2.

17. See David Archer, *The Global Carbon Cycle* (Princeton: Princeton University Press, 2010) for a readable, book-length description of the carbon cycle.

18. See the Kahn Academy page on atoms and molecules, https://www.khanacademy.org/science/biology/chemistry—of-life/chemical-bonds-and-reactions/a/chemical-bonds-article.

19. See https://scripps.ucsd.edu/programs/keelingcurve/ For a history of how these pivotal measurements came to be made, see Spencer R. Weart, *The Discovery of Global Warming* (Cambridge: Harvard University Press, 2008), 19-37.

20. For more information, see the Khan Academy page: https://www.khanacademy.org/science/biology/photosynthesis-in-plants/introduction-to-stages-of-photosynthesis/v/photosynthesis.

21. CB Field et al., "Primary Production of the Biosphere: Integrating Terrestrial and Oceanic Components," *Science* no. 281 (1998): 237-240, https://pubmed.ncbi.nlm.nih.gov/9657713/.

22. For an explanation of this chemistry, see https://www.khanacademy.org/science/biology/properties-of-carbon/carbon/a/carbon-and-hydrocarbons.

23. See Chapter 2 for the explanation of why we are so certain about this.

24. Climate scientist David Archer uses a similar analogy to describe environmental flows. Archer, *The Global Carbon Cycle*, Archer, *Global Warming*.

25. Nicolas Gruber et al., "The Oceanic Sink for Anthropogenic CO2 from 1994 to 2007," *Science* no. 363 (2019): 1193-1199, https://science.sciencemag.org/content/363/6432/1193?rss=1.

26. Chapter 10 discusses some land management practices that can help retain more carbon in farms, forests, and grasslands.

27. An excellent, readable description of the carbon cycle is by John Houghton, *Global Warming: The Complete Briefing*, 5th ed., (Cambridge: Cambridge University Press, 2015), 34-44.

28. The 37 Gt datum, and an enormous array of other statistics about the carbon cycle, are available from the Global Carbon Project, at https://www.globalcarbonproject.org.

29. Archer, *The Global Carbon Cycle*, 141-173.

30. Ciais, P., C. Sabine, G. Bala, L. Bopp, V. Brovkin, J. Canadell, A. Chhabra, R. DeFries, J. Galloway, M. Heimann, C. Jones, C. Le Quéré, R.B. Myneni, S. Piao and P. Thornton, 2013: Carbon and Other Biogeochemical Cycles. In: Climate Change 2013: The Physical Science Basis. Contribution of Working Group I to the Fifth Assessment Report of the Intergovernmental Panel on Climate Change [Stocker, T.F., D. Qin, G.-K. Plattner, M. Tignor, S.K. Allen, J. Boschung, A. Nauels, Y. Xia, V. Bex and P.M. Midgley (eds.)]. Cambridge University Press, Cambridge, United Kingdom and New York, NY, USA. https://www.ipcc.ch/site/assets/uploads/2018/02/WG1AR5_Chapter06_FINAL.pdf, 470-473. See also Bruhwiler, L., A. M. Michalak, R. Birdsey, J. B. Fisher, R. A. Houghton, D. N. Huntzinger, and J. B. Miller, 2018: Chapter 1: Overview of the global carbon cycle. In Second State of the Carbon Cycle Report (SOCCR2): A Sustained Assessment Report [Cavallaro, N., G. Shrestha, R. Birdsey, M. A. Mayes, R. G. Najjar, S. C. Reed, P. Romero-Lankao, and Z. Zhu (eds.)]. U.S. Global Change Research Program, Washington, DC, USA, 42-70, https://doi.org/10.7930/SOCCR2.2018.Ch1.

31. David Archer and Victor Brovkin, "The Millennial Anthropogenic Lifetime of Anthropogenic CO2," *Climatic Change* no. 90 (2008): 238-297, https://link.springer.com/article/10.1007/s10584-008-9413-1.

32. Fahey, D.W., S.J. Doherty, K.A. Hibbard, A. Romanou, and P.C. Taylor. 2017. Physical Drivers of Climate Change. In: Climate Science Special Report: Fourth National Climate Assessment, Volume I [Wuebbles D.J., D.W. Fahey, K.A. Hibbard, D.J. Dokken, B.C. Stewart, and T.K. Maycock (eds.)], U.S. Globa; Change Research Program, Washington DC, USA, 73-113, doi: 10.7930/J0513WCR.

33. Ciais, "Carbon and Other Biogeochemical Cycles," 470-473.

34. David Archer, *The Global Carbon Cycle*, 135.

35. See Chapter 10 for description of technologies for carbon removal from the atmosphere.

36. John Mason, "Understanding the Long-Term Carbon Cycle: Weathering of Rocks, a Vitally Important Carbon Sink," *Skeptical Science*, July 2, 2013, https://www.skepticalscience.com/weathering.html.

37. Ruddiman, *Earth's Climate*, 84-85, 98-100.

38. B.H. Samset et al., "Delayed Emergence of a Global Temperature Response After Emission Mitigation," *Nature Communications* (2020), 11(1):3261, www.nature.com/articles/s41467-020-17001-1.

39. The ocean liner analogy was made by climate scientist Marianne Lund, quoted in Chelsea Harvey, "Slash CO2—Then Wait, and Wait—For Temps to Drop," *E&E News*, July 9, 2020, https://www.scientificamerican.com/article/slash-co2-then-wait-and-wait-for-temperatures-to-drop/.

40. Granshaw, Frank, *Climate Toolkit: A Resource Manual for Science and Action*, PDXOpen: Open Educational Resources (2020): 28. https://pdxscholar.library.pdx.edu/pdxopen/28/0.15760/pdxopen-27.

Chapter 2 Notes

1. Al Gore, *Earth in the Balance: Ecology and the Human Spirit* (Boston: Houghton Mifflin Company, 1992).

2. Energy balances and natural drivers were introduced in Chapter 1.

3. IPCC, 2013: Summary for Policymakers. In: Climate Change 2013: *The Physical Science Basis. Contribution of Working Group I to the Fifth Assessment Report of the Intergovernmental Panel on Climate Change* [Stocker, T.F., D. Qin, G.-K. Plattner, M. Tignor, S.K. Allen, J. Boschung, A. Nauels, Y. Xia, V. Bex and P.M. Midgley (eds.)]. Cambridge University Press, Cambridge, United Kingdom and New York, NY, USA. https://www.ipcc.ch/site/assets/uploads/2018/02/WG1AR5_SPM_FINAL.pdf.

4. See https://www.ipcc.ch/about/.

5. "How Volcanoes Influence Climate," UCAR Center for Climate Education. https://scied.ucar.edu/shortcontent/how-volcanoes-influence-climate.

6. David Archer, Global Warming, 141.

7. See Figure 4 in Chapter 1 and the discussion therein.

8. "What is the Sun's Role in Climate Change?," *National Aeronautics and Space Administration*, https://climate.nasa.gov/blog/2910/what-is-the-suns-role-in-climate-change/.

9. The National Aeronautics and Space Administration (NASA) keeps detailed records of Earth's temperature, both regionally and as worldwide averages over the land and oceans. See NASA, "2020 Tied For Warmest Year on Record, NASA Analysis Shows," Release 21-005, January 14, 2021, https://www.nasa.gov/press-release/2020-tied-for-warmest-year-on-record-nasa-analysis-shows.

10. Rebecca Lindsey and LuAnn Dahlmann, "Climate Change: Global Temperature", *National Oceanic and Atmospheric Association*, Climate.gov, https://www.climate.gov/news-features/understanding-climate/climate-change-global-temperature.

11. Rebecca Lindsey and LuAnn Dahlmann, "Climate Change: Ocean Heat Content," *National Oceanic and Atmospheric Association*, Climate.gov, https://www.climate.gov/news-features/understanding-climate/climate-change-ocean-heat-content.

12. For an explanation of the ocean's high heat capacity, see https://www.khanacademy.org/science/high-school-biology/hs-biology-foundations/hs-water-and-life/v/hydrogen-bonding-in-water.

13. Zeke Hausfather, "Explainer: Will global warming 'stop' as soon as net-zero emissions are reached?" *CarbonBrief*, 29 April 2021, https://www.carbonbrief.org/explainer-will-global-warming-stop-as-soon-as-net-zero-emissions-are-reached

14. National Oceanic and Atmospheric Administration, "Arctic Report Card: Update for 2020." https://arctic.noaa.gov/Report-Card/Report-Card-2020.

15. See Chapter 1 for a discussion of Earth's albedo.

16. More examples of both positive and negative feedback loops are given in Chapter 3.

17. Chapter 4 describes impacts to Earth's cryosphere in more detail.

18. See Bibliography for Michael Mann, *The Hockey Stick and the Climate Wars*, and Naomi Oreskes and Eric Conway, *Merchants of Doubt*, for descriptions of the individuals and groups that are prominent in climate change denialism. Politics and implications for advocacy are discussed in Chapters 5 and 6.

19. The *Skeptical Science* website is a great resource for exploring climate deniers' arguments. For the arguments about the Sun as cause of climate change, see Skeptical Science, "Sun and Climate: Moving in Opposite Directions," https://www.skepticalscience.com/solar-activity-sunspots-global-warming-intermediate.htm.

20. Joseph Romm, Climate Change: *What Everyone Needs to Know.* (Oxford: Oxford University Press, 2016), 23-27.

21. Folland, C.K., T.R. Karl, J.R. Christy, R.A. Clarke, G.V. Gruza, J. Jouzel, M.E. Mann, J. Oerlemans, M.J. Salinger and S.-W. Wang, 2001: Observed Climate Variability and Change. In: *Climate Change 2001: The Scientific Basis. Contribution of Working Group I to the Third Assessment Report of the Intergovernmental Panel on Climate Change* [Houghton, J.T., Y. Ding, D.J. Griggs, M. Noguer, P.J. van der Linden, X. Dai, K. Maskell, and C.A. Johnson (eds.)]. Cambridge University Press, Cambridge, United Kingdom and New York, NY, USA, 881pp, 134.

22. See "Picture Climate: How Can We Learn From Tree Rings?," *NOAA*, https://www.ncdc.noaa.gov/news/picture-climate-how-can-we-learn-tree-rings.

23. Zeke Hausfather, "Explainer: How Data Adjustments Affect Global Temperature Records," *CarbonBrief*, https://www.carbonbrief.org/explainer-how-data-adjustments-affect-global-temperature-records.

24. See Mann, *The Hockey Stick and the Climate Wars*, for the full story by climate scientist Michael Mann, who has been the target of the most strident attacks.

25. "What Evidence is There for the Hockey Stick?," *Skeptical Science*, https://www.skepticalscience.com/broken-hockey-stick.htm.

26. For a detailed review of the state-of-the-art in paleoclimatology, see Jessica E. Tierney et al., "Past Climates Inform Our Future," *Science*, no. 370 (2020), 680, https://science.sciencemag.org/content/370/6517/eaay3701?rss=1.

27. Ruddiman, *Earth's Climate*, 95.

28. Ruddiman, *Earth's Climate*, 121-131.

29. Dr. Hansen coined this phrase in a lecture that he gave at a December, 2005 conference of the American Geophysical Union. Andrew C. Revkin, "Climate Scientist Says NASA Tried to Silence Him," *The New York Times*, January 29, 2006, https://www.nytimes.com/2006/01/29/science/earth/climate-expert-says-nasa-tried-to-silence-him.html.

30. See https://www.skepticalscience.com/climate-change-little-ice-age-medieval-warm-period-intermediate.htm.

31. Ruddiman, *Earth's Climate*, 137-140. For more details, see T. Westerhold et al., "An Astronomically Dated Record of Earth's Climate and its Predictability Over the Last 66 Million Years," *Science*, no. 369 (2020), https://science.sciencemag.org/content/369/6509/1383.

32. For a detailed map of the maximum ice sheet extent during the last glaciation, see https://www.arcgis.com/home/item.html?id=f1e7378b962d42168fdefec3b6eb8b5f.

33. For an account of climate scientist Lonnie Thompson's work to retrieve ice cores from high mountain glaciers, see Mark Bowen, *Thin Ice: Unlocking the Secrets of Climate in the World's Highest Mountains*, (New York: Henry Holt and Company, 2005).

34. Houghton, *Global Warming*, 80-84.

35. See Holli Riebeek, "Paleoclimatology: The Oxygen Balance," NASA Earth Observatory, https://earthobservatory.nasa.gov/features/Paleoclimatology_OxygenBalance. In Chapter 3, I give an explanation of carbon isotopes and how they are used to prove the fossil fuel origins of atmospheric carbon dioxide.

36. Alan Buis, "Milankovitch (Orbital) Cycles and Their Role in Earth's Climate," NASA's Jet Propulsion Laboratory, https://climate.nasa.gov/news/2948/milankovitch-orbital-cycles-and-their-role-in-earths-climate/.

37. Fiona Hibbert, Eelco Rohling and Katherine Grant, "Scientists Looked at Sea Levels 125,000 Years in the Past. The Results are Terrifying," *The Conversation*, http://theconversation.com/scientists-looked-at-sea-levels-125-000-years-in-the-past-the-results-are-terrifying-126017.

38. A Dutton & K Lambeck, "Ice Volume and Sea Level During the Last Interglacial," *Science*, no. 337 (2012): 216-219, https://science.sciencemag.org/content/337/6091/216.

39. See Chapter 4 for discussion of ice sheet melting. Carbon removal from the atmosphere is discussed in Chapter 10.

40. The uncertainty analysis in the IPCC reports is described in its Technical Summary document; see Stocker, T.F., D. Qin, G.-K. Plattner, L.V. Alexander, S.K. Allen, N.L. Bindoff, F.-M. Bréon, J.A. Church, U. Cubasch, S. Emori, P. Forster, P. Friedlingstein, N. Gillett, J.M. Gregory, D.L. Hartmann, E. Jansen, B. Kirtman, R. Knutti, K. Krishna Kumar, P. Lemke, J. Marotzke, V. Masson-Delmotte, G.A. Meehl, I.I. Mokhov, S. Piao, V. Ramaswamy, D. Randall, M. Rhein, M. Rojas, C. Sabine, D. Shindell, L.D. Talley, D.G. Vaughan and S.-P. Xie, 2013: Technical Summary. In: *Climate Change 2013: The Physical Science Basis. Contribution of Working Group I to the Fifth Assessment Report of the Intergovernmental Panel on Climate Change* [Stocker, T.F., D. Qin, G.-K. Plattner, M. Tignor, S.K. Allen, J. Boschung, A. Nauels, Y. Xia, V. Bex and P.M. Midgley (eds.)]. Cambridge University Press, Cambridge, United Kingdom and New York, NY, USA. https://www.ipcc.ch/site/assets/uploads/2018/02/WG1AR5_TS_FINAL.pdf.

41. Accelerated warming in recent years suggests that 1% may even be an underestimate. See "The Global Climate in 2015-2019," World Meteorological Organization, https://library.wmo.int/doc_num.php?explnum_id=9936.

42. The data supporting this statement is available from the International Energy Agency (IEA) World Energy Outlook, https://www.iea.org/weo/.

43. See Figure 3 in Chapter 1 for a depiction of the chemical structure of calcium carbonate.

44. Jocelyn Timperley, "Q&A: Why Cement Emissions Matter for Climate Change," *CarbonBrief*, https://www.carbonbrief.org/qa-why-cement-emissions-matter-for-climate-change. Chapter 9 describes carbon dioxide emissions from industry.

45. The effect of land use change on global warming is described later in this chapter.

46. For US emissions contributions, see "Overview of Greenhouse Gases," Environmental Protection Agency, https://www.epa.gov/ghgemissions/overview-greenhouse-gases. Global methane data are available from the Global Carbon Project, M. Sausnois et al., "The Global Methane Budget 2000-2017," *Earth System Science Data*, no. 12 (2020): 1-63, DOI:10.5194/essd-12-1561-2020, https://www.globalcarbonproject.org/methanebudget/index.htm.

47. A.J. Turner et al., "Interpreting Contemporary Trends in Atmospheric Methane," *Proceedings of the National Academy of Sciences of the. United States of America*, no. 116 (2019): 2805-2813, https://www.pnas.org/content/pnas/116/8/2805.full.pdf.

48. B. Himiel at al., "Preindustrial 14CH4 indicates greater anthropogenic fossil CH4 emissions," *Nature*, no. 578 (2020): 409, https://www.nature.com/articles/s41586-020-1991-8.pdf.

49. See Chapter 6 for further discussion of methane emissions from fossil fuel infrastructure.

50. The data on methane is from the Global Carbon Project, M. Sausnois et al., "The Global Methane Budget 2000-2017," *Earth System Science Data*, no. 12 (2020): 1-63, DOI:10.5194/essd-12-1561-2020, https://www.globalcarbonproject.org/methanebudget/index.htm.

51. Data for nitrous oxide is from the "World Meteorological Organization (WMO) Greenhouse Gas Bulletin (GHG Bulletin)– No. 14: The State of Greenhouse Gases in the Atmosphere Based on Global Observations Through 2017," World Meteorological Organization, available at: https://library.wmo.int/index.php?lvl=notice_display&id=20697>.

52. "Overview of Greenhouse Gases," Environmental Protection Agency, https://www.epa.gov/ghgemissions/overview-greenhouse-gases.

53. Sabrina Shankman, "What is Nitrous Oxide, and Why is it a Climate Threat?" *InsideClimate News*, https://insideclimatenews.org/news/11092019/nitrous-oxide-climate-pollutant-explainer-greenhouse-gas agriculture-livestock. More information about agricultural emissions and the industrialized food system is provided in Chapter 10.

54. UN environment programme, ozone secretariat, "20 Questions and Answers," United Nations Environment Programme, Ozone Secretariat, https://ozone.unep.org/20-questions-and-answers. See question 17 for a detailed discussion of how tropospheric and stratospheric ozone affect climate.

55. The sink for methane is described in the carbon cycle section of Chapter 1.

56. Thomas G. Spiro and William M. Stigliani, *Chemistry of the Environment*, 2nd ed. (Upper Saddle River: Prentice Hall, Inc., 2003), 226-231.

57. Read more about the ozone hole at this webpage from the American Chemical Society: https://www.acs.org/content/acs/en/education/whatischemistry/landmarks/cfcs-ozone.html.

58. "About Montreal Protocol," United Nations Environment Programme, https://www.unenvironment.org/ozonaction/who-we-are/about-montreal-protocol.

59. For the political part of the story, see the industry section of Chapter 9.

60. The nonprofit Center for Climate and Energy Solutions, C2ES, has written an informative fact sheet about black carbon, available at https://www.c2es.org/document/what-is-black-carbon/.

61. See the discussion of transportation solutions in Chapter 9.

62. Chelsea Harvey, "Cleaning up Air Pollution May Strengthen Global Warming," *ClimateWire*, https://www.scientificamerican.com/article/cleaning-up-air-pollution-may-strengthen-global-warming/.

63. "FactCheck: Is "Global Dimming" Shielding us From Catastrophe?," Climate Tipping Points, climatetippingpoints.info, https://climatetippingpoints.info/2019/04/15/fact-check-is-global-dimming-shielding-us-from-catastrophe/. Climate effects from human aerosols are discussed later in this chapter.

64. This is a form of solar geoengineering, which we will discuss in Chapter 10.

65. This description of cloud effects draws on Houghton, *Global Warming*, 53-59, and Archer, *Global Warming*, 78-81. For a review of recent findings suggesting that the overall cloud feedback may be positive, see Fred Pearce, "Why Clouds Are Key to New Troubling Projections on Warming," *Yale Environment*, no. 360 (2020), https://e360.yale.edu/features/why-clouds-are-the-key-to-new-troubling-projections-on-warming.

66. Ruddiman, *Earth's Climate*, 329-332.

67. ArchaeoGLOBE Project, "Archaeological Assessment Reveals Earth's Early Transformation Through Land Use," *Science*, no. 365 (2019): 897-902. https://science.sciencemag.org/content/365/6456/897.

68. Friedlingstein, P., O'Sullivan, M., Jones, M. W., Andrew, R. M., Hauck, J., Olsen, A., Peters, G. P., Peters, W., Pongratz, J., Sitch, S., Le Quéré, C., Canadell, J. G., Ciais, P., Jackson, R. B., Alin, S., Aragão, L. E. O. C., Arneth, A., Arora, V., Bates, N. R., Becker, M., Benoit-Cattin, A., Bittig, H. C., Bopp, L., Bultan, S., Chandra, N., Chevallier, F., Chini, L. P., Evans, W., Florentie, L., Forster, P. M., Gasser, T., Gehlen, M., Gilfillan, D., Gkritzalis, T., Gregor, L., Gruber, N., Harris, I., Hartung, K., Haverd, V., Houghton, R. A., Ilyina, T., Jain, A. K., Joetzjer, E., Kadono, K., Kato, E., Kitidis, V., Korsbakken, J. I., Landschützer, P., Lefèvre, N., Lenton, A., Lienert, S., Liu, Z., Lombardozzi, D., Marland, G., Metzl, N., Munro, D. R., Nabel, J. E. M. S., Nakaoka, S.-I., Niwa, Y., O'Brien, K., Ono, T., Palmer, P. I., Pierrot, D., Poulter, B., Resplandy, L., Robertson, E., Rödenbeck, C., Schwinger, J., Séférian, R., Skjelvan, I., Smith, A. J. P., Sutton, A. J., Tanhua, T., Tans, P. P., Tian, H., Tilbrook, B., van der Werf, G., Vuichard, N., Walker, A. P., Wanninkhof, R., Watson, A. J., Willis, D., Wiltshire, A. J., Yuan, W., Yue, X., and Zaehle, S.: "Global Carbon Budget 2020", *Earth Syst. Sci. Data*, 12, 3269–3340, https://doi.org/10.5194/essd-12-3269-2020, 2020.

69. See IPCC, 2019: Summary for Policymakers. In: *Climate Change and Land: an IPCC special report on climate change, desertification, land degradation, sustainable land management, food security, and greenhouse gas fluxes in terrestrial ecosystems* [P.R. Shukla, J. Skea, E. Calvo Buendia, V. Masson-Delmotte, H. – O. Pörtner, D. C. Roberts, P. Zhai, R. Slade, S. Connors, R. van Diemen, M. Ferrat, E. Haughey, S. Luz, S. Neogi, M. Pathak, J. Petzold, J. Portugal Pereira, P. Vyas, E. Huntley, K. Kissick, M. Belkacemi, J. Malley, (eds.)]. In press. https://www.ipcc.ch/site/assets/uploads/sites/4/2020/02/SPM_Updated-Jan20.pdf, 9.

70. Food and Agriculture Organization of the United Nations (FAO), *Global Forest Resources Assessment, 2015: How are the World's Forests Changing?*, http://www.fao.org/3/a-i4793e.pdf.

71. Houghton, *Global Warming*, 190-192.

72. See Chapter 10 for more discussion about emissions from the industrial food system.

73. R. Lal, "Carbon Emission From Farm Operations," *Environment International*, no. 30 (2004): 981-990, 982-984, 986-987.

74. See the IUCN webpage at https://www.iucn.org/resources/issues-briefs/blue-carbon.

75. The International Union for Conservation of Nature (IUCN) provides useful information and strategies for conserving peatlands. See https://www.iucn.org/resources/issues-briefs/peatlands-and-climate-change.

76. A good summary of urban heat islands is provided by the National Center for Atmospheric Research (NCAR) in Boulder, Colorado at https://scied.ucar.edu/longcontent/urban-heat-islands Also see Hibbard, K.A., F.M. Hoffmann, D. Huntzinger, and T.O. West, 2017. Changes in land cover and terrestrial biogeochemistry. In: Climate science Special report: Fourth National Climate assessment. Volume I[Wuebbles, D.J., D.W. Fahey, K.A. Hibbard, D.J. Dokken, B.C. Stewart, and T.K. Maycock (eds.)]. U.S. Global Change Research Program, Washington, DC, USA, pp. 277-302, doi: 10.7930/J0416V6X, Key Finding 4.

77. See Chapter 10 for discussion of beneficial land use practices in cities.

78. The nonprofit Greenhouse Gas Management Institute offers a very clear explanation of global warming potentials, at https://ghginstitute.org/2010/06/28/what-is-a-global-warming-potential/.

Chapter 3 Notes

1. An alternative (but mostly rejected) idea is the "abiotic" theory of fossil fuel formation, which holds that some oil is not of biological origin, but rather was formed from inorganic carbon and hydrogen-containing materials deep under the Earth's surface. See https://enviroliteracy.org/energy/fossil-fuels/abiotic-theory/.

2. The combustion reaction is described and compared to the respiration component of the photosynthesis cycle in Chapter 1.

3. Spiro and Stigliani, *Chemistry of the Environment*, 25-29.

4. The Carboniferous Period, https://ucmp.berkeley.edu/carboniferous/carboniferous.php.

5. Spiro and Stigliani, *Chemistry of the Environment*, 21-24, and J.M. Hunt et al., "Early Developments in Petroleum Geochemistry," *Organic Geochemistry*, no. 33 (2002): 1025-1052, https://www.sciencedirect.com/science/article/abs/pii/S0146638002000566.

6. For example, see David Goodstein, *Out of Gas: The End of the Age of Oil*, (New York: WW Norton and Company, 2004), and Matthew R. Simmons, *Twilight in the Desert: The Coming Saudi Oil Shock and the World Economy*. (Hoboken: John Wiley & Sons, 2005).

7. See Chapter 6 for a description of fracking technology.

8. The Rainforest Action Network has published an exhaustive analysis on fossil fuel expansion and how it might be financed. See https://www.ran.org/wpcontent/uploads/2019/03/Banking_on_Climate_Change_2019_vFINAL1.pdf.

9. Among the three fossil fuels, coal contributed 32 percent of total consumption in 2017, while oil contributed 40 percent and natural gas 28 percent. Ritchie and Roser, "Fossil Fuels," *Our World in Data*, https://ourworldindata.org/fossil-fuels. This resource is part of the highly regarded *Our World in Data* compilation, which measures and reports on a wide range of global problems and trends. See also the Global Change Data Lab, https://global-change-data-lab.org.

10. "Gas Hydrates," US Department of Energy, https://www.energy.gov/fe/science-innovation/oil-gas-research/methane-hydrate.

11. Exhaustive statistics on global fossil fuel use, prices, production, and reserves are available free from the *BP Statistical Review of World Energy*, (2020), https://www.bp.com/content/dam/bp/business-sites/en/global/corporate/pdfs/energy-economics/statistical-review/bp-stats-review-2020-full-report.pdf.

12. *BP Statistical Review of World Energy*, (2020), https://www.bp.com/content/dam/bp/business-sites/en/global/corporate/pdfs/energy-economics/statistical-review/bp-stats-review-2020-full-report.pdf. R/P for both oil and gas have remained nearly constant at about 50 years since at least 1988 (the earliest data included in BP's analysis). BP reports that the R/P for coal decreased from about 230 years to 130 years between 1998 and 2007, which is presumably due to large overestimates of proven reserves in 1998 and earlier. Since 2007 R/P for coal has remained nearly constant at 130 years.

13. For a definitive history of fossil fuel development, see Daniel Yergin, *The Prize: The Epic Quest for Oil, Money and Power*, (New York: Free Press, 2008).

14. *BP Statistical Review of World Energy*, (2020), https://www.bp.com/content/dam/bp/business-sites/en/global/corporate/pdfs/energy-economics/statistical-review/bp-stats-review-2020-full-report.pdf.

15. See the discussion of carbon cycle timeframes in Chapter 1.

16. The UNFCCC is an international environmental treaty among 197 member countries, which are known as Parties to the Convention. It was enacted in 1994. For a summary of its key elements and links to further information, see https://unfccc.int/process-and-meetings/the-convention/what-is-the-united-nations-framework-convention-on-climate-change. For the statement on dangerous anthropogenic interference, see *United Nations Framework Convention on Climate Change*, Article 2 (1992), https://unfccc.int/resource/docs/convkp/conveng.pdf.

17. Samuel Randalls, "History of the 2°C Climate Target," *WIREs Climate Change*, https://onlinelibrary.wiley.com/doi/abs/10.1002/wcc.62.

18. Since 1994, the Parties to the UNFCCC hold an annual conference. The Paris Agreement of 2015 was reached at the 21st Conference of the Parties (COP21). See *The Paris Agreement*, United Nations Climate Change, https://unfccc.int/process-and-meetings/the-paris-agreement/the-paris-agreement.

19. Climate impacts are discussed in detail in Chapter 4.

20. Some basic information about the carbon budget is given by Our World in Data, at https://ourworldindata.org/how-long-before-we-run-out-of-fossil-fuels. A useful summary in the form of an infographic is available from the World Resources Institute: https://www.wri.org/resources/data-visualizations/infographic-global-carbon-budget. This infographic shows the IPCC-based carbon budget from the 5th Assessment Report, which includes data through 2011. See the text for more recent analysis.

21. See Kerry A. Emanuel, *Climate Science and Climate Risk: A Primer*, ftp://texmex.mit.edu/pub/emanuel/PAPERS/Climate_Primer.pdf. Pages 8-12 provide a good introduction to the complexity of the climate system and development of climate models.

22. A readable and nicely detailed explanation of climate modeling is provided by CarbonBrief, at https://www.carbonbrief.org/qa-how-do-climate-models-work. CarbonBrief is an excellent journalism website based in the United Kingdom, which covers a wide range of topics on climate change and energy.

23. As discussed in Chapter 1, the parts of the climate system are the atmosphere, ocean, cryosphere, land, and biosphere. For more about general circulation models (GCMs), see https://www.ipcc-data.org/guidelines/pages/gcm_guide.html.

24. Houghton, Global Warming, 95-99. Also see Paul Voosen, "How Far Out Can We Forecast the Weather? Scientists Have a New Answer," *Science* (2019), doi:10.1126/science.aax0032. https://www.sciencemag.org/news/2019/02/how-far-out-can-we-forecast-weather-scientists-have-new-answer.

25. An informative video explaining the distinction between climate and weather, by climate scientist Katherine Hayhoe, is available at https://www.skepticalscience.com/weather-forecasts-vs-climate-models-predictions-intermediate.htm.

26. Senator Inhofe's performance can be viewed at http://time.com/3725994/inhofe-snowball-climate/.

27. Houghton, *Global Warming*, 99.

28. Houghton, *Global Warming*, 90-94.

29. Paul N. Edwards, *A Vast Machine, Computer Models, Climate Data, and the Politics of Global Warming*, (Cambridge: MIT Press, 2010), 139-140.

30. Archer, *Global Warming*, 23-26. In the improved model, the temperature at the ground comes out to be 19 percent warmer than the temperature high in the atmosphere, actually not a bad reproduction of reality.

31. The ice albedo feedback loop is discussed in the section on carbon dioxide and temperature in Chapter 2.

32. Brief explanations of 15 distinct climate feedback loops, both positive and negative, are available at https://earthhow. com/climate-feedback-loops/.

33. See Edwards, *A Vast Machine*, 143-149, for a very readable description of how climate scientists improved models over time, adding successively more complex features.

34. S. Sherwood et al., "An Assessment of Earth's Climate Sensitivity Using Multiple Lines of Evidence," *Reviews of Geophysics*, https://doi.org/10.1029/2019RG000678.

35. Paul Voosen, "Earth's Climate Destiny Finally Seen More Clearly," *Science*, no. 369 (2020):354-355, https://science. sciencemag.org/content/369/6502/354.

36. "How Reliable are Climate Models?," *Skeptical Science*, https://www.skepticalscience.com/climate-models.htm.

37. See http://ossfoundation.us/projects/environment/global-warming/models-v.-observations.

38. For a full development of these ideas, see Edwards, *A Vast Machine*, xiii-xxvii, 111-186.

39. For details on how general circulation models are used differently to model weather versus climate, see Edwards, *A Vast Machine*, 148-149.

40. Hayhoe, K., D.J. Wuebbles, D.R. Easterling, D.W. Fahey, S. Doherty, J. Kossin, W. Sweet, R. Vose, and M. Wehner, 2018: Our Changing Climate. In *Impacts, Risks, and Adaptation in the United States: Fourth National Climate Assessment, Volume II* [Reidmiller, D.R., C.W. Avery, D.R. Easterling, K.E. Kunkel, K.L.M. Lewis, T.K. Maycock, and B.C. Stewart (eds.)]. U.S. Global Change Research Program, Washington, DC, USA, pp. 72–144. doi: 10.7930/NCA4.2018.CH2, Figure 2.1.

41. The effects of climate change drivers to produce heating or cooling are discussed in Chapter 2.

42. Houghton, *Global Warming*, 116-120.

43. The uncertainty analysis in the IPCC reports is described in Stocker, "Carbon and Other Biogeochemical Cycles."

44. For a general discussion of the tendency toward conservative projections, see N. Oreskes, M. Oppenheimer & D. Jamieson, "Scientists Have Been Underestimating the Pace of Climate Change," https://blogs.scientificamerican.com/ observations/scientists-have-been-underestimating-the-pace-of-climate-change/.

45. Flato, G., J. Marotzke, B. Abiodun, P. Braconnot, S.C. Chou, W. Collins, P. Cox, F. Driouech, S. Emori, V. Eyring, C. Forest, P. Gleckler, E. Guilyardi, C. Jakob, V. Kattsov, C. Reason and M. Rummukainen, 2013: Evaluation of Climate Models. In: *Climate Change 2013: The Physical Science Basis. Contribution of Working Group I to the Fifth Assessment Report of the Intergovernmental Panel on Climate Change* [Stocker, T.F., D. Qin, G.-K. Plattner, M. Tignor, S.K. Allen, J. Boschung, A. Nauels, Y. Xia, V. Bex and P.M. Midgley (eds.)]. Cambridge University Press, Cambridge, United Kingdom and New York, NY, USA. https://www.ipcc.ch/site/assets/uploads/2018/02/WG1AR5_Chapter09_FINAL.pdf, 743-745.

46. Over 97 percent of peer-reviewed climate science papers find support for anthropogenic warming. See http://theconsensusproject.com.

47. See the discussion on land use change in Chapter 2.

48. Friedlingstein, "Global Carbon Budget 2020," Figure 9.

49. See Box 3 in Chapter 1 for how to interpret the gigaton units.

50. Friedlingstein, "Global Carbon Budget 2020," Figure 9.

51. See the discussion of the carbon cycle in Chapter 1.

52. T. DeVries et al., "Recent Increase in Oceanic Carbon Uptake Driven by Weaker Upper-Ocean Overturning," *Nature*, no. 542 (2017): 215-218.

53. See https://www.ipcc.ch/sr15/faq/faq-chapter-1/. Also see Rogelj, J., D. Shindell, K. Jiang, S. Fifita, P. Forster, V. Ginzburg, C. Handa, H. Kheshgi, S. Kobayashi, E. Kriegler, L. Mundaca, R. Séférian, and M.V. Vilariño, 2018: Mitigation Pathways Compatible with 1.5°C in the Context of Sustainable Development. In: *Global Warming of 1.5°C. An IPCC Special Report on the impacts of global warming of 1.5°C above pre-industrial levels and related global greenhouse gas emission pathways, in the context of strengthening the global response to the threat of climate change, sustainable development, and efforts to eradicate poverty* [Masson-Delmotte, V., P. Zhai, H.-O. Pörtner, D. Roberts, J. Skea, P.R. Shukla, A. Pirani, W. Moufouma-Okia, C. Péan, R. Pidcock, S. Connors, J.B.R. Matthews, Y. Chen, X. Zhou, M.I. Gomis, E. Lonnoy, T. Maycock, M. Tignor, and T. Waterfield (eds.)]. In Press. https://www.ipcc.ch/site/assets/uploads/sites/2/2019/05/SR15_Chapter2_Low_Res.pdf, section 2.2.2.2 and Table 2.2.

54. Global Carbon Project, "Carbon Budget 2020," https://www.globalcarbonproject.org/carbonbudget/index.htm.

55. The International Energy Agency (IEA) has calculated a carbon budget for a 2.0°C world that is similar to that provided in the IPCC1.5 report. See https://www.carbontracker.org/carbon-budgets-explained/.

56. Rogelj, "Mitigation Pathways Compatible with 1.5°C in the Context of Sustainable Development," section 2.2.2.2 and Table 2.2.

57. T. Mauritsen and Robert Pincus, "Committed Warming Inferred from Observations," *Nature Climate Change*, no. 7 (2017): 652.

58. A large release of methane and carbon dioxide from the Arctic permafrost represents a potential climate change tipping point. See Chapter 4 for discussion.

59. This is the author's estimate based on the amount of fossil fuel burned so far (Table 3.1), and the amount of carbon emitted from fossil fuel burning from 1850-2017, ignoring differences in carbon emissions among the three fossil fuels. Le Quere, C. et al., "Global Carbon Budget 2018," *Earth System Science Data*, no. 10 (2018): 2141-2194, https://www.earth-syst-sci-data.net/10/2141/2018/. Similar estimates of the fraction of burnable reserves have been made; see, for example, Ritchie, H. and Roser, M., "Fossil Fuels," *Our World in Data*, https://ourworldindata.org/fossil-fuels. (The authors estimate that 63 percent of proven reserves of fossil fuels must be stranded for a carbon budget that produces a 2.0°C world).

60. We will look at the economic and political implications of stranding fossil fuel assets in Chapter 6.

61. Ruddiman, *Earth's Climate*, 134-135.

62. R.E. Zeebe et al., "Anthropogenic Carbon Release Rate Unprecedented During the Past 66 Million Years," *Nature Geoscience*, no. 9 (2016): 325, https://www.nature.com/articles/ngeo2681.

63. See https://www.ipcc.ch/sr15/faq/faq-chapter-1/, FAQ1.2, Figure 1.

64. See Friedlingstein, "Global Carbon Budget 2020."

65. Houghton, *Global Warming*, 134-137.

66. We will take up the subject of climate change impacts in Chapter 4.

67. In 2011, the total anthropogenic forcing was estimated as 1.13–3.33 W/m². See Figure 1 in Chapter 2 and the related discussion on radiative forcing.

68. Hausfather, Z., "New Scenarios Show How the World Could Limit Warming to 1.5C in 2100," *CarbonBrief* (2018), https://www.carbonbrief.org/new-scenarios-world-limit-warming-one-point-five-celsius-2100.

69. GP Wayne, "The Beginners Guide to Representative Concentration Pathways," *Skeptical Science*, 14-18, https://skepticalscience.com/docs/RCP_Guide.pdf, https://skepticalscience.com/docs/RCP_Guide.pdf. Information in this review is drawn in part from DP van Vuuren et al., "The Representative Concentration Pathways: an Overview," *Climatic Change*, no. 109 (2011): 5-31.

70. See the Interlude.

71. IPCC, 2018: Summary for Policymakers. In: *Global Warming of 1.5°C. An IPCC Special Report on the impacts of global warming of 1.5°C above pre-industrial levels and related global greenhouse gas emission pathways, in the context of strengthening the global response to the threat of climate change, sustainable development, and efforts to eradicate poverty* [Masson-Delmotte, V., P. Zhai, H.-O. Pörtner, D. Roberts, J. Skea, P.R. Shukla, A. Pirani, W. Moufouma-Okia, C. Péan, R. Pidcock, S. Connors, J.B.R. Matthews, Y. Chen, X. Zhou, M.I. Gomis, E. Lonnoy, T. Maycock, M. Tignor, and T. Waterfield(eds.)] https://www.ipcc.ch/site/assets/uploads/sites/2/2019/05/SR15_SPM_version_report_LR.pdf.

72. IPCC, 2018, Summary for Policymakers. In: *Global Warming of 1.5°C.*, see especially Figures SPM.3a and SPM.3b. For criticism of the reliance of IPCC models on "carbon negative" approaches, see Benjamin Storrow and Chelsea Harvey, "IPCC Modelers' Secret Weapon: Negative Emissions Tech," E&E News, December 22, 2020, https://www.eenews.net/climatewire/2020/12/22/stories/1063721299.

73. See Chapter 10 for discussion of natural land management and industrial carbon sequestration.

74. See https://blogs.nicholas.duke.edu/citizenscientist/volcanic-carbon-dioxide/. For a more detailed discussion, see World Meterorological Organization, WMO Greenhouse Gas Bulletin, no. 15, November 25, 2019, https://library.wmo.int/doc_num.php?explnum_id=10100.

Chapter 4 Notes

1. For a broad overview of climate change impacts, see Jay, D.R. Reidmiller, C.W. Avery, D. Barrie, B.J. DeAngelo, A. Dave, M. Dzaugis, M. Kolian, K.L.M. Lewis, K. Reeves, and D. Winner, 2018: Overview. *In Impacts, Risks, and Adaptation in the United States: Fourth National Climate Assessment, Volume II*[Reidmiller, D.R., C.W. Avery, D.R. Easterling, K.E. Kunkel, K.L.M. Lewis, T.K. Maycock, and B.C. Stewart (eds.)]. U.S. Global Change Research Program, Washington, DC, USA, pp. 33–71. doi: 10.7930/NCA4.2018.CH1.

2. Michael Mann, "Doomsday Scenarios are as Harmful as Climate Change Denial," *The Washington Post*, last modified July 12, 2017, https://www.washingtonpost.com/opinions/doomsday-scenarios-are-as-harmful-as-climate-change-denial/2017/07/12/880ed002-6714-11e7-a1d7-9a32c91c6f40_story.html.

3. Guidance for advocacy is offered in Chapter 5.

4. IPCC, "Summary for Policymakers," *Global Warming of 1.5°C*," (2018). For a useful summary, also see O. Hoegh-Guldberg et al., "The Human Imperative of Stabilizing Global Climate Change at 1.5°C, *Science*, no. 365 (2019): 8 of 11, https://science.sciencemag.org/content/sci/365/6459/eaaw6974.full.pdf.

5. See the discussion in the last section of Chapter 3.

6. IPCC, "Summary for Policymakers," *Global Warming of 1.5°C*, (2018): 13-14.

7. B.C. O'Neill et al., "IPCC Reasons for Concern Regarding Climate Change Risks," *Nature Climate Change*, no. 7 (2017): 28-37, https://www.nature.com/articles/nclimate3179.

8. See the section on equity later in this chapter. Social justice is discussed further in the chapters on advocacy and fossil fuels, and the issue also arises in conjunction with many of the policies discussed in Chapters 7-10.

9. See https://climatetippingpoints.info.

10. W. Steffen et al., "Trajectories of the Earth System in the Anthropocene," *Proceedings of the National Academy of Sciences in the United States of America*, no. 115 (2018): 8252-8259, https://www.pnas.org/content/pnas/115/33/8252.full.pdf.

11. Sophie Yeo, "Anthropocene: The Journey to a New Geological Epoch," CarbonBrief, 2016, https://www.carbonbrief.org/anthropocene-journey-to-new-geological-epoch.

12. "Heinrich and Dansgaard-Oeschger Events," *National Centers for Environmental Information*, https://www.ncdc.noaa.gov/abrupt-climate-change/Heinrich%20and%20Dansgaard–Oeschger%20Events.

13. Ice cores are discussed in the paleoclimatology section of Chapter 2.

14. Robert McSweeney, "Explainer: Nine 'Tipping Points' That Could be Triggered by Climate Change," *CarbonBrief*, 2020, https://www.carbonbrief.org/explainer-nine-tipping-points-that-could-be-triggered-by-climate-change.

15. Gregory Cooper et al., "Regime Shifts Occur Disproportionately Faster in Larger Ecosystems," *Nature Communications*, no. 11 (2020), https://www.nature.com/articles/s41467-020-15029-x.

16. For a simple experiment that shows how heating water causes it to expand, see https://scied.ucar.edu/activity/thermal-expansion-water.

17. Houghton, *Global Warming*, 164-169. For more detail, see the NASA webpage on sea level change at https://sealevel.nasa.gov/understanding-sea-level/global-sea-level/ice-melt.

18. The National Snow and Ice Data Center is part of the Cooperative Institute for Research in Environmental Sciences (CIRES) at the University of Colorado, Boulder. See https://nsidc.org/cryosphere/quick-facts for details on the cryosphere and up-to-date information about ice melting.

19. For information about the iceberg cleavage event, see https://www.sciencenews.org/article/giant-iceberg-broke-antarctica-larsen-c-ice-shelf-stuck. Jeff Goodell, *The Water Will Come: Rising Seas, Sinking Cities, and the Remaking of the Civilized World* (New York: Little, Brown and Company, 2017) offers an excellent book-length reporter's account of sea level rise.

20. JC Ryan et al., "Greenland Ice Sheet Surface Melt Amplified by Snowline Migration and Bare Ice Exposure," *Science Advances* 5, no. 3 (2019), https://advances.sciencemag.org/content/5/3/eaav3738. For an informative narrative about ice sheet melting and sea level rise, emphasizing Greenland, see Jon Gertner, "The Secrets in Greenland's Ice Sheet," *The New York Times*, November 12, 2015.

21. Chelsea Harvey, "Historic Greenland Melt is 'Glimpse of the Future'," *E&E News*, August 2, 2019, https://www.eenews.net/climatewire/2019/08/02/stories/1060825907.

22. For up-to-date Arctic sea ice news and analysis, see the National Snow & Ice Data Center, http://nsidc.org/arcticseaicenews/. Also see Taylor, P.C., W. Maslowski, J. Perlwitz, and D.J. Wuebbles, 2017: Arctic changes and their effects on Alaska and the rest of the United States. In: *Climate Science Special Report: Fourth National Climate Assessment, Volume I* [Wuebbles, D.J., D.W. Fahey, K.A. Hibbard, D.J. Dokken, B.C. Stewart, and T.K. Maycock (eds.)]. U.S. Global Change Research Program, Washington, D.C., USA, pp. 303-332, doi: 10.7930/J00863GK, section 11.2.2.

23. Christina Hulbe, "Guest Post: How Close is the West Antarctic Ice Sheet to a 'Tipping Point'?," *CarbonBrief*, 2020, https://www.carbonbrief.org/guest-post-how-close-is-the-west-antarctic-ice-sheet-to-a-tipping-point.

24. E. Rignot et al., "Four Decades of Antarctic Ice Sheet Mass Balance From 1979-2017," *Proceedings of the National Academy of Sciences*, no. 116 (2019): 1095-1103, https://www.pnas.org/content/116/4/1095.

25. Chelsea Harvey, "World's 'Third Pole' is Melting Away," *E&E News*, February 4, 2019, https://www.eenews.net/climatewire/stories/1060119363.

26. "Melting Glaciers," *National Park Service*, https://www.nps.gov/glac/learn/nature/climate-change.htm.

27. For sea level rise data over the past five years, see "The Global Climate in 2015-2019," World Meteorological Organization (2019), https://library.wmo.int/doc_num.php?explnum_id=9936 For estimates of sea level rise later in the century, see Sweet, W.V., R. Horton, R.E.Kopp, A.N. LeGrande, and A. Romanou. 2017: Sea level rise. In: *Climate Science Special Report: Fourth National Climate Assessment, Volume I* [Wuebbles, D.J., D.W. Fahey, K.A. Hibbard, D.J. Dokken, B.C. Stewart, and T.K. Maycock (eds.)]. U.S. Global Change Research Program, Washington, D.C., USA, pp. 333-363, doi:10.7930/J0VM49F2, Key finding 2.

28. For sea level rise projections to the year 2100, see Rebecca Lindsey, "Climate Change: Global Sea Level," *Climate.gov*, November 19, 2019, https://www.climate.gov/news-features/understanding-climate/climate-change-global-sea-level.

29. A video showing the world with sea levels 230 feet higher than today is available; see "How the Earth Would Look if all the Ice Melted," https://www.youtube.com/watch?v=VbiRNT_gWUQ.

30. For basic information, see "What is High Tide Flooding?," *National Oceanic and Atmospheric Administration (NOAA)*, https://oceanservice.noaa.gov/facts/nuisance-flooding.html. For a summary of the 2020 NOAA report, see NOAA, "U.S. High Tide Flooding Continues to Increase," *NOAA* (2020), https://www.noaa.gov/media-release/us-high-tide-flooding-continues-to-increase.

31. "Storm surge," *U.S. Climate Resilience Toolkit*, last modified March 25, 2020, https://toolkit.climate.gov/topics/coastal/storm-surge.

32. For information about how the land rebounds from past glaciations, see "What is Glacial Isostatic Adjustment?" *NOAA*, https://oceanservice.noaa.gov/facts/glacial-adjustment.html. For details on the impact of land subsidies on the extremely rapid sea level rise in Louisiana, see https://sealevelrise.org/states/louisiana/.

33. This approach is similar to the modeling depicted in Color Plate 9 and described in Chapter 3, where the experimental data is the temperature record rather than sea level change or other aspects of the cryosphere.

34. Taylor, "Arctic Changes and Their Effects," Key Finding 4; Sweet, "Sea Level Rise," *Climate Science Special Report: Fourth National Climate Assessment, Volume I*, Key Finding 1.

35. See Table 1 in Chapter 2 for a summary of IPCC assessments that global warming is attributable to human activities.

36. Brad Plumer, "How the Weather gets Weaponized in Climate Change Messaging," *The New York Times*, March 4, 2019, https://www.nytimes.com/2019/03/01/climate/weather-climate-change.html.

37. James Hansen, Makiko Sato, and Reto Ruedy, "The New Climate Dice: Public Perception of Climate Change," *NASA Goddard Institute for Space Studies*, 2012, https://www.giss.nasa.gov/research/briefs/hansen_17/.

38. For a clear description of the basic principles of attribution science, see https://www.sciline.org/evidence-blog/climate-attribution. For a detailed study, see National Academies of Science, Engineering and Medicine, "Attribution of Extreme Weather Events in the Context of Climate Change," *National Academies of Science, Engineering and Medicine* (2016), https://doi.org/10.17226/21852.

39. Vose, R.S., D.R. Easterling, K.E. Kunkel, A.N. LeGrande, and M.F. Wehner, 2017: Temperature changes in the United States. In: *Climate Science Special Report: Fourth National Climate Assessment, Volume I* [Wuebbles, D.J., D.W. Fahey, K.A. Hibbard, D.J. Dokken, B.C. Stewart, and T.K. Maycock (eds.)]. U.S. Global Change Research Program, Washington, D.C., USA, pp. 185-206, doi:10.7930/J0N29V45, Table 6.3.

40. John Schwartz, "That Siberian Heat Wave? Yes, Climate Change was a Big Factor," *The New York Times*, July 15, 2020, https://www.nytimes.com/2020/07/15/climate/siberia-heat-wave-climate-change.html.

41. E.C.J. Oliver et al., "Longer and More Frequent Marine Heatwaves Over the Past Century," *Nature Communications* no. 9 (2018), https://www.nature.com/articles/s41467-018-03732-9.

42. JE Walsh et al., "The High Latitude Marine Heat Wave of 2016 and its Impacts on Alaska," *Explaining Extreme Events of 2016 From a Climate Perspective, Bulletin of the American Meteorological Society*, no. 99 (2018): 34-38, https://journals.ametsoc.org/doi/pdf/10.1175/BAMS-ExplainingExtremeEvents2016.1.

43. Vose, "Temperature changes in the United States," *Climate Science Special Report: Fourth National Climate Assessment, Volume I*, Key Finding 2.

44. The Stanford Earth website provides an explanation and short video depicting the polar vortex; see https://earth.stanford.edu/news/polar-vortex-science-behind-cold#gs.x8b12a.

45. R. Blackport et al., "Minimal Influence of Reduced Arctic Sea Ice on Coincident Cold Winters in Mid-Latitudes," *Nature Climate Change* (2019), https://doi.org/10.1038/s41558-019-0551-4. Also see Taylor, "Arctic Changes and Their Effects," supporting information for Key Finding 5.

46. Houghton, *Global Warming*, 143-145.
47. See Chapter 2 for discussion of how clouds affect Earth's radiation balance.
48. For further discussion of Hadley cells, see North Carolina Climate Office, General Circulation of the Atmosphere, https://climate.ncsu.edu/edu/AtmosCirculation.
49. For a primer on desertification, see Robert McSweeney, "Explainer: 'Desertification' and the Role of Climate Change," *CarbonBrief*, 2019, https://www.carbonbrief.org/explainer-desertification-and-the-role-of-climate-change.
50. Wehner, M.F., J.R. Arnold, T. Knutson, K.E. Kunkel, and A.N. LeGrande, 2017: Droughts, floods and wildfires. In: *Climate Science Special Report: Fourth National Climate Assessment, Volume I* [Wuebbles, D.J., D.W. Fahey, K.A. Hibbard, D.J. Dokken, B.C. Stewart, and T.K. Maycock (eds.)]. U.S. Global Change Research Program, Washington, D.C., USA, pp. 231-256, doi: 10.7930/J0CJ8BNN, Key Finding 4.
51. N.S. Diffenbaugh, D.L. Swain, and D. Touma, "Anthropogenic Warming has Increased Drought Risk in California," *Proceedings of the National Academy of Sciences of the United States of America*, no. 112 (2015): 3931-3936. See also K. Marvel et al., "Twentieth-Century Hydroclimate Changes Consistent with Human Influence." *Nature*, no. 569 (2019): 59-65, https://www.pnas.org/content/112/13/3931.
52. A. Park Williams et al., "Large Contribution from Anthropogenic Warming to an Emerging North American Megadrought," *Science*, no. 368 (2020): 314-318, https://science.sciencemag.org/content/368/6488/314.
53. J.A. Vano et al., "Hydroclimatic Extremes as Challenges for the Water Management Community: Lessons from Oroville Dam and Hurricane Harvey," *American Meteorological Society*, 2018, http://www.ametsoc.net/eee/2017a/ch3_EEEof2017_Vano.pdf. For a list of studies investigating human influence that revealed mixed outcomes, see Easterling, D.R., K.E. Kunkel, J.R. Arnold, T. Knutson, A.N. LeGrande, L.R. Leung, R.S. Vose, D.E. Waliser, and M.F. Wehner, 2017: Precipitation change in the United States, In: Climate Science Special Report: Fourth National Climate Assessment, Volume I [Wuebbles, D.J., D.W. Fahey, K.A. Hibbard, D.J. Dokken, B.C. Stewart, and T.K. Maycock (eds.)]. U.S. Global Change Research Program, Washington, D.C., USA, pp. 207-230, doi: 10.7930/J0H993CC, Table 7.1.
54. Wehner, "Droughts, Floods and Wildfires," *Climate Science Special Report: Fourth National Climate Assessment, Volume I*, Key Finding 1. For a nuanced discussion of the role of climate change in the 2019 Midwest floods, see Daniel Cusick, "Record Floods Worsened by Warming and Levees. 'How Idiotic'," *E&E News*, May 7, 2019, https://www.eenews.net/climatewire/2019/05/07/stories/1060287345. See also Phil McKenna, "Extreme Weather Flooding the Midwest Looks a Lot Like Climate Change," *InsideClimate News*, last modified May 6, 2017, https://insideclimatenews.org/news/03052017/flooding-storms-climate-change-extreme-weather-missouri-arkansas.
55. See Adam J.P. Smith et al., "Climate Change Increases the Risk of Wildfires," ScienceBrief, last modified September 2020, https://sciencebrief.org/uploads/reviews/ScienceBrief_Review_WILDFIRES_Sep2020.pdf.
56. Wehner, "Droughts, Floods and Wildfires," *Climate Science Special Report: Fourth National Climate Assessment, Volume I*, Key Finding 6.
57. Abatzoglou JT and Willams AP, "Impact of Anthropogenic Climate Change on Wildfire across Western US Forests," *Proceedings of the National Academy of Sciences of the United States of America*, no. 113 (2016), https://www.pnas.org/content/113/42/11770. A digestible summary of this article is available: https://www.pnas.org/content/113/42/11649.
58. *World Weather Attribution*, "Attribution of the Australian Bushfire Risk to Anthropogenic Climate Change," World Weather Attribution, 2020, https://www.worldweatherattribution.org/bushfires-in-australia-2019-2020/.
59. See "Evapotranspiration and the Water Cycle," *United States Geological Survey*, https://www.usgs.gov/special-topic/water-science-school/science/evapotranspiration-and-water-cycle?qt-science_center_objects=0#qt-science_center_objects.
60. Wehner, "Droughts, Floods and Wildfires," *Climate Science Special Report: Fourth National Climate Assessment, Volume I*, Key Finding 2.
61. Zoe Cormier, "Why the Arctic is Smoldering," *BBC*, last modified August 27, 2019, http://www.bbc.com/future/story/20190822-why-is-the-arctic-on-fire.
62. Chelsea Harvey, "Soot on Snow Speeds up Warming. Here's Where it Comes From," *ClimateWire*, last modified February 15, 2019, https://www.eenews.net/climatewire/stories/1060121165.
63. Vaclav Smil, *Harvesting the Biosphere: What We Have Taken From Nature* (Cambridge: MIT Press, 2013) 5-14, 221-252.
64. The Intergovernmental Science-Policy Platform on Biodiversity and Ecosystem Services, (IPBES), 2019: Summary for policymakers of the global assessment report on biodiversity and ecosystem services of the Intergovernmental Science-Policy Platform on Biodiversity and Ecosystem Services. S. Díaz, J. Settele, E. S. Brondízio E.S., H. T. Ngo, M. Guèze, J. Agard, A. Arneth, P. Balvanera, K. A. Brauman, S. H. M. Butchart, K. M. A. Chan, L. A. Garibaldi, K. Ichii, J. Liu, S. M. Subramanian, G. F. Midgley, P. Miloslavich, Z. Molnár, D. Obura, A. Pfaff, S. Polasky, A. Purvis, J. Razzaque, B. Reyers, R. Roy Chowdhury, Y. J. Shin, I. J. Visseren-Hamakers, K. J. Willis, and C. N. Zayas (eds.). IPBES secretariat, Bonn, Germany. 56 pages. https://doi.org/10.5281/zenodo.3553579. For an overall description, see Brad Plumer, "Humans are Speeding Extinction and Altering the Natural World at an 'Unprecedented' Pace," *The New York Times*, last modified May 6, 2019, https://www.nytimes.com/2019/05/06/climate/biodiversity-extinction-united-nations.html?action=click&module=Top%20Stories&pgtype=Homepage.
65. Meera Subramanian, "Humans Versus Earth: The Quest to Define the Anthropocene," *Nature*, no. 572 (2019): 168-170, https://www.nature.com/articles/d41586-019-02381-2.
66. IPBES, *Summary for Policymakers of the Global Assessment Report on Biodiversity and Ecosystem Services of the Intergovernmental Science-Policy Platform on Biodiversity and Ecosystem Services*, section D, 7. See also "Global Biodiversity Outlook 5," Convention on Biological Diversity, 2020, https://www.cbd.int/gbo5.
67. Houghton, *Global Warming*, 190.
68. IPBES, *Summary for Policymakers of the Global Assessment Report on Biodiversity and Ecosystem Services of the Intergovernmental Science-Policy Platform on Biodiversity and Ecosystem Services*, sections A5 and A6.

69. For a beautiful and terrifying narrative of the sixth extinction, see Elizabeth Kolbert, *The Sixth Extinction* (New York: Henry Holt & Company, 2014). However, we should note that some experts question whether the term "sixth extinction" is appropriate to the present crisis, as well as the extent to which climate change contributes in addition to human extractive activities. See Peter Brannen, "Earth is not in the Midst of a Sixth Mass Extinction," *The Atlantic*, June 13, 2017, https://www.theatlantic.com/science/archive/2017/06/the-ends-of-the-world/529545/.

70. IPBES, *Summary for Policymakers of the Global Assessment Report on Biodiversity and Ecosystem Services of the Intergovernmental Science-Policy Platform on Biodiversity and Ecosystem Services*.

71. "Crop diversity: Why it Matters," *Crop Trust*, https://www.croptrust.org/our-mission/crop-diversity-why-it-matters/.

72. See IPBES, 2019, "Summary for policymakers of the global assessment report on biodiversity and ecosystem services of the Intergovernmental Science-Policy Platform on Biodiversity and Ecosystem Services", sections A5 through A7, and Kolbert, *The Sixth Extinction*, for first-hand reports of the threats to a wide variety of contemporary species. For further information on biodiversity impacts, see IPCC, 2019, Summary for Policymakers. In: IPCC Special Report on the Ocean and Cryosphere in a Changing Climate [H.-O. Pörtner, D.C. Roberts, V. Masson-Delmotte, P. Zhai, M. Tignor, E. Poloczanska, K. Mintenbeck, A. Alegría, M. Nicolai, A. Okem, J. Petzold, B. Rama, N.M. Weyer (eds.)]. In press. https://www.ipcc.ch/site/assets/uploads/sites/3/2019/09/SROCC_SPM_HeadlineStatements.pdf.

73. Julie Shaw, "Why is Biodiversity Important?" *Conservation International*, last modified November 15, 2018, https://www.conservation.org/blog/why-is-biodiversity-important.

74. Louis Bergeron, "Discovering Mammals Cause for Worry," *Stanford Report*, last modified February 9, 2009, https://news.stanford.edu/news/2009/february11/numa-021109.html.

75. See Chapter 10 for actions that can be taken to sequester carbon on the land.

76. R.L. Thoman, J. Richter-Menge, and M. L. Druckenmiller, eds., *Arctic Report Card 2020*, https://doi.org/10.25923/mn5p-t549. For a short description of Arctic fragility and distinctions with the Antarctic, see Robin McKie, "Nature's Last Refuge: Climate Change Threatens our Most Fragile Ecosystem," *The Guardian*, last modified August 29, 2015, https://www.theguardian.com/world/2015/aug/30/my-arctic-journey-fragile-ecosystem-northwest-passage.

77. Kolbert, *The Sixth Extinction*, discusses ocean acidification and its effects on corals and other marine organisms. Information is also available at the International Coral Reef Initiative website: https://www.icriforum.org/about-coral-reefs/what-are-corals. The Reef Resilience Network provides information on the economic and medicinal value of corals; see https://reefresilience.org/value-of-reefs/. Another good resource is "CO2 and Ocean Acidification: Causes, Impacts, Solutions," *Union of Concerned Scientists* https://www.ucsusa.org/global-warming/global-warming-impacts/co2-ocean-acidification#bf-toc-2.

78. See the depiction of a crystalline form of limestone in Figure 1.3.

79. Hoegh-Guldberg, "The Human Imperative of Stabilizing Global Climate Change at 1.5°C," 3-5.

80. "Basic Information About Coral Reefs," *United States Environmental Protection Agency (US EPA)*, https://www.epa.gov/coral-reefs/basic-information-about-coral-reefs See also "Value of Reefs," *Reef Resilience Network*, https://reefresilience.org/value-of-reefs/.

81. Atmospheric carbon removal, by improved natural land management and industrial-scale technologies, is discussed in Chapter 10.

82. A. Witze, "Acidic Oceans Linked to Greatest Extinction Ever," *Nature*, 2015, https://www.nature.com/news/acidic-oceans-linked-to-greatest-extinction-ever-1.17276.

83. Lucas Joel, "The Dinosaur-Killing Asteroid Acidified the Ocean in a Flash," *The New York Times*, last modified October 21, 2019, https://www.nytimes.com/2019/10/21/science/chicxulub-asteroid-ocean-acid.html.

84. X-P Song et al., "Global land change from 1982 to 2016," Nature, no. 560 (2018): 639-643,https://pubmed.ncbi.nlm.nih.gov/30089903/.

85. Vose, "Temperature Changes in the United States," *Climate Science Special Report: Fourth National Climate Assessment, Volume I*, Key Message 2: Ecosystem Services.

86. Jim Robbins, "The Rapid and Startling Decline of World's Vast Boreal Forests," *Yale Environment*, no. 360 (2015), https://e360.yale.edu/features/the_rapid_and_startling_decline_of_worlds_vast_boreal_forests.

87. Ignacio Amigo, "When Will the Amazon Hit a Tipping Point?," *Nature*, no. 578 (2020): 505-507, https://www.nature.com/articles/d41586-020-00508-4. See also E. Aparecido Trondoli Matricardi et al., "Long-Term Forest Degradation Surpasses Deforestation in the Brazilian Amazon," Science 369, no. 6509 (2020): 1378-1382, https://science.sciencemag.org/content/369/6509/1378.

88. Carbon dioxide fertilization refers to improved growth of vegetation from higher levels of carbon dioxide in the atmosphere. See Chapter 1.

89. Anja Rammig, "Tropical Carbon Sinks are Saturating at Different Times on Different Continents," *Nature*, 2020, https://www.nature.com/articles/d41586-020-00423-8.

90. IPCC "Summary for Policymakers," *Global Warming of 1.5°C*, 13.

91. Hoegh-Guldberg, "The Human Imperative of Stabilizing Global Climate Change at 1.5°C," 4-5.

92. A brief synopsis of the environmental justice movement is offered by Renee Skelton & Vernice Miller, "The Environmental Justice Movement," National Resources Defense Council (NRDC), 2016, https://www.nrdc.org/stories/environmental-justice-movement. For evidence that climate change today is already having distributional consequences, see Noah S. Diffenbaugh & Marshall Burke, "Global Warming Has Increased Global Social Inequality," *Proceedings of the National Academies of Science of the United States of America*, vol. 116 (2019): 9808-9813, https://www.pnas.org/content/116/20/9808. For a study showing that a climate-stabilized 1.5°C world would alleviate inequality, see M. Burke, W.M. Davis, and N.S. Diffenbaugh, "Large Potential Reduction in Economic Damages Under UN Mitigation Targets," *Nature*, no. 557 (2018):549-553, https://www.nature.com/articles/s41586-018-0071-9.

93. K Song et al., "Scale, Distributions and Variations of Global Greenhouse Gas Emissions Driven by US Households," *Environment International*, vol. 133 (2019), part A, https://www.sciencedirect.com/science/article/pii/S0160412019315752?via%3Dihub.

94. "Transforming our World: The 2030 Agenda for Sustainable Development," *United Nations General Assembly*, 2015, https://www.unfpa.org/resources/transforming-our-world-2030-agenda-sustainable-development. For a 2019 IPCC report on how climate change affects land management and food security worldwide, see IPCC, 2019: Climate Change and Land.

95. Kim Severson, "From Apples to Popcorn, Climate Change is Altering the Foods America Grows," *The New York Times*, last modified April 30, 2019, https://www.nytimes.com/2019/04/30/dining/farming-climate-change.html.

96. Gowda, P., J.L. Steiner, C. Olson, M. Boggess, T. Farrigan, and M.A. Grusak, 2018: Agriculture and Rural Communities. In: *Impacts, Risks, and Adaptation in the United States: Fourth National Climate Assessment, Volume II* [Reidmiller, D.R., C.W. Avery, D.R. Easterling, K.E. Kunkel, K.L.M. Lewis, T.K. Maycock, and B.C. Stewart (eds.)]. U.S. Global Change Research Program, Washington, DC, USA, pp. 391–437. doi: 10.7930/NCA4.2018.CH10.

97. Angel, J., C. Swanston, B.M. Boustead, K.C. Conlon, K.R. Hall, J.L. Jorns, K.E. Kunkel, M.C. Lemos, B. Lofgren, T.A. Ontl, J. Posey, K. Stone, G. Takle, and D. Todey, 2018: Midwest. In: *Impacts, Risks, and Adaptation in the United States: Fourth National Climate Assessment, Volume II* [Reidmiller, D.R., C.W. Avery, D.R. Easterling, K.E. Kunkel, K.L.M. Lewis, T.K. Maycock, and B.C. Stewart (eds.)]. U.S. Global Change Research Program, Washington, DC, USA, pp. 872–940. doi: 10.7930/NCA4.2018.CH21, Key Message 1.

98. See Chapter 10 for discussion of climate-friendly approaches to agricultural land management.

99. Scott Shigeoka, "Think Rural America Doesn't Care About the Climate? Think Again," *Grist*, October 29, 2019, https://grist.org/article/think-rural-america-doesnt-care-about-the-climate-think-again/.

100. Gowda, "Agriculture and Rural Communities," *Impacts, Risks, and Adaptation in the United States: Fourth National Climate Assessment, Volume II*, Key Message 4.

101. See https://tribalclimate.uoregon.edu.

102. See https://www.usdn.org/uploads/cms/documents/glcan_network_info_may2016.pdf.

103. Rudy Baum, "Climate Change and Drought in the American Southwest," *Climate Institute*, November 2015, https://climate.org/effects-of-21st-century-climate-change-on-drought-disk-in-the-american-southwest/. For an explanation of different kinds of droughts and their links to climate change, see "Causes of Drought: What's the Climate Connection?," *Union of Concerned Scientists*, https://www.ucsusa.org/global-warming/science-and-impacts/impacts/causes-of-drought-climate-change-connection.html.

104. Lall, U., T. Johnson, P. Colohan, A. Aghakouchak, C. Brown, G. McCabe, R. Pulwarty, and A. Sankarasubramanian, 2018: Water. In: *Impacts, Risks, and Adaptation in the United States: Fourth National Climate Assessment, Volume II* [Reidmiller, D.R., C.W. Avery, D.R. Easterling, K.E. Kunkel, K.L.M. Lewis, T.K. Maycock, and B.C. Stewart (eds.)]. U.S. Global Change Research Program, Washington, DC, USA, pp. 145–173. doi: 10.7930/NCA4.2018.CH3, Regional Summary.

105. For more information, see Colorado River District, Water Banking Option, https://www.coloradoriverdistrict.org/water-banking/.

106. "Resilience Strategies for Drought," Center for Climate and Energy Solutions, October 2018, https://www.c2es.org/site/assets/uploads/2018/10/resilience-strategies-for-drought.pdf.

107. Ebi, K.L., J.M. Balbus, G. Luber, A. Bole, A. Crimmins, G. Glass, S. Saha, M.M. Shimamoto, J. Trtanj, and J.L. White-Newsome, 2018: Human Health. In: *Impacts, Risks, and Adaptation in the United States: Fourth National Climate Assessment, Volume II* [Reidmiller, D.R., C.W. Avery, D.R. Easterling, K.E. Kunkel, K.L.M. Lewis, T.K. Maycock, and B.C. Stewart (eds.)]. U.S. Global Change Research Program, Washington, DC, USA, pp. 539–571. doi: 10.7930/NCA4.2018.CH14, Figure 14.1; Key Message 2.

108. Ebi, "Human Health," *Impacts, Risks, and Adaptation in the United States: Fourth National Climate Assessment, Volume II*, Key Message 1.

109. Kevin Krajick, "Humidity May Prove Breaking Point for Some Areas as Temperatures Rise," *Earth Institute, Columbia University*, 2017, https://blogs.ei.columbia.edu/2017/12/22/humidity-may-prove-breaking-point-for-some-areas-as-temperatures-rise-says-study/.

110. Sarofim, M.C., S. Saha, M.D. Hawkins, D.M. Mills, J. Hess, R. Horton, P. Kinney, J. Schwartz, and A. St. Juliana, 2016: Ch. 2: Temperature-Related Death and Illness. The Impacts of Climate Change on Human Health in the United States: A Scientific Assessment. U.S. Global Change Research Program, Washington, D.C., 43-68. http://dx.doi.org/10.7930/J0MG7MDX, 51.

111. Christopher Flavelle, "Climate Change Tied to Pregnancy Risks, Affecting Black Mothers Most," *The New York Times*, June 18, 2020, https://www.nytimes.com/2020/06/18/climate/climate-change-pregnancy-study.html.

112. See the Center for Disease Control (CDC) Social Vulnerability Index, https://svi.cdc.gov, and the National Integrated Heat Health Information System, https://nihhis.cpo.noaa.gov/vulnerability-mapping.

113. Ebi, "Human Health," *Impacts, Risks, and Adaptation in the United States: Fourth National Climate Assessment, Volume II*, Key Message 1.

114. Beard, C.B., R.J. Eisen, C.M. Barker, J.F. Garofalo, M. Hahn, M. Hayden, A.J. Monaghan, N.H. Ogden, and P.J. Schramm. 2016: Ch. 5: Vectorborne Diseases. The Impacts of Climate Change on Human Health in the United States: A Scientific Assessment. U.S. Global Change Research Program, Washington, D.C., 129-156, http://dx.doi.org/10.7930/J0765C7V, 130.

115. Sherry Towers et al., "Climate Change and Influenza: the Likelihood of Early and Severe Influenza Seasons Following Warmer than Average Winters," *Public Library of Science (PLOS)*, January 28, 2013, https://currents.plos.org/influenza/article/climate-change-and-influenza-the-likelihood-of-early-and-severe-influenza-seasons-following-warmer-than-average-winters/.

116. Marlene Cimons, "Climate Change Could Mean Shorter Winters, but Longer Flu Seasons," *NexusMedia*, November 26, 2019, https://nexusmedianews.com/climate-change-could-mean-shorter-winters-but-longer-flu-seasons-bd632a8b3ba0.

117. Nolte, C.G., P.D. Dolwick, N. Fann, L.W. Horowitz, V. Naik, R.W. Pinder, T.L. Spero, D.A. Winner, and L.H. Ziska, 2018: Air Quality. In: *Impacts, Risks, and Adaptation in the United States: Fourth National Climate Assessment, Volume II* [Reidmiller, D.R., C.W. Avery, D.R. Easterling, K.E. Kunkel, K.L.M. Lewis, T.K. Maycock, and B.C. Stewart (eds.)]. U.S. Global Change Research Program, Washington, DC, USA, pp. 512–538. doi: 10.7930/NCA4.2018.CH13, Key Messages 1, 3, 4.

118. Drew Shindell, *Health and Economic Benefits of a 2°C Climate Policy: Testimony to the House Committee on Oversight and Reform Hearing on "The Devastating Impacts of Climate Change on Health,* 2020, https://oversight.house.gov/sites/democrats.oversight.house.gov/files/Testimony%20Shindell.pdf.

119. Connie Roser-Renouf and Edward Wile Maibach, "Strategic Communication Research to Illuminate and Promote Public Engagement with Climate Change," *Change and Maintaining Change, (Nebraska Symposium on Motivation 65),* (New York, Springer, 2018): 167-218. For health professionals advocacy, see https://medsocietiesforclimatehealth.org/become-champion-climate-health/.

120. See Chapter 5 for discussion of the interplay between adaptation and mitigation.

121. For a list of individual weather events exceeding $1 billion in financial damages, see https://www.ncdc.noaa.gov/billions/events.pdf. For a summary of statistics by category of events, see https://www.ncdc.noaa.gov/billions/events.

122. Regional chapters in the National Climate Assessment are an excellent resource to investigate these impacts; see https://nca2018.globalchange.gov.

123. Robert Rich, "The Great Recession," Federal Reserve History, 2013 https://www.federalreservehistory.org/essays/great_recession_of_200709.

124. Sverre LeRoy and Richard Wiles, "High Tide Tax: Sea-Level Rise Cost Study," *The Center for Climate Integrity, and Resilient Analytics,* June 2019, https://www.climatecosts2040.org/files/ClimateCosts2040_Report.pdf.

125. United States Government Accountability Office, "Climate Change: A Climate Migration Pilot Program Could Enhance the Nation's Resilience and Reduce Federal Fiscal Exposure," United States Government Accountability Office, July 2020, https://www.gao.gov/assets/710/707961.pdf.

126. Abrahm Lustgarden, "How Climate Migration Will Reshape America," *The New York Times,* September 15, 2020, https://www.nytimes.com/interactive/2020/09/15/magazine/climate-crisis-migration-america.html.

127. See Chapter 6 for discussion of the business and politics of climate change denial.

128. Thomas Frank, "'This Program is Sick.' Fight Brews Over Flood Insurance," *E&E News,* March 14, 2019, https://www.eenews.net/climatewire/stories/1060127249/feed.

129. Christopher Flavelle, "Climate Risk in the Housing Market has Echoes of Subprime Crisis, Study Finds," *The New York Times,* September 27, 2019, https://www.nytimes.com/2019/09/27/climate/mortgage-climate-risk.html; Christopher Flavelle, "Rising Seas Threaten an American Institution: the 30 Year Mortgage, *The New York Times,* June 19, 2020, https://www.nytimes.com/2020/06/19/climate/climate-seas-30-year-mortgage.html.

130. Simon Evans and Zeke Hausfather, "Q&A: How 'Integrated Assessment Models' Are Used to Study Climate Change," *Carbon Brief,* February 10, 2018, https://www.carbonbrief.org/qa-how-integrated-assessment-models-are-used-to-study-climate-change.

131. Hoegh-Guldberg, O., D. Jacob, M. Taylor, M. Bindi, S. Brown, I. Camilloni, A. Diedhiou, R. Djalante, K.L. Ebi, F. Engelbrecht, J. Guiot, Y. Hijioka, S. Mehrotra, A. Payne, S.I. Seneviratne, A. Thomas, R. Warren, and G. Zhou, 2018: Impacts of 1.5°C Global Warming on Natural and Human Systems. In: *Global Warming of 1.5°C. An IPCC Special Report on the impacts of global warming of 1.5°C above pre-industrial levels and related global greenhouse gas emission pathways, in the context of strengthening the global response to the threat of climate change, sustainable development, and efforts to eradicate poverty* [Masson-Delmotte, V., P. Zhai, H.-O. Pörtner, D. Roberts, J. Skea, P.R. Shukla, A. Pirani, W. Moufouma-Okia, C. Péan, R. Pidcock, S. Connors, J.B.R. Matthews, Y. Chen, X. Zhou, M.I. Gomis, E. Lonnoy, T. Maycock, M. Tignor, and T. Waterfield (eds.)]. In Press, 3-155.

132. Burke, Davis, and Diffenbough, "Large Potential Reduction in Economic Damages Under UN Mitigation Targets," 549.

133. R DeFries et al., "The Missing Economic Risks in Assessments of Climate Change Impacts," *Grantham Research Institute, Earth Institute, and Potsdam Institute,* September 2019, http://www.lse.ac.uk/GranthamInstitute/wp-content/uploads/2019/09/The-missing-economic-risks-in-assessments-of-climate-change-impacts-2.pdf.

134. A readable introduction to the SCC is "The Social Cost of Carbon," *CarbonBrief,* February 14, 2017, https://www.carbonbrief.org/qa-social-cost-carbon. See also William Nordhaus, *The Climate Casino: Risk, Uncertainty and Economics for a Warming World* (New Haven: Yale University Press, 2013), 15-19.

135. To calculate present-day equivalent dollars from an estimated future climate cost, multiply the estimated climate cost by $(1+r)^{-n}$, where r is the discount rate (use 0.05 for 5 percent) and n is the number of years in the future. So, for a $1 trillion dollar climate change cost that is 100 years in the future, 5 percent discounting gives a present-day equivalent of $7.6 billion. Nordhaus, 2013, 160.

136. For a detailed discussion of the discounting problem, see D.A. Weisbach and C.R. Sunstein, "Climate Change and Discounting the Future: A Guide for the Perplexed," *27 Yale Law and Policy Review,* 2009: 433-458, https://papers.ssrn.com/sol3/papers.cfm?abstract_id=1223448.

137. "Social Cost of Carbon," United States Government Accountability Office, June 2020, https://www.gao.gov/assets/710/707776.pdf.

138. The ethical view is most prominently associated with the well-known economist Nicholas Stern. See Nicholas Stern, *Why Are We Waiting? The Logic, Urgency and Promise of Tackling Climate Change* (Cambridge: MIT Press, 2015), 151-184.

139. N. Oreskes and N. Stern, "Climate Change Will Cost Us Even More Than We Think," *The New York Times,* October 23, 2019, https://www.nytimes.com/2019/10/23/opinion/climate-change-costs.html.

140. To learn more about the chemistry of acids and bases, see https://www.khanacademy.org/science/high-school-biology/hs-biology-foundations/hs-ph-acids-and-bases/v/introduction-to-ph. For more details about the chemistry of ocean acidification, see https://www.khanacademy.org/science/biology/biodiversity-and-conservation/threats-to-biodiversity/v/ocean-acidification-and-biodiversity-impacts.

141. See https://www.khanacademy.org/science/biology/biodiversity-and-conservation/threats-to-biodiversity/v/ocean-acidification-and-biodiversity-impacts.

142. Elizabeth Kolbert, "Louisiana's Disappearing Coast," *The New Yorker,* March 25, 2019, https://www.newyorker.com/magazine/2019/04/01/louisianas-disappearing-coast; Nathaniel Rich, "Destroying a Way of Life to Save Louisiana," *The New York Times,* July 21, 2020, https://www.nytimes.com/interactive/2020/07/21/magazine/louisiana-coast-engineering.html.

143. To read the comprehensive Master Plan, see *2017 Coastal Master Plan,* State of Louisiana Coastal Protection and Restoration Authority (CPRA), https://coastal.la.gov/our-plan/2017-coastal-master-plan/.

144. See Chapter 6 for discussion of the basis for lawsuits against oil and gas companies.

145. "Two NREL Studies find Gulf of Mexico Well Positioned for Offshore Wind Development," *National Renewable Energy Laboratory*, May 6, 2020, https://www.nrel.gov/news/program/2020/studies-find-gulf-of-mexico-well-positioned-for-offshore-wind-development.html.

Interlude Notes

1. Several international organizations have also advanced scenarios for how the energy transition might unfold globally. See Sustainable Development Solutions Network and Fondazione Eni Enrico Mattei, *Roadmap to 2050. A Manual for Nations to Decarbonize by Mid-Century*, September 2019, https://roadmap2050.report/static/files/roadmap-to-2050.pdf. The International Energy Agency (IEA) has created sustainable development scenarios for every country; see IEA, "World Energy Model, Sustainable Development Scenario," International Energy Agency, https://www.iea.org/reports/world-energy-model/sustainable-development-scenario.

2. "Nationally Determined Contributions (NDCs)," *United Nations Climate Change*, https://unfccc.int/process-and-meetings/the-paris-agreement/nationally-determined-contributions-ndcs. See Chapter 3 for further information and links to the UNFCCC.

3. "What is the Kyoto Protocol?," United Nations, https://unfccc.int/kyoto_protocol. Also, see Brad Plumer, "Past Climate Treaties Failed. So the Paris Deal Will Try Something Radically Different," *Vox*, December 14, 2015, https://www.vox.com/2015/12/14/10105422/paris-climate-deal-history.

4. "Pathways to 2050: Alternative Scenarios for Decarbonizing the US Economy," Center for Climate and Energy Solutions (C2ES), May 2019, https://www.c2es.org/site/assets/uploads/2019/05/pathways-to-2050-scenarios-for-decarbonizing-the-us-economy-final.pdf.

5. R.B. Jackson et al., "Global Energy Growth is Outpacing Decarbonization," *Global Carbon Project*, September 2019, https://www.globalcarbonproject.org/global/pdf/GCP_2019_Global%20energy%20growth%20outpace%20decarbonization_UN%20Climate%20Summit_HR.pdf.

6. See Chapters 2 and 3.

7. "Emissions Gap Report 2019," United Nations Environment Programme, https://www.unenvironment.org/interactive/emissions-gap-report/2019/.

8. Zeke Hausfather, "UNEP: 1.5°C Climate Target 'Slipping Out of Reach'," *CarbonBrief*, November 26, 2019, https://www.carbonbrief.org/unep-1-5c-climate-target-slipping-out-of-reach.

9. Global Carbon Project, "Global Carbon Budget 2020," December 11, 2020, https://www.globalcarbonproject.org/carbonbudget/20/files/GCP_CarbonBudget_2020.pdf. For details on how US greenhouse gas emissions from different economic sectors dropped in 2020, see Kate Larsen, Hannah Pitt, and Alfredo Rivera, "Preliminary US Greenhouse Gas Emissions Estimates for 2020," Rhodium Group, January 12, 2021, https://rhg.com/research/preliminary-us-emissions-2020/.

10. For an overall description of the far-reaching changes required to stabilize climate for a 1.5°C world, see World Energy Outlook 2020, "Achieving Net-Zero Emissions by 2050," International Energy Agency, https://www.iea.org/reports/world-energy-outlook-2020/achieving-net-zero-emissions-by-2050#abstract.

11. Carbon sequestration is discussed in Chapter 10.

12. "Target Atmospheric CO2: Where Should Humanity Aim?," NASA, December 2008, https://www.giss.nasa.gov/research/briefs/hansen_13/. Paleoclimatology and climate tipping points are discussed in Chapters 2-4.

13. Z. Hausfather and G.P. Peters, "Emissions—the 'Business as Usual' Story is Misleading," *Nature*, no. 577 (2020): 618-620, https://www.nature.com/articles/d41586-020-00177-3.

14. UN environment programme, *Cut Global Emissions by 7.6 Percent Every Year for Next Decade to Meet 1.5 °C Paris target—UN report*, November 26, 2019, https://www.unenvironment.org/news-and-stories/press-release/cut-global-emissions-76-percent-every-year-next-decade-meet-15degc.

15. *Kieran Mulvaney, "Climate Change Report Card: These Countries are Reaching Targets," National Geographic*, September 19, 2019, https://www.nationalgeographic.com/environment/2019/09/climate-change-report-card-co2-emissions/. For individual countries' progress toward meeting Paris pledges, see "The Truth Behind the Climate Pledges," *Universal Ecological Fund*, https://feu-us.org/behind-the-climate-pledges/.

16. The US withdrew from the Paris Agreement on November 4, 2020 but rejoined when President Biden assumed office.

17. Meehan Crist, "What the Coronavirus Means for Climate Change," *The New York Times*, March 27, 2020, https://www.nytimes.com/2020/03/27/opinion/sunday/coronavirus-climate-change.html.

18. Kate Larsen et al., "Taking Stock 2020: The COVID-19 Edition," *Rhodium Group*, July 9, 2020, https://rhg.com/research/taking-stock-2020/.

19. For a description of SSPs, see Z. Hausfather, "Explainer: How "Shared Socioeconomic Pathways" Explore Future Climate Change," *CarbonBrief*, April 19, 2018, https://www.carbonbrief.org/explainer-how-shared-socioeconomic-pathways-explore-future-climate-change.

20. The RCPs were introduced in the last section of Chapter 3. In Chapter 4, they were used to illustrate climate impacts in different possible futures corresponding to 1.5°C, 2°C, or warmer worlds.

21. Z. Hausfather, 2018, 5; 19.

22. See Chapter 5 for discussion of how some US groups link climate advocacy to social justice concerns.

23. See https://usddpp.org and http://deepdecarbonization.org/about/. For the US report, see J.H. Williams et al., "Pathways to Deep Decarbonization in the United States," *The U.S. Report of the Deep Decarbonization Pathways Project of the Sustainable Development Solutions Network and the Institute for Sustainable Development and International Relations*, last modified November 16, 2015, https://usddpp.org/downloads/2014-technical-report.pdf.

24. IPCC, 2018. Summary for Policymakers, Global Warming of 1.5°C, 12.

25. J.H. Williams, et al., "Pathways to Deep Decarbonization in the United States," 2015. For a later US report that also addresses non-CO_2 greenhouse gases, see "United States Mid-Century Strategy for Deep Decarbonization," *The White House*, 2016, https://unfccc.int/files/focus/long-term_strategies/application/pdf/mid_century_strategy_report-final_red.pdf.

26. Chapter 8 is devoted to exploring carbon-free power sources and policies to promote a carbon-free electricity grid by 2050.

27. Vaclav Smil, *Energy Transitions: Global and National Perspectives*. 2nd ed., (Santa Barbara, CA: Praeger, 2017), 83.

28. Rewiring America's plan is available for download at https://www.rewiringamerica.org/hand-book. For an interview with founder Saul Griffith, see Ezra Klein, "How to Decarbonize America—and Create 25 Million Jobs," Vox, August 27, 2020, https://www.vox.com/podcasts/2020/8/27/21403184/saul-griffith-ezra-klein-show-solve-climate-change-green-new-deal-rewiring-america.

29. For review and analysis of *Rewiring America's Plan*, see David Roberts, "How to Drive Fossil Fuels Out of the US Economy, Quickly," *Vox*, August 6, 2020, https://www.vox.com/energy-and-environment/21349200/climate-change-fossil-fuels-rewiring-america-electrify.

30. For a summary of Princeton's effort, see Molly Seltzer, "Big But Affordable Effort Needed for America to Reach Net-Zero Emissions by 2050, Princeton Study Shows," *Princeton Environmental Research*, December 15, 2020, https://environmenthalfcentury.princeton.edu/research/2020/big-affordable-effort-needed-america-reach-net-zero-emissions-2050-princeton-study. For the full interim report, see https://environmenthalfcentury.princeton.edu/sites/g/files/toruqf331/files/2020-12/Princeton_NZA_Interim_Report_15_Dec_2020_FINAL.pdf.

31. Smil, *Energy Transitions*.

32. David Roberts, "Many Technologies Needed to Solve the Climate Crisis are Nowhere Near Ready," *Vox*, July 14, 2020, https://www.vox.com/energy-and-environment/2020/7/14/21319678/climate-change-renewable-energy-technology-innovation-net-zero-emissions.

33. In Chapter 6, we discuss the schemes of the fossil fuel companies to maintain their grip on the energy system, including funding the network of climate change denialists.

34. International Energy Agency, *Special Report on Clean Energy Innovation*, Energy Technology Perspectives 2020, July 2020, https://www.iea.org/reports/clean-energy-innovation. For a comprehensive review of all technologies needed for the clean energy transition, see International Energy Agency, Energy Technology Perspectives 2020, September 2020, https://www.iea.org/reports/energy-technology-perspectives-2020.

35. Varun Sivaram, Julio Friedmann, and Colin Cunliff, "To Confront the Climate Crisis, the US Should Launch a National Energy Innovation Mission," *Columbia SIPA Center on Global Energy Policy*, September 15, 2020, https://www.energy-policy.columbia.edu/research/op-ed/confront-climate-crisis-us-should-launch-national-energy-innovation-mission.

36. Evolved Energy Research, "350 ppm Pathways for the United States," *Deep Decarbonization Pathways Project*, May 8, 2019, https://docs.wixstatic.com/ugd/294abc_95dfdf602afe4e11a184ee65ba565e60.pdf.

37. Renewable hydrogen and the circular carbon economy are discussed in Chapter 9. Atmospheric carbon removal is discussed in Chapter 10.

38. Saul Griffith and Sam Calisch, *Mobilizing for a Zero-Carbon America—Jobs, Jobs, Jobs and More Jobs*, Rewiring America, July 2020, https://www.rewiringamerica.org/jobs-report.

39. See Chapter 7 for discussion of carbon pricing at state and national levels.

40. See Chapters 6 and 8, respectively, for discussion of advocacy against fossil fuels and engagement in decision making for the electricity grid.

41. J. C. Minx et al., "Negative Emissions—Part 1: Research Landscape and Synthesis," *Environmental Research Letters*, no. 13 (2018), https://iopscience.iop.org/article/10.1088/1748-9326/aabf9b/meta.

42. See https://electrifynow.net.

43. See https://climatecrisis.house.gov/report.

44. The Green New Deal was introduced into the House of Representatives on February 7, 2019, by freshman legislator Alexandria Ocasio-Cortez and 94 of her Democratic colleagues. See "Recognizing the Duty of the Federal Government to Create a Green New Deal," H.R. Res. 109, 116th Congress (2019-2020), https://www.congress.gov/bill/116th-congress/house-resolution/109/text. The companion Senate Resolution 59 was cosponsored on the same day by 14 Senate Democrats (*Congressional Record Vol. 165, No. 24*).

45. A summary of the House plan is available. See "Solving the Climate Crisis," *House Select Committee on the Climate Crisis*, June 2020, https://climatecrisis.house.gov/sites/climatecrisis.house.gov/files/4%20Pager_compressed.pdf.

46. For the full House plan, see "Solving the Climate Crisis, The Congressional Action Plan for a Clean Energy Economy and a Healthy, Resilient and Just America," *House Select Committee on the Climate Crisis*, June 2020, https://climatecrisis.house.gov/sites/climatecrisis.house.gov/files/Climate%20Crisis%20Action%20Plan.pdf.

47. Megan Mahajan, Robbie Orvis, and Sonia Aggarwal, "Modeling the Climate Crisis Action Plan," *Energy Innovation*, June 2020, https://energyinnovation.org/wp-content/uploads/2020/07/Modeling-the-Climate-Crisis-Action-Plan_FOR-RELEASE.pdf.

Chapter 5 Notes

1. Shannon Osaka, "The New York Times and the Super-Wicked Problem of Climate Change," *Grist*, August 2, 2018, https://grist.org/article/what-the-new-york-times-got-right-and-wrong-about-the-super-wicked-problem-of-climate-change/.

2. See the discussion on future discounting, in the Economy section of Chapter 4.

3. Sara Gorman and Jack M. Gorman, "Climate Change Denial: Facing a Reality Too Big to Believe," *Psychology Today*, January 12, 2019, https://www.psychologytoday.com/us/blog/denying-the-grave/201901/climate-change-denial.

4. See the discussion in the last section of Chapter 3.

5. William Nordhaus, "Climate Clubs to Overcome Free-Riding," *Issues in Science and Technology*, 31 no. 4 (Summer 2015), https://issues.org/climate-clubs-to-overcome-free-riding/.

6. The bible of this movement is E.F. Schumacher, *Small is Beautiful: Economics as if People Mattered*, (Vancouver, BC: Hartley & Marks, 1999). For quotes from this 25th anniversary edition of the original text, see https://centerforneweconomics.org/envision/legacy/ernst-friedrich-schumacher-small-is-beautiful-quotes/.

7. P.R. Ehrlich and J.P. Holdren, "Critique," and B. Commoner, "Response" on "A Bulletin Dialogue on 'The Closing Circle," *Bulletin of the Atomic Scientists*, vol. 28. no. 5 (1972), https://e4anet.files.wordpress.com/2014/09/610_wk1-ehrlich-and-holdren-one-dimensional-ecology.pdf.

8. Hans Rosling, *Factfulness* (New York: Flatiron Books, 2018), 77-92.

9. See https://www.worldometers.info/world-population/world-population-by-year/.

10. Rosling, *Factfulness*, 86.

11. See https://population.un.org/wpp/Graphs/Probabilistic/POP/TOT/900.

12. See https://www.un.org/en/sections/issues-depth/population/index.html.

13. Lyman Stone, "Why You Shouldn't Obsess About 'Overpopulation'," *Vox*, last modified July 11, 2018, https://www.vox.com/the-big-idea/2017/12/12/16766872/overpopulation-exaggerated-concern-climate-change-world-population.

14. Amit Kapoor and Bibek Debroy, "GDP is Not a Measure of Human Well-Being," *Harvard Business Review*, October 4, 2019, https://hbr.org/2019/10/gdp-is-not-a-measure-of-human-well-being.

15. See the discussion in the last section of Chapter 3.

16. A.E. Raftery et al., "Less Than 2°C Warming by 2100 Unlikely," *Nature Climate Change*, no. 7 (2017): 637-641, https://www.nature.com/articles/nclimate3352.pdf.

17. "Gross Domestic Product, 2nd Quarter 2020 (Advance Estimate) and Annual Update," *United States Bureau of Economic Analysis*, July 30, 2020, https://www.bea.gov/news/2020/gross-domestic-product-2nd-quarter-2020-advance-estimate-and-annual-update.

18. Milena Buchs and Max Koch, "Challenges for the Degrowth Transition: The Debate About Wellbeing," *Futures* no. 105 (2019): 155-165, https://www.sciencedirect.com/science/article/pii/S0016328718300715.

19. D. Saha and M. Muro, "Growth, Carbon and Trump: States are 'Decoupling' Economic Growth from Emissions Growth," *Brookings Institute*, December 8, 2016. https://www.brookings.edu/blog/the-avenue/2016/12/08/decoupling-economic-growth-from-emissions-growth/.

20. See https://www.sunrisemovement.org. See also Ella Nilsen, "The New Face of Climate Activism is Young, Angry—and Effective," *Vox*, September 17, 2019, https://www.vox.com/the-highlight/2019/9/10/20847401/sunrise-movement-climate-change-activist-millennials-global-warming.

21. David Roberts, "The Green New Deal, Explained," *Vox*, March 30, 2019, https://www.vox.com/energy-and-environment/2018/12/21/18144138/green-new-deal-alexandria-ocasio-cortez.

22. "Impact of Climate Policy on 2020 Election," *All Things Considered*, NPR, September 1, 2020, https://www.npr.org/2020/09/01/908456852/impact-of-climate-policy-on-2020-presidential-election.

23. See https://350.org.

24. See https://www.climaterealityproject.org.

25. See https://citizensclimatelobby.org.

26. CCL's efforts bore fruit with the introduction of the Energy Innovation and Carbon Dividend Act in Congress. See Chapter 7 and https://energyinnovationact.org.

27. See https://fridaysforfuture.org/what-we-do/who-we-are/, and Alleen Brown, "Can Extinction Rebellion Build a U.S. Climate Movement Big Enough to Save the Earth?," *The Intercept*, October 12, 2019, https://theintercept.com/2019/10/12/extinction-rebellion-climate-movement-direct-action/.

28. See https://time.com/person-of-the-year-2019-greta-thunberg/. For inspiration, try Greta Thunberg, *No One Is Too Small to Make a Difference* (London: Penguin Books, 2018).

29. "Big Green," *SourceWatch*, The Center for Media and Democracy, https://www.sourcewatch.org/index.php?title=Big_Green.

30. Jason Mark, "Naomi Klein: 'Big Green Groups are More Damaging Than Climate Deniers'," *The Guardian*, September 10, 2013, https://www.theguardian.com/environment/2013/sep/10/naomi-klein-green-groups-climate-deniers.

31. Ryan Grim and Briahna Gray, "Alexandria Ocasio-Cortez Joins Climate Activists in Protest at Democratic Leader Nancy Pelosi's Office," *The Intercept*, November 13, 2018, https://theintercept.com/2018/11/13/alexandria-ocasio-cortez-sunrise-activists-nancy-pelosi/.

32. Joseph Robertson, "Bipartisan Carbon Fee and Dividend Bill Now Before US Congress," *Carbon Pricing Leadership Coalition*, February 4, 2019, https://www.carbonpricingleadership.org/blogs/2019/2/3/bipartisan-carbon-fee-and-dividend-bill-now-before-us-congress.

33. It is unclear what forms the noncooperation would take. See https://www.sunrisemovement.org/about.

34. Ella Nilsen, "The New Face of Climate Activism is Young, Angry—and Effective."

35. David Bornstein, "Cracking Washington's Gridlock to Save the Planet," *The New York Times*, May 19, 2017, https://www.nytimes.com/2017/05/19/opinion/cracking-washingtons-gridlock-to-save-the-planet.html.

36. The bill is the Energy Innovation and Carbon Dividend Act, which will be discussed at length in Chapter 7. See Joseph Robertson, "Bipartisan Carbon Fee and Dividend Bill Now Before US Congress," *Carbon Pricing Leadership Coalition*, February 4, 2019, https://www.carbonpricingleadership.org/blogs/2019/2/3/bipartisan-carbon-fee-and-dividend-bill-now-before-us-congress.

37. C. Jones et al., "Project 1: Briefing Paper on Climate Countermovement Coalitions," in T. Roberts et al., "Brown's Climate and Development Lab Begins New Chapter to Uncover Networks of Denial," *Climate and Development Lab*, http://www.climatedevlab.brown.edu/home/browns-climate-and-development-lab-begins-new-chapter-to-uncover-networks-of-denial.

38. Scott Waldman, "Research Finds Broad Array of Groups Fighting Climate Policy," *E&E News*, October 22, 2019, https://www.eenews.net/stories/1061341211. See also Hiroko Tabuchi, "How One Firm Drove Influence Campaigns Nationwide for Big Oil," *The New York Times*, November 11, 2020, https://www.nytimes.com/2020/11/11/climate/fti-consulting.html.

39. Jennifer R. Marlon et al., "How Hope and Doubt Affect Climate Change Mobilization," *Frontiers in Communication*, May 21, 2019, https://climateadvocacylab.org/system/files/Marlon%20et%20al_2019_How%20Hope%20and%20Doubt%20Affect%20Climate%20Change%20Mobilization.pdf.

40. The politics and organizations of climate change denialism are discussed in Chapter 6.

41. "Climate Unplugged," *Niskanen Center*, https://www.climateunplugged.com.

42. "Climate Science 101," *republicEn*, https://republicen.org/climate-science.

43. See https://clcouncil.org.

44. For information about Young Conservatives for Carbon Dividends, see https://www.yccdaction.org. For information about Students for Carbon Dividends, see https://www.s4cd.org.

45. "Where We Stand on Important Issues," *American Conservation Coalition*, https://www.acc.eco/platform.

46. See Chapter 7 for discussion of how high carbon prices should be to achieve climate goals.

47. See Chapter 10.

48. Marilynne Robinson, "What Kind of Country Do We Want?" *The New York Review of Books*, June 11, 2020, https://www.nybooks.com/articles/2020/06/11/what-kind-of-country-do-we-want/.

49. A brief synopsis of the environmental justice movement is offered by Renee Skelton and Vernice Miller, "The Environmental Justice Movement," NRDC, March 17, 2016, https://www.nrdc.org/stories/environmental-justice-movement. For evidence that climate change today is already having distributional consequences, see Diffenbaugh and Burke, "Global Warming Has Increased Global Social Inequality." For a study showing that a climate-stabilized 1.5°C world would alleviate inequality, see Burke, Davis, and Diffenbaugh, "Large Potential Reduction in Economic Damages Under UN Mitigation Targets," 549–553.

50. Kaihui Song et al., "Scale, Distributions and Variations of Global Greenhouse Gas Emissions Driven by US Households," *Environment International*, vol. 133, Part A (December 2019), https://www.sciencedirect.com/science/article/pii/S0160412019315752?via%3Dihub.

51. "Jemez Principles for Democratic Organizing," *Southwest Network for Environmental and Economic Justice*, (Jemez, New Mexico), December 1996, https://www.ejnet.org/ej/jemez.pdf.

52. Colby Itkowitz, Dino Grandoni, and Jeff Stein, "AFL-CIO Criticizes Green New Deal, Calling It 'Not Achievable or Realistic'," *Washington Post*, March 12, 2019, https://www.washingtonpost.com/politics/afl-cio-criticizes-green-new-deal-calling-it-not-achievable-or-realistic/2019/03/12/842784fe-44dd-11e9-aaf8-4512a6fe3439_story.html.

53. Aviva Chomsky, "Labor and Environment Movement's Complex Stance on Green New Deal," *Consortium News*, December 21, 2019, https://consortiumnews.com/2019/08/12/labor-and-environment-movements-complex-stance-on-green-new-deal/.

54. See Stern, *Why are we Waiting?*, 276-281, 291-299.

55. Stern, *Why are we Waiting?*, 294.

56. Michael Tomasky, "The Party Cannot Hold," *The New York Review of Books*, March 26, 2020, https://www.nybooks.com/articles/2020/03/26/democratic-party-cannot-hold/.

57. The meaning of this phrase is elaborated in the preamble of the GND resolution. See "Recognizing the Duty of the Federal Government to Create a Green New Deal," H. Res. 109, 116th Congress (2019-2020), https://www.congress.gov/bill/116th-congress/house-resolution/109/text.

58. Ann Pettifor, *The Case for the Green New Deal* (London: Verso, 2019), 8-15.

59. The three essential elements were gleaned from extensive interviews with key players. Roberts, "The Green New Deal, Explained." Another good summary is Joel Benjamin, "The Urgent Case for a Green New Deal," *Carbon Tracker*, November 27, 2019, https://carbontracker.org/the-urgent-case-for-a-green-new-deal/. Detailed formulation of policies is being done by the group Data for Progress, who mobilize "state of the art techniques in data science to support progressive activists and causes." See https://www.dataforprogress.org/about.

60. "Mountains and Molehills: Achievements and Distractions on the Road to Decarbonization," *J.P. Morgan Private Bank*, March 2019, https://www.jpmorgan.com/jpmpdf/1320746971070.pdf.

61. Emily Atkin, "Why Won't These Democrats Reject Fossil Fuel Money?" *The New Republic*, March 19, 2019, https://newrepublic.com/article/153336/democrats-refuse-sign-no-fossil-fuel-money-pledge.

62. For the House Green New Deal resolution, see "Recognizing the Duty of the Federal Government to Create a Green New Deal," H.Res. 109, 116th Congress (2019-2020), https://www.congress.gov/bill/116th-congress/house-resolution/109/text. For the Senate resolution, see "A Resolution Recognizing the Duty of the Federal Government to Create a Green New Deal," S. Res. 59, 116th Congress (2019-2020), https://www.congress.gov/bill/116th-congress/senate-resolution/59?q=%7B%22search%22%3A%5B%22green+new+deal%22%5D%7D&a=1&r=2.

63. Nate Cohn, "Moderate Democrats Fared Best in 2018," *The New York Times*, September 11, 2019, https://www.nytimes.com/2019/09/10/upshot/2020-North-Carolina-moderate-democrats.html.

64. David Roberts, "Democratic Hopes for Climate Policy May Come Down to this One Weird Senate Trick," *Vox*, February 20, 2020, https://www.vox.com/energy-and-environment/2019/5/28/18636759/climate-change-budget-reconciliation-democrats.

65. For a detailed criticism of the political feasibility of the US Green New Deal proposal, see Jerry Taylor, "An Open Letter to Green New Dealers," Niskanen Center, March 31, 2019, https://www.niskanencenter.org/an-open-letter-to-green-new-dealers/.

66. See https://climatecommunication.yale.edu/news-events/act-on-climate-change/. For a thorough compilation of public polling on climate change, see Anthony Leiserowitz et al., "Climate Change in the American Mind: March 2018," *Yale Program on Climate Change Communication*, April 17, 2018, https://climatecommunication.yale.edu/publications/climate-change-american-mind-march-2018/.

67. See https://www.ucsusa.org/resources/engaging-policymakers.

68. See https://www.climateadvocacylab.org.

69. See https://www.audubon.org/climate-action-guide.

70. Rick Moore, "Climate Action Plans: What Are They, and Do They Work?" *Grand Canyon Trust*, January 18, 2018, https://www.grandcanyontrust.org/blog/climate-action-plans-what-are-they-and-do-they-work.

71. The USDDPP project is discussed in the Interlude. See https://usddpp.org/local-action/.

72. "Climate Action Plans," *Institute for Local Government*, https://www.ca-ilg.org/climate-action-plans.

73. See https://www.holyoke.org/news/holyoke-coalition-initiates-clean-energy-transition-plan-aided-by-400000-in-grants/. Information about the Rocky Mountain Institute is available at https://rmi.org/about/.

74. Jochen Hinkel et al., "Transformative Narratives for Climate Action," *Climatic Change*, no. 160 (2020): 495-506, https://link.springer.com/article/10.1007/s10584-020-02761-y.

75. Franzisca Funke and David Klenert, "Climate Change After COVID-19: Harder to Defeat Politically, Easier to Tackle Economically," *World Economic Forum*, August 19, 2020, https://www.weforum.org/agenda/2020/08/climate-change-after-covid-19-harder-to-defeat-politically-easier-to-tackle-economically.

76. Nick Sobczyk, "How the Pandemic Upended Climate Politics," *E&E News*, April 20, 2020, https://www.eenews.net/stories/1062901321.

77. Adam Corner et al., "Principles for Effective Communication and Public Engagement on Climate Change: A Handbook for IPCC Authors," *Climate Outreach*, January 2018, https://climateoutreach.org/resources/ipcc-communications-handbook/.

78. Jonathan Haidt, *The Righteous Mind: Why Good People are Divided by Politics and Religion* (New York, NY: Pantheon, 2012), 144-149.

79. See the Evangelical Environmental Network, https://www.creationcare.org. Also see Katherine Hayhoe, "I'm a Climate Scientist Who Believes in God. Hear Me Out," *The New York Times*, October 31, 2019, https://www.nytimes.com/2019/10/31/opinion/sunday/climate-change-evangelical-christian.html. See also J.L. Dickinson et al., "Which Moral Foundations Predict Willingness to Make Lifestyle Changes to Avert Climate Change in the USA?" *PLOS One*, October 19, 2016, https://journals.plos.org/plosone/article?id=10.1371/journal.pone.0163852.

80. Available at https://laudatosi.com/watch.

81. See https://climatecommunication.yale.edu.

82. See https://climatecommunication.yale.edu/about/projects/global-warmings-six-americas/.

83. See https://climatecommunication.yale.edu/wp-content/uploads/2014/03/Global_Warmings_Six_Americas_book_chapter_2014.pdf.

84. Leiserowitz, A., Maibach, E., Rosenthal, S., Kotcher, J., Bergquist, P., Ballew, M., Goldberg, M. & Gustafson, A. (2019). *Climate change in the American mind*: November 2019. Yale University and George Mason University. New Haven, CT: Yale Program on Climate Change Communication.

85. See https://www.pbs.org/newshour/science/support-for-the-endangered-species-act-remains-high-as-trump-administration-and-congress-try-to-gut-it. For lists of endangered and threatened species, see https://www.fws.gov/endangered/. For good stories about disappearing wildlife, see IPBES, "Summary for Policymakers of the Global Assessment Report," and Kolbert, *The Sixth Extinction*.

86. Ezra Markowitz and Julie Sweetland, "Entering Climate Change Communications Through the Side Door," *Stanford Social Innovation Review*, July 10, 2018, https://ssir.org/articles/entry/entering_climate_change_communications_through_the_side_door?utm_source=Enews&utm_medium=Email&utm_campaign=SSIR_Now&utm_content=Title.

87. "States are Learning What Happens to COVID-19 Cases If You Reopen Too Early," *Healthline*, June 30, 2020, https://www.healthline.com/health-news/covid19-cases-rising-states-reopened.

88. Mark Muro, David G. Victor, and Jacob Whiton, "How the Geography of Climate Damage Could Make the Politics Less Polarizing," *Brookings*, January 29, 2019, https://www.brookings.edu/research/how-the-geography-of-climate-damage-could-make-the-politics-less-polarizing/. For state adaptation rankings, see Dena P. Adler and Emma Gosliner, "State Hazard Mitigation Plans & Climate Change: Rating the States 2019 Update," Sabin Center for Climate Change Law, Columbia Law School, September 2019, http://columbiaclimatelaw.com/files/2019/09/Adler-Gosliner-SHMP-Report-Sept-10-2019.pdf.

89. See the discussion in Chapter 4.

90. See https://www.climateinteractive.org/about/.

91. Lempert, R., J. Arnold, R. Pulwarty, K. Gordon, K. Greig, C. Hawkins Hoffman, D. Sands, and C. Werrell, 2018: Reducing Risks Through Adaptation Actions. In *Impacts, Risks, and Adaptation in the United States: Fourth National Climate Assessment, Volume II* [Reidmiller, D.R., C.W. Avery, D.R. Easterling, K.E. Kunkel, K.L.M. Lewis, T.K. Maycock, and B.C. Stewart (eds.)]. U.S. Global Change Research Program, Washington, DC, USA, pp. 1309–1345. doi: 10.7930/NCA4.2018.CH28, Key Message 4. Detailed information on specific adaptation measures in particular US regions and economic sectors is provided throughout the fourth National Climate Assessment; see https://nca2018.globalchange.gov.

92. See Chapter 10 for discussion of carbon storage on agricultural lands.

93. See https://archive.thinkprogress.org/these-deep-red-states-are-going-green-a885cfa040ca/.

94. Easterling, "Precipitation change in the United States," *Climate Science Special Report: Fourth National Climate Assessment, Volume I*, Key Finding 4; Wehner, "Droughts, Floods and Wildfires", *Climate Science Special Report: Fourth National Climate Assessment, Volume I*, Key Finding 4.

95. See https://protectourwinters.org.

96. Porter Fox, "Why Can't Rich People Save Winter?" *The New York Times*, February 2, 2019, https://www.nytimes.com/2019/02/02/opinion/sunday/winter-snow-ski-climate.html.

97. See https://protectourwinters.org.
98. See http://www.nsaa.org/environment/outdoor-business-climate-partnership/.
99. See https://www.wltx.com/article/news/olympics-and-winter-sports-are-at-risk-due-to-climate-change/101-516805937.
100. "Ski Federation Joins UN Climate Change Initiative," *E&E News*, September 27, 2019, https://www.eenews.net/climatewire/2019/09/27/stories/1061172747.
101. See http://www.usclimatealliance.org.
102. See http://www.usclimatealliance.org/publications/2019/12/9/us-climate-alliance-states-within-reach-of-their-commitment-to-the-paris-agreement.
103. See https://www.americaspledgeonclimate.com/fulfilling-americas-pledge/. For information on how states in the Alliance are performing toward climate goals, see Environmental Defense Fund, *Turning Climate Commitments into Results: Progress on State-led Climate Action*, https://www.edf.org/sites/default/files/documents/FINAL_State%20Emission%20Gap%20Analysis.pdf.
104. See https://www.wearestillin.com.
105. To access the En-ROADS simulator, go to https://www.climateinteractive.org/tools/en-roads/. For the En-ROADS Climate Ambassador program, see https://www.climateinteractive.org/tools/en-roads/en-roads-ambassador-program/.

Chapter 6 Notes

1. Michael Mobilia, "Even as Renewables Increase, Fossil Fuels Continue to Dominate US Energy Mix," *Today in Energy*, US Energy Information Administration, July 3, 2017, https://www.eia.gov/todayinenergy/detail.php?id=31892.
2. The four largest US banks are JP Morgan Chase, Wells Fargo, Citi, and Bank of America. See "Banking on Climate Change 2020: Fossil Fuel Finance Report Card," *Oil Change International*, March 18, 2020, http://priceofoil.org/2020/03/18/banking-on-climate-change-report-card-2020/.
3. See "How the IPCC's 1.5ºC Report Demonstrates the Risks of Overinvestment in Oil and Gas," *Global Witness*, Executive Summary, April 23, 2019, https://www.globalwitness.org/en/campaigns/oil-gas-and-mining/overexposed/. For a list of the top 10 companies by forecast capital expenditure, see https://www.statista.com/chart/17835/top-10-companies-by-forecast-capital-expenditure-in-new-fields/.
4. By 2030, governments plan to produce over twice the fossil fuel consistent with a 1.5°C world. See SEI, IISD, ODI, E3G, and UNEP. (2020). The Production Gap Report: 2020 Special Report. https://productiongap.org/2020report/. See also Steven Leahy, "We Have Too Many Fossil Fuel Power Plants to Meet Climate Goals," National Geographic, July 1, 2019, https://www.nationalgeographic.com/environment/2019/07/we-have-too-many-fossil-fuel-power-plants-to-meet-climate-goals/.
5. See Chapter 8 for discussion of electricity generation.
6. International Energy Agency, Data and Statistics, 2020, https://www.iea.org/data-and-statistics?country=USA&fuel=Energy%20supply&indicator=Electricity%20generation%20by%20source. See Chapter 8 for discussion of the electricity power sector.
7. Silvio Marcacci, "The Coal Cost Crossover: 74% of US Coal Plants Now More Expensive Than New Renewables, 86% by 2025," *Forbes*, March 26, 2019, https://www.forbes.com/sites/energyinnovation/2019/03/26/the-coal-cost-crossover-74-of-us-coal-plants-now-more-expensive-than-new-renewables-86-by-2025/#30950dd222d9. For a description of how the economics of coal is becoming unfavorable worldwide, see David Roberts, "4 Astonishing Signs of Coal's Declining Economic Viability," *Vox*, March 14, 2020, https://www.vox.com/energy-and-environment/2020/3/14/21177941/climate-change-coal-renewable-energy.
8. See Chuck Jones, "Even Trump Can't Keep Coal Companies From Declaring Bankruptcies," Forbes, November 9, 2019, https://www.forbes.com/sites/chuckjones/2019/11/09/even-trump-cant-keep-coal-companies-from-declaring-bankruptcy/#4fd0c2c910c4.
9. See Unfriend Coal, "Insuring Coal No More. The 2019 Scorecard on Insurance, Coal and Climate Change," December 2019, https://unfriendcoal.com/2019scorecard/.
10. Brad Plumer, "In a First, Renewable Energy is Poised to Eclipse Coal in US," *The New York Times*, May 13, 2020, https://www.nytimes.com/2020/05/13/climate/coronavirus-coal-electricity-renewables.html.
11. Stephen York and Mark Morey, "More Power Generation Came From Natural Gas in First Half of 2020 Than First Half of 2019," *Today in Energy*, August 12, 2020, https://www.eia.gov/todayinenergy/detail.php?id=44716#.
12. See Chapter 8 for discussion of how electric utilities make decisions about which power sources to use.
13. In Chapter 8 we will discuss the sources of power for the electricity grid, and ways that the intermittent nature of solar and wind power can be accommodated on the grid.
14. Carbon capture and storage technology is discussed in Chapter 10. See Jeff Goodell, *Big Coal: The Dirty Secret Behind America's Energy Future*, (New York: Houghton Mifflin Company, 2007), 210-250, for a discussion of coal industry technology, pollution, finance, and politics at a time when the industry was at its peak. For a recent assessment of how a still-existing "clean coal" tax break is climate-unfriendly, see Brian C Prest and Alan Krupnick, "How Clean is 'Refined Coal'? An Empirical Assessment of a Billion Dollar Tax Break," *Resources for the Future*, last modified February 2020, https://www.eenews.net/assets/2020/03/02/document_daily_01.pdf.
15. Bell Terence, "What You Should Know About Metallurgical Coal," *Thought Co.*, July 15, 2019, https://www.thebalance.com/what-is-metallurgical-coal-2340012. Metallurgical coal (also termed "coking coal") use is detailed in "Data and Statistics," *International Energy Agency*, https://www.iea.org/data-and-statistics?country=USA&fuel=Energy%20supply&indicator=Coal%20production%20by%20type. Most US coal exports are metallurgical.
16. See Chapter 9 for discussion of greenhouse gas emissions from steelmaking.
17. For statistics and mapping of the world's coal plants, see *CarbonBrief*, https://www.carbonbrief.org/mapped-worlds-coal-power-plants. Another good resource is Christine Shearer et al., "Boom and Bust 2019: Tracking the Global Coal Plant Pipeline," March 2019, https://endcoal.org/wp-content/uploads/2019/03/BoomAndBust_2019_r6.pdf.
18. See the IEA report, "Coal 2019, Analysis and Forecasts to 2024," December 2019, https://www.iea.org/reports/coal-2019. See also Lauri Myllivirta, Dave Jones, and Tim Buckley, "Analysis: Global Coal Power Set for Record Fall in 2019," *CarbonBrief*, November 25, 2019, https://www.carbonbrief.org/analysis-global-coal-power-set-for-record-fall-in-2019.

19. For the IPCC's analysis of how fast coal use must shrink to meet climate targets, see P.A. Yanguas Parra et al., "Global and Regional Coal Phase-Out Requirements of the Paris Agreement: Insights from the IPCC Special Report on 1.5°C," September 23, 2019, https://climateanalytics.org/publications/2019/coal-phase-out-insights-from-the-ipcc-special-report-on-15c-and-global-trends-since-2015/. See also Jason Bordoff, "Yes, We Can Get Rid of the World's Dirtiest Fuel," Foreign Policy, August 26, 2020, https://foreignpolicy.com/2020/08/26/coal-mining-electricity-climate-change/?utm_source=Center+on+Global+Energy+Policy+Mailing+List&utm_campaign=85cf68c2f6-EMAIL_CAMPAIGN_2019_09_18_12_40_COPY_01&utm_medium=email&utm_term=0_0773077aac-85cf68c2f6-102337809.

20. For an analysis of the US coal sector, see Howard Gruenspecht, "The U.S. Coal Sector: Recent and Continuing Challenges," Brookings Institute, January 2019, https://www.brookings.edu/research/the-u-s-coal-sector/.

21. Goodell, Big Coal, 120–146, gives a harrowing account of the environmental damage and human health impacts from coal mining.

22. See Chapter 7 for discussion of carbon pricing.

23. See Chapter 8 for discussion of how electricity production is regulated by state government, and how advocates can become involved in this process.

24. For a list of large US coal power plants, see https://www.worldatlas.com/articles/the-largest-coal-power-stations-in-the-united-states.html. For a summary of EIA data on US coal plant retirements, see https://ieefa.org/u-s-coal-plant-retirements-to-top-10gw-in-2019-eia/.

25. Adele Morris, Noah Kaufman, and Siddhi Doshi, "The Risk of Fiscal Collapse in Coal-Reliant Communities," Columbia | SIPA Center on Global Energy Policy, July 15, 2019, https://energypolicy.columbia.edu/research/report/risk-fiscal-collapse-coal-reliant-communities.

26. Fossil fuel uses are described by the US Energy Information Administration (EIA); see https://www.eia.gov/energyexplained/. For an overall look at the oil and gas business, see Bob Iaccino, "How Much Does Oil and Gas Drive US GDP?" The Street, June 4, 2019, https://www.thestreet.com/markets/how-much-does-oil-and-gas-drive-u-s-gdp-14981567. Oil and gas statistics are compiled by M. Garside, "U.S. Oil and Gas Industry—Statistics and Facts," Statista, September 11, 2019, https://www.statista.com/topics/1706/oil-and-gas/#dossierSummary__chapter3.

27. Iaccino, "How Much Does Oil and Gas Drive U.S. GDP?," 2019.

28. For lists of oil and gas firms, see http://blog.evaluateenergy.com/list-of-u-s-oil-gas-companies.

29. Jack Perrin and Kristen Tsai, "US Crude Oil and Natural Gas Production Increased in 2018, with 10% Fewer Wells," Today in Energy, US Energy Information Administration, February 3, 2020, https://www.eia.gov/todayinenergy/detail.php?id=42715.

30. See https://www.eia.gov/todayinenergy/detail.php?id=26112, and https://www.eia.gov/todayinenergy/detail.php?id=25372.

31. See https://www.eia.gov/dnav/pet/pet_crd_crpdn_adc_mbbl_m.htm.

32. Mark Gillespie, "Ethane Storage Seen as Key to Revitalization of Appalachia," AP News, April 29, 2019, https://apnews.com/8873394dd33647ddb369978bcfdc4e2d.

33. "America Lifts its Ban on Oil Exports," The Economist, December 18, 2015, https://www.economist.com/finance-and-economics/2015/12/18/america-lifts-its-ban-on-oil-exports.

34. US natural gas production hit a new record high in 2019. See "U.S. Natural Gas Production Grew Again in 2019, Increasing by 10%," US Energy Information Administration, March 10, 2020, https://www.eia.gov/todayinenergy/detail.php?id=43115. For information about the US liquified natural gas industry, see https://www.eia.gov/energyexplained/natural-gas/liquefied-natural-gas.php.

35. Emily Geary, "US Natural Gas Production Hit a New High in 2018," Today in Energy, US Energy Information Administration, March 14, 2019, https://www.eia.gov/todayinenergy/detail.php?id=38692.

36. Environmental Protection Agency, Inventory of U.S. Greenhouse Gas Emissions and Sinks: 1990–2018, Chapter 2: Trends in Greenhouse Gas Emissions, https://www.epa.gov/sites/production/files/2020-04/documents/us-ghg-inventory-2020-chapter-2-trends.pdf.

37. Benjamin Storrow, "Is Gas Really Better Than Coal for the Climate?" E&E News, May 4, 2020, https://www.eenews.net/stories/1063041299.

38. The World Bank, "Global Gas Flaring Tracker Report," July 22, 2020, https://www.eenews.net/assets/2020/07/22/document_ew_06.pdf.

39. Ryan Dezember, "Energy Producers' New Years Resolution: Pay the Tab for the Shale Drilling Bonanza," The Wall Street Journal, January 1, 2020, https://www.wsj.com/articles/energy-producers-new-years-resolution-pay-the-tab-for-the-shale-drilling-bonanza-11577880001. See also David Roberts, "Coronavirus Stimulus Money Will be Wasted on Fossil Fuels," Vox, June 29, 2020, https://www.vox.com/2020/4/20/21224659/coronavirus-stimulus-money-oil-prices-fossil-fuels-bailout.

40. Nick Cunningham, "Why Big Oil May Not Recover From the Pandemic," Sierra Club, May 5, 2020, https://www.sierraclub.org/sierra/why-big-oil-may-not-recover-pandemic. See also Clifford Krauss, "Oil Industry Turns to Mergers and Acquisitions to Survive," The New York Times, October 19, 2020, https://www.nytimes.com/2020/10/19/business/energy-environment/conocophillips-concho-oil-merger.html.

41. "Emissions and Coal Have Peaked as Covid-19 saves 2.5 Years of Emissions, Accelerates Energy Transition," Bloomberg NEF, October 27, 2020, https://about.bnef.com/blog/emissions-and-coal-have-peaked-as-covid-19-saves-2-5-years-of-emissions-accelerates-energy-transition/.

42. "McKinsey: Global Oil Demand to Peak Earlier Than Forecasted," Fuels and Lubes Daily, July 31, 2019, https://www.fuelsandlubes.com/mckinsey-report-global-oil-demand-slow-0-5-annum-2018-2035/.

43. "Handbrake Turn. The Cost of Failing to Anticipate an Inevitable Policy Response to Climate Change," Carbon Tracker, January 31, 2020, https://www.carbontracker.org/reports/handbrake-turn/.

44. For a description of how much crude oil is dedicated to various end products, see https://energyeducation.ca/encyclopedia/In_a_barrel_of_oil.

45. Laura Parker, "The World's Plastics Pollution Crisis Explained," National Geographic, June 7, 2019, https://www.nationalgeographic.com/environment/habitats/plastic-pollution/. For advocacy work on plastics, see https://www.plasticpollutioncoalition.org/the-coalition.

46. Jean-Marc Ollagnier, "Oil and Petrochemical Companies: Run With or Be Run Over by Plastics Recycling," *Forbes*, August 1, 2019, https://www.forbes.com/sites/jeanmarcollagnier/2019/08/01/oil-and-petrochemical-companies-run-with-or-be-run-over-by-plastics-recycling/#454b82ef5632.

47. *Carbon Tracker*, "The Future's Not in Plastics: Why Plastics Demand Won't Rescue the Oil Sector," September 4. 2020, https://carbontracker.org/reports/the-futures-not-in-plastics/.

48. See https://oilandgasclimateinitiative.com/about-us/#members.

49. See https://oilandgasclimateinitiative.com/action-and-engagement/improve-energy-efficiency/#carbon-target.

50. Benjamin Storrow and Mike Lee, "BP Swerves From Renewables in Seismic Shift From Oil, Governor's Wind & Solar Energy Coalition," August 5, 2020, https://governorswindenergycoalition.org/bp-swerves-toward-renewables-in-seismic-shift-from-oil/. See also Heather Richards, "In a Quest for Carbon-Zero, BP Buys into U.S. Offshore Wind," *E&E News*, September 10, 2020, https://www.eenews.net/stories/1063713389.

51. Stephen Naimoli and Sarah Ladislaw, "Oil and Gas Industry Engagement on Climate Change," Center for Strategic & International Studies, October 2019, https://www.csis.org/analysis/oil-and-gas-industry-engagement-climate-change.

52. Simon Dietz et al., "A Survey of the Net Zero Positions of the World's Largest Energy Companies," Transition Pathway Initiative, University of Oxford, December 11, 2019, https://www.transitionpathwayinitiative.org/tpi/publications/41.pdf?type=Publication. Also see "Absolute Impact: Why Oil Majors' Climate Ambitions Fall Short of Paris Limits," *Carbon Tracker*, June 24, 2020, https://carbontracker.org/reports/absolute-impact/.

53. Clifford Krauss, "U.S. and European Oil Giants Go Different Ways on Climate Change," *The New York Times*, September 21, 2020, https://www.nytimes.com/2020/09/21/business/energy-environment/oil-climate-change-us-europe.html.

54. ClimateWire, "Chevron Has Reality Check for 'Aspirational' Carbon Targets," *E&E News*, March 5, 2020, https://www.eenews.net/climatewire/2020/03/05/stories/1062517909.

55. EcoRight advocacy groups such as the CLC are described in Chapter 5. The CLC plan is available at https://clcouncil.org/Bipartisan-Climate-Roadmap.pdf.

56. Benjamin Storrow, "Conservative Group Drops Proposed Legal Shield for Big Oil," E&E News, September 12, 2019, https://www.eenews.net/climatewire/stories/1061113823?t=https%3A%2F%2Fwww.eenews.net%2Fstories%2F1061113823.

57. Valerie Volcovici, "Business Roundtable CEO Group Announces its Support for Carbon Pricing to Help Fight Climate Change," *Business Insider*, September 16, 2020, https://www.businessinsider.com/us-ceo-group-says-it-supports-carbon-pricing-to-fight-climate-change-2020-9.

58. Kate Wheeling, "Here's How the Oil Industry Plans to Solve Climate Change," *Pacific Standard*, May 24, 2019, https://psmag.com/environment/heres-how-the-oil-industry-plans-to-solve-climate-change.

59. See Chapter 7 for a thorough description of carbon pricing. A good summary of climate policies supported by fossil fuel companies is Karl Evers-Hillstrom and Raymond Arke, "Fossil Fuel Companies Lobby Congress On Their Own Solutions to Curb Climate Change," *OpenSecrets.org*, Center for Responsive Politics, May 17, 2019, https://www.opensecrets.org/news/2019/05/fossil-fuel-lobby-congress-on-climate-change/.

60. See Chapter 7 for discussion of carbon pricing politics.

61. J.F. Mercure et al., "Macroeconomic Impact of Stranded Fossil Fuel Assets," *Nature Climate Change*, no. 8 (2018): 588-593, https://www.nature.com/articles/s41558-018-0182-1.

62. "It's Closing Time: The Huge Bill to Abandon Oilfields Comes Early," *Carbon Tracker*, June 2020, https://carbontracker.org/reports/its-closing-time/.

63. Jeffrey Ball, "Climate Change is Hitting the Insurance Industry Hard. Here's How Swiss Re is Adapting," *Fortune*, October 24, 2019, https://fortune.com/longform/insurance-industry-climate-change-swiss-re-reinsurance/.

64. Christopher Flavelle, "Global Financial Giants Swear Off Funding an Especially Dirty Fuel," *The New York Times*, February 12, 2020, https://www.nytimes.com/2020/02/12/climate/blackrock-oil-sands-alberta-financing.html. For a description of tar sands deposits, see "What are Tar Sands?" Union of Concerned Scientists, last modified February 23, 2016, https://www.ucsusa.org/resources/what-are-tar-sands.

65. "Addressing Climate Change as a Systemic Risk: A Call to Action for U.S. Financial Regulators," Ceres, June 1, 2020, https://www.ceres.org/resources/reports/addressing-climate-systemic-risk. Also see Christopher Flavelle, "Climate Change Poses 'Systemic Threat' to the Economy, Big Investors Warn," *The New York Times*, July 21, 2020, https://www.nytimes.com/2020/07/21/climate/investors-climate-threat-regulators.html.

66. See https://www.cftc.gov/About/CFTCReports/index.htm.

67. Nick Sobczyk, "Hill Takes Notice as Big Banks Shun Fossil Fuels," *E&E News*, June 26, 2020, https://www.eenews.net/stories/1063453415.

68. Avery Ellfeldt, "3 Wall Street Giants Can Force Action on Climate. Will They?" *E&E News*, May 4, 2020, https://www.eenews.net/climatewire/2020/05/04/stories/1063013889.

69. See the Climate Action 100+ 2019 Progress Report, available at http://www.climateaction100.org/?mc_cid=13f0d6d001&mc_eid=17061e3674.

70. See https://www.cdp.net/en/.

71. Somini Sengupta and Veronica Penney, "Big Tech has a Big Climate Problem. Now, It's Being Forced to Clean Up," *The New York Times*, July 21, 2020, https://www.nytimes.com/2020/07/21/climate/apple-emissions-pledge.html.

72. Justine Calma, "Democrats are Pushing a National Climate Bank," The Verge, January 29, 2020, https://www.theverge.com/2020/1/29/21113300/democrats-green-bank-national-climate-change-capital-greenhouse-gases.

73. See Sarah Dougherty and Ann Shikany, "Introduction: National Climate Bank," Natural Resources Defense Council, July 15, 2019, https://www.nrdc.org/experts/sarah-dougherty/national-climate-bank.

74. IEA and the Centre for Climate Finance and Investment, *Energy Investing: Exploring Risk and Return in the Capital Markets*, 2nd ed. (London, UK: Imperial College Business School, June 2020), https://imperialcollegelondon.app.box.com/s/f1r832z4apqypwofakk1k4ya5w30961g.

75. See Avery Ellfeldt, "Bank Regulators Could Prevent a Climate Crisis—Report," *E&E News*, February 4, 2020. https://www.eenews.net/climatewire/2020/02/04/stories/1062257007.

76. See Bill McKibben, *Falter: Has the Human Game Begun to Play Itself Out?*, (New York, NY: Henry Holt and Company, 2019), 196, and https://stopthemoneypipeline.com.

77. Brad Plumer and Henry Fountain, "Trump Administration Finalizes Plan to Open Arctic Refuge to Drilling," *The New York Times*, August 17, 2020, https://www.nytimes.com/2020/08/17/climate/alaska-oil-drilling-anwr.html?action=click&module=News&pgtype=Homepage.

78. Parker M. Shea, "Why isn't the Ivy League Divesting From Fossil Fuels?" *E&E News*, March 2, 2020, https://www.eenews.net/climatewire/stories/1062493481.

79. See https://stopthemoneypipeline.com/tools/move-your/ and Corbin Hiar, "Investment Indices Focus on Companies with Falling CO2," *E&E News*, April 21, 2020, https://www.eenews.net/stories/1062927487.

80. Karin Kirk, "Fossil Fuel Political Giving Outdistances Renewables 13 to One," *Yale Climate Connections*, January 6, 2020, https://www.yaleclimateconnections.org/2020/01/fossil-fuel-political-giving-outdistances-renewables-13-to-one/. Information on donations accepted by politicians at all levels of government is maintained by the Center for Responsive Politics, which operates the OpenSecrets.org website: https://www.opensecrets.org.

81. "Trade Associations and the Public Relations Industry," Climate Investigations Center, https://climateinvestigations.org/trade-association-pr-spending/.

82. Myles Martin, *Citizens United v. FEC* (Supreme Court), Federal Election Commission, February 1, 2010, https://www.fec.gov/updates/citizens-united-v-fecsupreme-court/.

83. Jane Mayer, *Dark Money. The Hidden History of the Billionaires Behind the Rise of the Radical Right* (New York: Anchor Books, 2016), 278-294.

84. Benjamin Storrow, "How Utilities Use Secret Campaigns Against Climate Action," *E&E News*, August 14, 2020, https://www.eenews.net/stories/1063711675.

85. Matt Kaspar, "Final Campaign Finance Reports Filed Before Election Shows APS Spending $30 Million to Defeat Prop 127," Energy and Policy Institute, October 31, 2018, https://www.energyandpolicy.org/aps-spending-30-million-to-defeat-prop-127/. Also see Ted Macdonald, "National Right-Wing Media Outlets Bash Renewable Energy Ballot Initiative in Arizona," *Salon*, November 4, 2018, https://www.salon.com/2018/11/04/national-right-wing-media-outlets-bash-renewable-energy-ballot-initiative-in-arizona_partner/. For further examples and detailed analysis of clean energy policy and politics, see Leah Cardamore Stokes, *Short Circuiting Policy: Interest Groups and the Battle Over Clean Energy and Climate Policy in the American States* (New York: Oxford University Press, 2020).

86. Oil Change International, *Dirty Energy Dominance: Dependent upon Denial*, October 2017, http://priceofoil.org/content/uploads/2017/10/OCI_US-Fossil-Fuel-Subs-2015-16_Final_Oct2017.pdf.

87. Environmental and Energy Study Institute, "Fact Sheet: Fossil Fuel Subsidies: A Closer Look at Tax Breaks and Societal Costs," July 29, 2019, https://www.eesi.org/papers/view/fact-sheet-fossil-fuel-subsidies-a-closer-look-at-tax-breaks-and-societal-costs#1.

88. P. Erickson et al., "Why Fossil Fuel Producer Subsidies Matter," *Nature*, no. 578 (2020): E1-E3, https://www.nature.com/articles/s41586-019-1920-x.epdf?author_access_token=SjUTjrtDbCtKRGtQgVANitRgNojAjWel9jnR3ZoT-voMv5JGKl3gtKAeEYJVTQuBQ6wZJirmD9bocEKcS34GxBxr6Ea9gZYL3E_Q6ivEj7kg4BNlEaLi4zToksif-Tllk_KEi7aCEkEkEVkhgisHyrg%3D%3D.

89. See John Feldmann et al., "What the Clean Energy for America Act Gets Right - And How it Can Improve," World Resources Institute, April 30, 2021, https://www.wri.org/insights/what-clean-energy-america-act-gets-right-and-how-it-can-improve.

90. See Oreskes and Conway, *Merchants of Doubt*, for a definitive treatment of the history of science denialism in the US.

91. Koch Industries is the second largest private company in the US, with heavy investments in fossil fuels and related areas: https://www.kochind.com/about/what-we-do. For details on the role of General Motors and Ford in fostering climate change denialism, see Maxine Joselow, "Exclusive: GM, Ford Knew About Climate Change 50 Years Ago," *E&E News*, October 26, 2020, https://www.eenews.net/stories/1063717035.

92. ExxonMobil's funding of denialist groups has been investigated by The Royal Society, the independent scientific academy of the United Kingdom. For a letter from The Royal Society to ExxonMobil, and for ExxonMobil's response, see https://royalsociety.org/topics-policy/publications/2006/royal-society-exxonmobil/.

93. John Cook et al., *America Misled: How the Fossil Fuel Industry Deliberately Misled Americans About Climate Change* (Fairfax, VA: George Mason University Center for Climate Change Communication, 2019). Available at https://www.climatechangecommunication.org/america-misled/.

94. *Examining the Oil Industry's Efforts to Suppress the Truth About Climate Change*, 116th Cong. (October 23, 2019), (statements by Ed Garvey and Martin Hoffert, House Oversight Committee), https://oversight.house.gov/legislation/hearings/examining-the-oil-industry-s-efforts-to-suppress-the-truth-about-climate-change.

95. Geoffrey Supran and Naomi Oreskesi, "Assessing ExxonMobil's Climate Change Communications (1977-2014)," *Environmental Research Letters* 12, no. 8 (August 23, 2017), https://iopscience.iop.org/article/10.1088/1748-9326/aa815f#back-to-top-target.

96. See the Consensus Project, at http://theconsensusproject.com.

97. See "The Consensus Project," *Skeptical Science*, https://www.skepticalscience.com/tcp.php.

98. Oreskes and Conway, *Merchants of Doubt*, 266–274.

99. For a lucid description of Trump era political psychology, see Brian Resnick, "9 Essential Lessons from Psychology to Understand the Trump Era," *Vox*, January 10, 2019, https://www.vox.com/science-and-health/2018/4/11/16897062/political-psychology-trump-explain-studies-research-science-motivated-reasoning-bias-fake-news.

100. See Nadja Popovich, "Climate Change Rises as a Public Priority. But It's More Partisan Than Ever," *The New York Times*, February 20, 2020, https://www.nytimes.com/interactive/2020/02/20/climate/climate-change-polls.html. For YouTube data, see Carly Cassella, "Conspiracy Theorists are 'Hijacking' Climate Change Concepts on YouTube," *ScienceAlert*, July 25, 2019, https://www.sciencealert.com/beware-of-getting-climate-info-from-youtube-a-study-shows-it-s-riddled-with-conspiracies.

101. For a historical narrative of how things got this bad, see Coral Davenport and Eric Lipton, "How G.O.P. Leaders Came to View Climate Change as Fake Science," *The New York Times*, June 3, 2017, https://www.nytimes.com/2017/06/03/us/politics/republican-leaders-climate-change.html.

102. David Roberts, "This One Weird Trick Will Not Convince Conservatives to Fight Climate Change," *Vox*, December 28, 2016, https://www.vox.com/science-and-health/2016/12/28/14074214/climate-denialism-social.

103. Mark Bowen, *Censoring Science. Inside the Political Attack on Dr. James Hansen and the Truth of Global Warming* (New York, NY: Penguin Group Inc., 2008)

104. See Chapter 2 for a description of the "hockey stick" temperature record. See Mann, *The Hockey Stick and the Climate Wars*, for a book-length description of how his work has been attacked.

105. For a list of dozens of denialist organizations, see "Koch-Funded Climate Denial Front Groups," Greenpeace, https://www.greenpeace.org/usa/global-warming/climate-deniers/front-groups/. For a study of the financial resources of organizations making up the climate change counter-movement, see Robert J. Brulle, "Institutionalizing Delay: Foundation Founding and the Creation of U.S. Climate Change Counter-Movement Organizations," *Climatic Change* 122 (January 2014): 681-694, doi: 10.1007/s10584-013-1018-7.

106. Examining the Oil Industry's Efforts to Suppress the Truth about Climate Change, 116th Cong. (October 23, 2019) (statement by Naomi Oreskes, Professor of the History of Science and Affiliated Professor of Earth and Planetary Sciences), https://www.democrats.senate.gov/imo/media/doc/Naomi_Oreskes_Testimony.pdf.

107. Jessica Lee and Neela Banerjee, "Science Teachers Respond to Climate Materials Sent by Heartland Institute," *Inside Climate News*, December 22, 2017, https://insideclimatenews.org/news/22122017/science-teachers-heartland-institute-anti-climate-booklet-survey. See also Katie Worth, "Climate Change Skeptic Group Seeks to Influence 200,000 Teachers," PBS Frontline, PBS | OPB , March 28, 2017, https://www.pbs.org/wgbh/frontline/article/climate-change-skeptic-group-seeks-to-influence-200000-teachers/.

108. Meehan Crist, "How the New Climate Denial is Like the Old Climate Denial," The Atlantic, February 10, 2017, https://www.theatlantic.com/science/archive/2017/02/the-new-rhetoric-of-climate-denial/516198/. See also Michael Mann, *The New Climate War: The Fight to Take Back Our Planet* (New York, NY: Hachette Book Group, 2021).

109. See the discussion of EcoRight groups in Chapter 5.

110. Climate change cases are collected by the Sabin Center for Climate Change Law. See http://climatecasechart.com.

111. Public Health Law Center, *The Master Settlement Agreement: An Overview*, January 2019, https://publichealthlawcenter.org/sites/default/files/resources/MSA-Overview-2019.pdf.

112. Jennifer Hijazi, "Climate Nuisance Cases: Where Things Stand," *E&E News*, September 11, 2019, https://www.eenews.net/stories/1061111983.

113. Jennifer Hijazi, "States Test New Climate Strategies in Big Oil Showdowns," *E&E News,* June 29, 2020, https://www.eenews.net/stories/1063470853.

114. David Hasemyer, "Fossil Fuels on Trial: Where the Major Climate Change Lawsuits Stand Today," *Inside Climate News*, January 17, 2020, https://insideclimatenews.org/news/04042018/climate-change-fossil-fuel-company-lawsuits-timeline-exxon-children-california-cities-attorney-general. For a description of ExxonMobil's tactics in the lawsuit, see David Hasemyer, "With Bare Knuckles and Big Dollars, Exxon Fights Climate Probe to a Legal Stalemate," *Inside Climate News*, June 5, 2017, https://insideclimatenews.org/news/05062017/exxon-climate-change-fraud-investigation-eric-schneiderman-rex-tillerson-exxonmobil.

115. See https://www.ourchildrenstrust.org.

116. Mary Christina Wood, *Nature's Trust: Environmental Law for a New Ecological Age* (Cambridge, UK: Cambridge University Press, 2014).

117. For a timeline of events in the case, see https://www.youthvgov.org/our-case. For an illustrated description suitable for young readers, see https://www.dropbox.com/s/tvuup6iwdwekfym/OCT.PathwayToClimateRecovery1.pdf?dl=0.

118. For information on actions in all fifty states, see https://www.ourchildrenstrust.org/other-proceedings-in-all-50-states. For actions abroad, see https://www.ourchildrenstrust.org/global-legal-actions.

119. For a timeline of events in the Juliana case, see https://www.ourchildrenstrust.org/juliana-v-us.

120. All proceedings in the *Juliana* case are available at http://climatecasechart.com/case/juliana-v-united-states/.

121. See Michael P. Joy and Sashe D. Dimitroff, "Oil and Gas Regulation in the US: Overview," *Thomson Reuters Practical Law*, https://content.next.westlaw.com/Document/I466099551c9011e38578f7ccc38dcbee/View/FullText.html?transitionType=Default&contextData=(sc.Default)&firstPage=true&bhcp=1.

122. National Constitution Center, "Executive Orders 101: What Are They and How Do Presidents Use Them?," January 23, 2017, https://constitutioncenter.org/blog/executive-orders-101-what-are-they-and-how-do-presidents-use-them/.

123. Tara Golshan, "Bernie Sanders and AOC Want Congress to Declare a National Emergency Over Climate Change," Vox, July 9, 2019, https://www.vox.com/2019/7/9/20687526/bernie-sanders-aoc-national-climate-change-emergency. For a description of the broadening scope of climate emergency declarations in the US and worldwide, see Justine Calma, "2019 was the Year of 'Climate Emergency' Declarations," *The Verge*, December 27, 2019, https://www.theverge.com/2019/12/27/21038949/climate-change-2019-emergency-declaration.

124. National Emergencies Act, 50 U.S.C. §1621 (1976).

125. Brennan Center for Justice, *A Guide to Emergency Powers and Their Use*, last modified September 4, 2019, https://www.brennancenter.org/our-work/research-reports/guide-emergency-powers-and-their-use. The initial few pages give a good readable summary of the laws.

126. "Congress Reaches 100 Supporters for Climate Emergency Bill," *The Climate Mobilization*, December 20, 2019, https://www.theclimatemobilization.org/blog/2019/12/20/congress-reaches-100-sponsors-for-climate-emergency-bill/.

127. Dan Farber, "Using Emergency Powers to Fight Climate Change," *American Constitution Society*, January 15, 2019, https://www.acslaw.org/expertforum/using-emergency-powers-to-fight-climate-change/.

128. Youngstown Sheet & Tube Co. v. Sawyer, 343 U.S. 579 (1952), https://www.law.cornell.edu/supremecourt/text/343/579.

129. Congressional Research Service, *U.S. Crude Oil and Natural Gas Production in Federal and Nonfederal Areas* last modified October 23, 2018, https://crsreports.congress.gov/product/pdf/R/R42432.

130. See "Coal," Tennessee Valley Authority, https://www.tva.gov/Energy/Our-Power-System/Coal.

131. See "History of Reducing Air Pollution from Transportation in the United States," United States Environmental Protection Agency, https://www.epa.gov/transportation-air-pollution-and-climate-change/accomplishments-and-success-air-pollution-transportation.

132. For a book-length overview of environmental law for nonspecialists, see James Salzman and Barton H. Thompson, Jr., *Environmental Law and Policy*, 5th ed. (New York, NY: Foundation Press, 2019).

133. For a very brief overview of agency functions, see https://www.justia.com/administrative-law/. A description of the regulatory process at the Environmental Protection Agency is available at https://www.epa.gov/laws-regulations/basics-regulatory-process.

134. See https://www.justia.com/administrative-law/rulemaking-writing-agency-regulations/notice-and-comment/. State administrative agencies follow similar procedures.

135. Excellent guidance for writing comments may be found in Adam Looney, "How to Effectively Comment on Regulations," Center on Regulation and Markets, Brookings Institution, August 2018, https://www.brookings.edu/wp-content/uploads/2018/08/ES_20180809_RegComments.pdf.

136. Nadja Popovich, Livia Albeck-Ripka, and Kendra Pierre-Louis, "The Trump Administration is Reversing More Than 100 Environmental Rules. Here's the Full List," *The New York Times*, last modified November 10, 2020, https://www.nytimes.com/interactive/2019/climate/trump-environment-rollbacks.html.

137. See Massachusetts v. EPA, 549 U.S. 497 (2007).

138. For a good brief description, see Natural Resources Defense Council, *EPA's Endangerment Finding: The Legal and Scientific Foundation for Climate Action*, May 2017, https://www.nrdc.org/sites/default/files/epa-endangerment-finding-fs.pdf.

139. Brad Plumer, "How Obama's Clean Power Plan Actually Works—a Step-by-Step Guide," *Vox*, August 5, 2015, https://www.vox.com/2015/8/4/9096903/clean-power-plan-explained.

140. For a description of the ACE rule, see Umair Irfan, "Trump's EPA Just Replaced Obama's Signature Climate Policy with a Much Weaker Rule," *Vox*, June 19, 2019, https://www.vox.com/2019/6/19/18684054/climate-change-clean-power-plan-repeal-affordable-emissions. For the decision overturning the ACE rule, see Lisa Friedman, "Court Voids a 'Tortured' Trump Climate Rollback," The New York Times, January 19, 2021, https://www.nytimes.com/2021/01/19/climate/trump-climate-change.html.

141. See David Roberts, "Obama's Carbon Rule Hangs on this One Legal Question," *Grist*, February 9, 2015, https://grist.org/climate-energy/obamas-carbon-rule-hangs-on-this-one-legal-question/.

142. See https://www.leahy.senate.gov/press/leahy-carper-and-others-introduce-the-clean-economy-act-of-2020.

143. See Mark Hafstead, "Carbon Pricing 101," *Resources for the Future*, June 6, 2019, https://www.rff.org/publications/explainers/carbon-pricing-101/.

144. See Chapters 7 and 8 for discussion of carbon pricing and carbon-free electricity standards, respectively.

145. Global Methane Initiative, *Global Methane Emissions and Mitigation Opportunities*, 2020, https://www.globalmethane.org/documents/gmi-mitigation-factsheet.pdf. See Chapter 2 for a discussion of natural and anthropogenic methane sources.

146. Stephen Lee, "EPA Estimate Undercounts Methane Emissions, Environmentalists Say," *Bloomberg Green*, April 17, 2020, https://www.bloomberg.com/news/articles/2020-04-17/epa-estimate-undercounts-methane-emissions-environmentalists.

147. Benjamin Himiel at al., "Preindustrial ¹⁴CH4 Indicates Greater Anthropogenic Fossil CH4 Emissions," *Nature*, no. 578 (2020): 409-412, https://www.nature.com/articles/s41586-020-1991-8.pdf.

148. Dan Utech, "A Strategy to Cut Methane Emissions," *The White House*, March 28, 2014, https://obamawhitehouse.archives.gov/blog/2014/03/28/strategy-cut-methane-emissions.

149. Romany M. Webb, "The Status of Methane Regulation in the US," *Climate Law Blog*, Sabin Center for Climate Change Law, January 31, 2020, http://blogs.law.columbia.edu/climatechange/2020/01/31/the-status-of-methane-regulation-in-the-u-s/.

150. Kimberly Brubeck, "Interior Department Announces Final Rule to Reduce Methane Emissions & Wasted Gas on Public, Tribal Lands," US Department of the Interior, November 15, 2016, https://www.doi.gov/pressreleases/interior-department-announces-final-rule-reduce-methane-emissions-wasted-gas-public.

151. Mark Olalde, "The US' Hidden Methane Problem," *Climate Home News*, August 13, 2018, https://www.climatechangenews.com/2018/08/13/us-methane-problem/.

152. See California v. Bernhardt, Case No. 4:18-cv-05712-YGR (N.D. Cal. 2020), http://climatecasechart.com/case/california-v-zinke/.

153. Coral Davenport, "EPA to Lift Obama-Era Controls on Methane, a Potent Greenhouse Gas," *The New York Times*, August 10, 2020, https://www.nytimes.com/2020/08/10/climate/trump-methane-climate-change.html.

154. Devashree Saha, "As U.S. Government Retreats on Reducing Climate-Warming Methane, 4 States Step Up," *World Resources Institute*, September 18, 2019, https://www.wri.org/blog/2019/09/us-government-retreats-reducing-climate-warming-methane-4-states-step-up. For a description of the parts of the December 2020 omnibus budget bill that concern energy and climate, see "13 ways the massive omnibus hits energy," *E&E News*, December 23, 2020, https://www.eenews.net/stories/1063721401.

155. See Chapter 2 for discussion of the properties of these greenhouse gases, and Chapters 9 and 10 for policies to reduce nitrous oxide and hydrofluorocarbon emissions.

156. See https://www.ncel.net/offshore-drilling/.

157. See Rob Friedman, "Five Years After Ban, New York Can Do More Against Fracking," National Resources Defense Council, December 17, 2019, https://www.nrdc.org/experts/rob-friedman/5-years-after-ban-new-york-can-do-more-against-fracking. Important work is being done by the Fractracker Alliance, which addresses human health and community concerns from fracking across the US. See https://www.fractracker.org.

158. See https://www.arctictoday.com/alaskas-new-governor-has-dismissed-the-states-climate-team-and-scrapped-climate-policy-and-plan/.

159. Costs of climate change in Alaska are from Markon, C., S. Gray, M. Berman, L. Eerkes-Medrano, T. Hennessy, H. Huntington, J. Littell, M. McCammon, R. Thoman, and S. Trainor, 2018: *Alaska. In Impacts, Risks, and Adaptation in the United States: Fourth National Climate Assessment, Volume II* [Reidmiller, D.R., C.W. Avery, D.R. Easterling, K.E. Kunkel, K.L.M. Lewis, T.K. Maycock, and B.C. Stewart (eds.)]. U.S. Global Change Research Program, Washington, DC, USA, pp. 1185–1241. doi: 10.7930/NCA4.2018.CH26, Key Message 5. For oil revenues, see the Oil and Gas Industry webpage by Alaska's Resource Development Council, https://www.akrdc.org/oil-and-gas.

160. Adele C. Morris, "The Challenge of State Reliance on Revenue From Fossil Fuel Production," Climate and Energy Economics Project, Brookings Institute, August 9, 2016, https://www.brookings.edu/wp-content/uploads/2016/08/state-fiscal-implications-of-fossil-fuel-production-0809216-morris.pdf. For an insightful analysis of the political problems in a leading fossil fuel state, see Heather Richards, "Beyond Coal and Oil: Wyo. Faces Crisis," *E&E News*, November 19, 2019, https://www.eenews.net/stories/1061587029.

161. Zahra Hirji, "Portland Bans New Fossil Fuel Infrastructure in Stand Against Climate Change," *Inside Climate News*, December 15, 2016, https://insideclimatenews.org/news/14122016/portland-oregon-ban-fossil-fuels-oil-and-gas-pipelines-coal-global-warming. Also see Blair Stenvick, "Portland City Council Renews Ban on New Fossil Fuel Terminals," *Portland Mercury*, December 18, 2019, https://www.portlandmercury.com/blogtown/2019/12/18/27669176/portland-city-council-renews-ban-on-new-fossil-fuel-terminals.

162. See the last section of Chapter 9 for further discussion about energy savings in buildings and cities.

163. Mike Baker, "To Fight Climate Change, One City May Ban Heating Homes with Natural Gas," *The New York Times*, January 5, 2020, https://www.nytimes.com/2020/01/05/us/bellingham-natural-gas-ban.html?action=click&module=News&pgtype=Homepage. For federal initiatives, see David Iaconangelo, "As Natural Gas Bans Go National, Can Cities Fill the Gap?" *E&E News*, August 3, 2020, https://www.eenews.net/stories/1063674561.

164. For a description of legal preemption, see https://www.law.cornell.edu/wex/preemption.

165. Jason Plautz, "Arizona Set to Preempt Local Natural Gas Bans," *Smart Cities Dive*, February 21, 2020, https://www.smartcitiesdive.com/news/arizona-bill-preempt-local-natural-gas-bans/572693/. Also see Ari Phillips, "Texas Governor Signs Bill That Makes Local Fracking Bans Illegal," *ThinkProgress*, May 19, 2015, https://thinkprogress.org/texas-governor-signs-bill-that-makes-local-fracking-bans-illegal-bccd73b6046/.

166. "Coal Explained. Mining and Transportation of Coal," US Energy Information Administration, reviewed December 10, 2020, https://www.eia.gov/energyexplained/coal/mining-and-transportation.php. See also "Key Facts," *Coal Train Facts*, http://www.coaltrainfacts.org/key-facts.

167. Brandon J. Murrill, *Pipeline Transportation of Natural Gas and Crude Oil: Federal and State Regulatory Authority*, Congressional Research Service, March 28, 2016, https://fas.org/sgp/crs/misc/R44432.pdf.

168. See https://www.sightline.org/research/thin-green-line/ for a short video on the coal, oil shale, and Canadian tar sand resources whose export has been blocked from Pacific NW ports.

169. Earthfix, "Coal Scorecard: Your Guide to Coal in the Northwest," Jefferson Public Radio, October 24, 2017, https://www.klcc.org/post/coal-scorecard-your-guide-to-coal-northwest.

170. See Hal Bernton, "Vancouver Energy Ends Bid to Build Nation's Biggest Oil Train Terminal Along Columbia River," *The Seattle Times*, February 27, 2018, https://www.seattletimes.com/seattle-news/no-oil-train-terminal-on-the-columbia-river-vancouver-energy-gives-up-plan/.

171. See https://www.ferc.gov/about/ferc-does.asp.

172. For an overview of NEPA, see "What is the National Environmental Policy Act?," United States Environmental Protection Agency, https://www.epa.gov/nepa/what-national-environmental-policy-act.

173. Sharon Buccino, "Understanding Trump's Harmful Attack on NEPA," Natural Resources Defense Council, July 15, 2020, https://www.nrdc.org/experts/sharon-buccino/understanding-trumps-harmful-attack-nepa.

174. For a good basic explanation of fracking, see Brad Plumer, "Fracking, Explained," *Vox*, July 30, 2015, https://www.vox.com/2014/4/14/18076690/fracking.

175. "Does Fracking Cause Earthquakes?," United States Geological Survey, https://www.usgs.gov/faqs/does-fracking-cause-earthquakes?qt-news_science_products=0#qt-news_science_products.

176. "The Halliburton Loophole," Earthworks, https://earthworks.org/issues/inadequate_regulation_of_hydraulic_fracturing/.

177. Kristina Marusic, "After a Decade of Research, Here's What Scientists Know About the Health Impacts of Fracking," *Environmental Health News*, April 15, 2019, https://www.ehn.org/health-impacts-of-fracking-2634432607.html.

178. See https://www.ferc.gov/final-environmental-impact-statement-jordan-cove-energy-project for the final EIS and supporting information.

Chapter 7 Notes

1. The analogy between tobacco and fossil fuels is described by Barry Rabe, *Can We Price Carbon?* (Cambridge, MA: MIT Press, 2018), 1-4.

2. "Where Carbon is Taxed," Carbon Tax Center, https://www.carbontax.org/where-carbon-is-taxed/.

3. For a summary, see Garrett Hardin, "The Tragedy of the Commons," The Library of Economics and Liberty, https://www.econlib.org/library/Enc/TragedyoftheCommons.html. For a fuller description, see Garrett Hardin, "The Tragedy of the Commons," Science 162, no.3859 (1968): 1243-1248, https://science.sciencemag.org/content/162/3859/1243.

4. Hal Harvey, *Designing Climate Solutions: A Policy Guide for Low-Carbon Energy* (Washington, D.C.: Island Press, 2018), 254-255.

5. See the discussion in Chapter 6.

6. Rogelj, "Mitigation Pathways Compatible with 1.5°C in the Context of Sustainable Development," *Global Warming of 1.5°C: An IPCC Special Report on the Impacts of Global Warming of 1.5°C*, 95. Professor William Nordhaus, the 2018 Nobel Prize winner in Economics, garnered prominent support for carbon pricing by showing that "...the most efficient remedy for problems caused by greenhouse gases is a global scheme of universally imposed carbon taxes." See https://www.nobelprize.org/prizes/economic-sciences/2018/summary/.

7. J. Larsen et al., "Energy and Environmental Implications of a Carbon Tax in the United States," Columbia | SIPA Center on Global Energy Policy (2018): 19-26, https://energypolicy.columbia.edu/research/report/energy-and-environmental-implications-carbon-tax-united-states.

8. See the discussion in Chapters 8 and 9.

9. Shi-Ling Hsu, *The Case for a Carbon Tax* (Washington, D.C.: Island Press, 2011), 39-40.

10. Hsu, *The Case for a Carbon Tax*, 72.

11. For a good primer on cap and trade that explains these points in more detail, see https://www.c2es.org/content/cap-and-trade-basics/.

12. See https://ww3.arb.ca.gov/cc/ab32/ab32.htm.

13. Brad Plumer, "California is About to Find Out What a Truly Radical Climate Policy Looks Like," *Vox*, September 9, 2016, https://www.vox.com/2016/8/29/12650488/california-climate-law-sb-32. For a summary of changes introduced in SB32, see Jason Ye, "Summary of California's Extension of its Cap and Trade Program," Center for Climate and Energy Solutions, https://www.c2es.org/site/assets/uploads/2017/09/summary-californias-extension-its-cap-trade-program.pdf.

14. *California's 2017 Climate Change Scoping Plan* (California Air Resources Board, 2017) https://ww3.arb.ca.gov/cc/scopingplan/scoping_plan_2017_es.pdf.

15. We will look at some of these other initiatives in Chapters 8-10.

16. See https://ww2.arb.ca.gov/ghg-inventory-graphs.

17. The California Supreme Court upheld the legality of the cap and trade law in 2017; see "California's Landmark Cap-and-Trade Program Upheld by California Supreme Court," Environmental Defense Fund, June 28, 2017, https://www.edf.org/media/californias-landmark-cap-and-trade-program-upheld-california-supreme-court.

18. For details on how funds were spent, see http://www.caclimateinvestments.ca.gov and https://ww2.arb.ca.gov/news/california-climate-investments-track-set-annual-record-more-900-million-projects-statewide. For the 2020 report on cap and trade revenue spending, see https://ww2.arb.ca.gov/sites/default/files/classic//cc/capandtrade/auctionproceeds/2020_cci_annual_report.pdf.

19. For an overview of the main provisions, see "California Cap and Trade," Center for Climate and Energy Solutions, https://www.c2es.org/content/california-cap-and-trade/. For a more thorough description, see *The 2017-18 Budget: Cap and Trade*, Legislative Analyst's Office, February 13, 2017, https://lao.ca.gov/publications/report/3553. Regulations are available at https://ww3.arb.ca.gov/cc/capandtrade/capandtrade/ct_reg_unofficial.pdf.

20. See Chapter 3 for a discussion of carbon budgets.

21. *California's Cap-and-Trade Program, Step by Step*, Environmental Defense Fund, https://www.edf.org/sites/default/files/californias-cap-and-trade-program-step-by-step.pdf.

22. Hsu, *The Case for a Carbon Tax*, 62-63.

23. When cap and trade prices are constrained by both a floor and a ceiling, then the program design resembles a fixed carbon tax. See Hsu, *The Case for a Carbon Tax*, 104-114.

24. For a good brief description of secondary carbon markets, see https://www.c2es.org/site/assets/uploads/2016/04/secondary-carbon-markets.pdf.

25. See *California Greenhouse Gas Emissions for 2000 to 2017* (California Air Resources Board, 2019), https://ww3.arb.ca.gov/cc/inventory/pubs/reports/2000_2017/ghg_inventory_trends_00-17.pdf for detailed statistics on emissions through 2017.

26. David Roberts, "California's Cap-and-Trade System May Be Too Weak to Do Its Job," *Vox*, December 13, 2018, https://www.vox.com/energy-and-environment/2018/12/12/18090844/california-climate-cap-and-trade-jerry-brown. Also see Anne Mulkern, "Bleak Cap-and-Trade Results Raise Doubts About Program," *E&E News*, June 16, 2020, https://www.eenews.net/climatewire/2020/06/16/stories/1063394847.

27. For the basic elements of RGGI, see https://www.rggi.org/program-overview-and-design/elements.

28. The Model Rule is available at https://www.rggi.org/sites/default/files/Uploads/Design-Archive/Model-Rule/2017-Program-Review-Update/2017_Model_Rule_revised.pdf.

29. Rabe, *Can We Price Carbon?*, 125-161.

30. Rabe, *Can We Price Carbon?*, 129.

31. *The Regional Greenhouse Gas Initiative: Background, Impacts and Selected Issues* (Congressional Research Service, 2019), 14, https://fas.org/sgp/crs/misc/R41836.pdf.

32. For future RGGI target levels, see https://www.rggi.org/sites/default/files/Uploads/Program-Review/12-19-2017/Principles_Accompanying_Model_Rule.pdf. For data on emissions reductions in RGGI states, see *Regional Greenhouse Gas Initiative*, 7.

33. *Regional Greenhouse Gas Initiative*, 15.

34. Carbon capture and storage (CCS) is described in Chapter 10.

35. The WCI is a private, nonprofit corporation that administers emissions trading programs, linking independent jurisdictions such as California and Quebec. See Alex Green and Robert Jackel, "Will Other States Join California's International Climate Pact?" *The Atlantic*, August 10, 2017, https://www.theatlantic.com/politics/archive/2017/08/california-emissions-cap-trade/536430/.

36. For further reading on cap-and trade-programs, see Rabe, *Can We Price Carbon?*, and https://onclimatechangepolicydotorg.wordpress.com/carbon-pricing/. For a detailed manual for policymakers, see *Emissions Trading in Practice: A Handbook on Design and Implementation* (World Bank Group, 2016) http://documents.worldbank.org/curated/en/353821475849138788/Emissions-trading-in-practice-a-handbook-on-design-and-implementation.

37. To access the exhaustive hearings and documents held and considered by the legislature in developing the law (HB2020), visit https://olis.leg.state.or.us/liz/2019R1/Measures/Overview/HB2020.

38. Ted Sickinger, "Oregon Climate Change: Governor Signs Executive Order to Reduce Greenhouse Gas Emissions," *OregonLive*, March 11, 2020, https://www.oregonlive.com/politics/2020/03/oregon-climate-change-governor-signs-executive-order-to-reduce-greenhouse-gas-emissions.html.

39. See https://www.rggi.org/sites/default/files/Uploads/Participation/RGGI_New_State_Participation_Overview.pdf.

40. See https://climateadvocacylab.org/system/files/GuideToCommunicatingCarbonPricing.pdf.

41. Charles Komanoff & Matthew Gordon, "British Columbia's Carbon Tax: By the Numbers," Carbon Tax Center, December 2015, https://www.carbontax.org/wp-content/uploads/CTC_British_Columbia's_Carbon_Tax_By_The_Numbers.pdf.

42. See "Where Carbon is Taxed," Carbon Tax Center, https://www.carbontax.org/where-carbon-is-taxed/.

43. Hsu, *The Case for a Carbon Tax*, 48-52.

44. Rabe, *Can We Price Carbon?*, 95.

45. Hsu, *The Case for a Carbon Tax*, 77-83.

46. Marc Hafstead., "The Year of the Carbon Pricing Proposal," *Resources for the Future*, August 2, 2019, https://www.resourcesmag.org/common-resources/the-year-of-the-carbon-pricing-proposal/.

47. See https://energyinnovationact.org.

48. The text of the EICDA is available at https://energyinnovationact.org/wp-content/uploads/2019/01/Energy-Innovation-and-Carbon-Dividend-Act-2019.pdf.

49. Noah Kaufman et al., *An Assessment of the Energy Innovation and Carbon Dividend Act* (Columbia | SIPA Center on Global Energy Policy, and Rhodium Group, 2019): 25, https://energypolicy.columbia.edu/sites/default/files/file-uploads/FICDA_CGEP-Report.pdf.

50. Carbon Pricing Leadership Coalition, *Report of the High-Level Commission on Carbon Prices*, (Carbon Pricing Leadership Coalition, 2017) https://static1.squarespace.com/static/54ff9c5ce4b0a53decccfb4c/t/59b7f2409f-8dce5316811916/1505227332748/CarbonPricing_FullReport.pdf.

51. Rogelj, "Mitigation Pathways Compatible," 152. See Chapter 3 for the discussion of carbon budgets.

52. The social cost of carbon is discussed in the intergenerational equity section of Chapter 4.

53. Noah Kaufman et al, "A Near-Term to Net Zero Alternative to the Social Cost of Carbon for Setting Carbon Prices," *Nature. Climate Change*, August 17, 2020, https://doi.org/10.1038/s41558-020-0880-3.

54. Kaufman et al., *An Assessment of the Energy Innovation*, 13-17.

55. John Larsen et al., *Energy and Environmental Implications of a Carbon Tax in the United States*, 52.

56. For discussion of the CLC proposal, see Chapters 5 and 6.

57. New regulations triggered by the EICDA would no longer come under the authority of the Clean Air Act, which could avoid legal issues. See the discussion of greenhouse gas regulations in Chapter 6.

58. See Ted Halstead, *Unlocking the Climate Puzzle* (Climate Leadership Council, 2017) https://www.clcouncil.org/wp-content/uploads/2017/02/Unlocking_The_Climate_Puzzle.pdf.

59. See EU Emissions Trading System (EU ETS), https://ec.europa.eu/clima/policies/ets_en.

60. Kevin Ummel, *Household Impact Study II, The Impact of a Carbon Fee and Dividend Policy on the Finances of US Households* (2020), https://citizensclimatelobby.org/wp-content/uploads/2018/06/HIS2-Working-Paper-v1.1.pdf.

61. "How We're Putting a Price on Carbon Pollution," Government of Canada, last modified June 28, 2019,https://www.canada.ca/en/environment-climate-change/services/climate-change/pricing-pollution-how-it-will-work/putting-price-on-carbon-pollution.html.

62. Jake Cigainero, "Who Are France's Yellow Vest Protesters, and What Do They Want?" *NPR*, December 3, 2018, https://www.npr.org/2018/12/03/672862353/who-are-frances-yellow-vest-protesters-and-what-do-they-want.

63. Emily Mazzacurati, "What Would a Federal Carbon Tax Mean for California Cap and Trade?" *Four Twenty Seven*, March 14, 2013, http://427mt.com/2013/03/14/what-would-a-federal-carbon-tax-mean-for-california-cap-and-trade/.

64. The dark money network and the oil and gas industries' support for the CLC's carbon pricing plan are explained in Chapter 6.

65. David Roberts, "Oregon Republicans Are Subverting Democracy by Running Away. Again," *Vox*, February 29, 2020, https://www.vox.com/energy-and-environment/2020/2/29/21157246/oregon-republicans-walk-out-climate-change-cap-trade-democracy.

66. Kate Aronoff, "BP Claims to Support Taxing Carbon, but It's Spending $13 Million Against an Initiative That Would Do Just That," *The Intercept*, November 1, 2018, https://theintercept.com/2018/11/01/bp-washington-state-carbon-tax-initiative/.

67. See Nathaniel Johnson, "So What Did California Do with That $1.4 Billion in Cap-and-Trade Money?" *Grist*, March 27, 2019, https://grist.org/article/so-what-did-california-do-with-that-1-4-billion-in-cap-and-trade-money/, and *The Investment of RGGI Proceeds in 2018* (Regional Greenhouse Gas Initiative, 2020) https://www.rggi.org/sites/default/files/Uploads/Proceeds/RGGI_Proceeds_Report_2018.pdf.

68. Kaufman et al., *An Assessment of the Energy Innovation*, https://www.energypolicy.columbia.edu/research/report/assessment-energy-innovation-and-carbon-dividend-act.

69. Hafstead., "The Year of the Carbon Pricing Proposal."

70. The progressive political stance, as embodied in the Green New Deal, is discussed in Chapter 5.

71. For a critique of carbon pricing from the progressive Left, see Matto Mildenberger and Leah Stokes, "The Trouble With Carbon Pricing," *Boston Review*, September 24, 2020, http://bostonreview.net/science-nature-politics/matto-mildenberger-leah-c-stokes-trouble-carbon-pricing. For a rebuttal, see Joseph Majkut, "The Immediate Case for a Carbon Price," Niskanen Center, October 26, 2020, https://www.niskanencenter.org/the-immediate-case-for-a-carbon-price/.

72. See https://www.sierraclub.org/washington/sierra-club-position-carbon-washington-ballot-initiative-732.

73. See https://350pdx.org/board-statement-on-hb2020/.

74. For a list of the 86 cosponsors of the EICDA, see https://www.congress.gov/bill/116th-congress/house-bill/763/cosponsors?searchResultViewType=expanded&KWICView=false. For a list of the 99 cosponsors of the Green New Deal resolution in the House of Representatives, see https://www.congress.gov/bill/116th-congress/house-resolution/109/cosponsors?searchResultViewType=expanded&KWICView=false.

75. See https://www.carbontax.org/carbon-tax-vs-the-alternatives/offsets/.

76. For allowed California offsets, see https://ww3.arb.ca.gov/cc/capandtrade/offsets/offsets.htm. For offsets in the RGGI program, see https://www.rggi.org/allowance-tracking/offsets.

Chapter 8 Notes

1. For an excellent overall description, see "How The Electricity Grid Works," Union of Concerned Scientists, February 17, 2015, https://www.ucsusa.org/resources/how-electricity-grid-works. For a bit more technical detail, see "Electricity Explained," US Energy Information Administration, https://www.eia.gov/energyexplained/electricity/electricity-in-the-us-generation-capacity-and-sales.php. For all the nitty gritty, see Department of Energy, *United States Electricity Industry Primer* (Department of Energy, 2015), https://www.energy.gov/sites/prod/files/2015/12/f28/united-states-electricity-industry-primer.pdf.

2. "What is U.S. Electricity Generation by Energy Source?" US Energy Information Administration, last modified February 27, 2020, https://www.eia.gov/tools/faqs/faq.php?id=427&t=3.

3. Nadja Popovich and Brad Plumer, "How Does Your State Make Electricity?" *New York Times*, October 28, 2020, https://www.nytimes.com/interactive/2020/10/28/climate/how-electricity-generation-changed-in-your-state-election.html.

4. See David Roberts, "Transmission Week: Why We Need More Big Power Lines," Volts, January 25, 2021, https://www.volts.wtf/p/transmission-week-why-we-need-more, and David Roberts, "Transmission Week: How to Start Building More High-Voltage Power Lines," Volts, January 27, 2021, https://www.volts.wtf/p/transmission-week-how-to-start-building.

5. "Demand Response," US Department of Energy, https://www.energy.gov/oe/activities/technology-development/grid-modernization-and-smart-grid/demand-response.

6. "Understanding Energy Capacity and Capacity Factor," NMPP Energy, http://www.nmppenergy.org/feature/capacity_factor.

7. The generation capacities and capacity factors for today's grid are given in Table 8.2. The contributions that these sources make to electrical energy output are shown in Figure 1. The total energy output depends on both the total capacity and how much of that capacity is used. See Chapter 2 for the distinction between energy and power.

8. In 2018, the national average for the number of homes powered by 1 MW of PV solar was 190. See Solar Energies Industries Association, *What's in a Megawatt? Calculating the Number of Homes Powered by Solar Energy*, (Solar Energy Industries Association), https://www.seia.org/initiatives/whats-megawatt.

9. "Solar Explained," US Energy Information Administration, https://www.eia.gov/energyexplained/solar/photovoltaics-and-electricity.php.

10. Stephen Battersby, "News Feature: The Solar Cell of the Future," *Proceedings of the National Academy of Sciences of the United States of America* 116 no. 1 (2019): 7-10, https://www.pnas.org/content/116/1/7.

11. See "What is the Levelized Cost of Electricity?" Corporate Finance Institute, https://corporatefinanceinstitute.com/resources/knowledge/finance/levelized-cost-of-energy-lcoe/.

12. "Dispatchable Source of Electricity," Energy Education, University of Calgary, https://energyeducation.ca/encyclopedia/Dispatchable_source_of_electricity.

13. See "Why Does the Electricity Grid Have to Stay in Balance?" Energuide.be, https://www.energuide.be/en/questions-answers/why-does-the-electricity-grid-have-to-stay-in-balance/2136/.

14. Ker Than, "Critical Minerals Scarcity Could Threaten Renewable Energy Future," *Stanford Earth*, January 17, 2018, https://earth.stanford.edu/news/critical-minerals-scarcity-could-threaten-renewable-energy-future#gs.7hzybe.

15. See "Solar Energy Development Environmental Considerations," Solar Energy Development Programmatic EIS, http://solareis.anl.gov/guide/environment/.

16. Krisztina Pjeczka, "Reducing the Land Use Impact of Solar Energy—a Triple Win for Climate, Agriculture and Biodiversity," *Yale Environment Review*, September 14, 2018, https://environment-review.yale.edu/reducing-land-use-impact-solar-energy-triple-win-climate-agriculture-and-biodiversity.

17. See https://www.eia.gov/energyexplained/solar/solar-thermal-power-plants.php. Another good overview of CSP is "Concentrating Solar Power Plants," Union of Concerned Scientists, December 23, 2015, https://www.ucsusa.org/resources/concentrating-solar-power-plants.

18. "Solar Desalination," United States Department of Energy Office of Energy Efficiency and Renewable Energy, https://www.energy.gov/sites/prod/files/2018/09/f55/Concentrating-Solar-Thermal-Power-FactSheet.pdf.

19. "Concentrating Solar Power," National Renewable Energy Laboratory, https://atb.nrel.gov/electricity/2018/index.html?t=sc.

20. "Research Spotlight: Concentrating Solar-Thermal Power," US Department of Energy Office of Energy Efficiency and Renewable Energy, https://www.energy.gov/sites/prod/files/2018/09/f55/Concentrating-Solar-Thermal-Power-FactSheet.pdf.

21. Nathan Bracken et al., *Concentrating Solar Power and Water Issues in the U.S. Southwest* (Joint Institute for Strategic Energy Analysis, 2015), https://www.nrel.gov/docs/fy15osti/61376.pdf.

22. For a description of windmill components and designs, see *Wind Power Technology Brief* (International Renewable Energy Agency, 2016), 9-13, https://irena.org/-/media/Files/IRENA/Agency/Publication/2016/IRENA-ETSAP_Tech_Brief_Wind_Power_E07.pdf.

23. Shaun Campbell, "Clash of the Titans—Top 5 Biggest Onshore Turbines," *Wind Power Monthly*, August 2, 2019, https://www.windpowermonthly.com/article/1592000/clash-titans-top-5-biggest-onshore-turbines.

24. P. Veers et al., "Grand Challenges in the Science of Wind Energy," *Science* (2019), https://science.sciencemag.org/content/sci/early/2019/10/09/science.aau2027.full.pdf.

25. For more information on the US wind energy resource, see National Renewable Energy Laboratory, https://www.nrel.gov/gis/wind.html.

26. Russell McClendon, "6 Ways to Protect Bats and Birds From Wind Turbines," *Mother Nature Network*, October 22, 2019, https://www.mnn.com/earth-matters/energy/blogs/6-ways-to-protect-bats-and-birds-from-wind-turbines.

27. Jeffrey Tomich, "End of Iowa's Wind Boom? Renewable Rules Spark Fears," E&E News, July 8, 2020, https://www.eenews.net/stories/1063523833.

28. John Fialka, "Massive Turbines Designed for US Offshore Wind Farms," *E&E News*, June 10, 2020, https://www.eenews.net/climatewire/2020/06/10/stories/1063360375?utm_medium=email&utm_source=eenews%3Aclimatewire&utm_campaign=edition%2BiZ%2B%2FftFV%2B2LxUfHtN5bxJQ%3D%3D. For a look at offshore wind turbine design, see https://www.ge.com/renewableenergy/wind-energy/offshore-wind/haliade-x-offshore-turbine.

29. See Wind Europe, *Offshore Wind in Europe*, Wind Europe, https://windeurope.org/about-wind/statistics/offshore/european-offshore-wind-industry-key-trends-statistics-2019/.

30. World Wildlife Foundation, *Environmental Impacts of Offshore Wind Power Production in the North Sea* (World Wildlife Foundation, 2014), https://www.wwf.no/assets/attachments/84-wwf_a4_report___havvindrapport.pdf.

31. Ross Davies, "Vineyard Wind: Delayed Project Reveals Bluster in US's Offshore Wind Ambitions," *Power Technology*, December 4, 2019, https://www.power-technology.com/features/vineyard-wind-delayed-project-reveals-bluster-in-uss-offshore-wind-ambitions/.

32. John Fialka, "The U.S. Has 7 Ocean Turbines. Companies see Hundreds Soon," *E&E News*, July 30, 2020, https://www.eenews.net/stories/1063653141. See also John Fialka, "16 Projects Are on the Way. Will the Grid Fail?" *E&E News*, August 3, 2020, https://www.eenews.net/stories/1063676089.

33. National Renewable Energy Laboratory, "Two NREL Studies Find Gulf of Mexico Well Positioned for Offshore Wind Development," May 6, 2020, https://www.nrel.gov/news/program/2020/studies-find-gulf-of-mexico-well-positioned-for-offshore-wind-development.html. For reporting on the pandemic's impact on the Louisiana oil and gas industry, see Nathaniel Rich, "In Louisiana, Covid-19 Has Achieved What Big Oil Protesters Could Not," *New York Times*, May 26, 2020, https://www.nytimes.com/interactive/2020/05/26/magazine/coronavirus-louisiana-unemployment-jobs.html.

34. US Geological Survey, *Hydroelectric Power: How It Works*, https://www.usgs.gov/special-topic/water-science-school/science/hydroelectric-power-how-it-works?qt-science_center_objects=0#qt-science_center_objects.

35. Emily Grubert, "Conventional Hydroelectricity and the Future of Energy: Linking National Inventory of Dams and Energy Information Administration Data to Facilitate Analysis of Hydroelectricity," *Electricity Journal* 33, no. 1 (2020): https://www.sciencedirect.com/science/article/abs/pii/S1040619019302970.

36. US Department of Energy, *Hydropower Vision: A New Chapter for America's 1st Renewable Energy Source*, (US Department of Energy, 2016):,11,https://www.energy.gov/eere/water/articles/hydropower-vision-new-chapter-america-s-1st-renewable-electricity-source.

37. Ilissa B. Ocko and Steven P. Hamburg, "Climate Impacts of Hydropower: Enormous Differences Among Facilities and Over Time," *Environmental Science & Technology* 53, no. 23 (2019): 14070-14082, https://pubs.acs.org/doi/10.1021/acs.est.9b05083.

38. "Water Quality Degradation from Hydropower," Energy Education, https://energyeducation.ca/encyclopedia/Water_quality_degradation_from_hydropower.

39. US Department of Energy, *Hydropower Vision*.

40. "Pumped Storage," University of Calgary, Energy Education, https://energyeducation.ca/encyclopedia/Pumped_storage.

41. "Hydropower Explained," US Energy Information Administration, last modified March 30, 2020, https://www.eia.gov/energyexplained/hydropower/where-hydropower-is-generated.php.

42. National Hydropower Association, *2018 Pumped Storage Report* (National Hydropower Association), https://www.hydro.org/wp-content/uploads/2018/04/2018-NHA-Pumped-Storage-Report.pdf.

43. "Geothermal Heat Pumps," Department of Energy, https://www.energy.gov/energysaver/heat-and-cool/heat-pump-systems/geothermal-heat-pumps.

44. US Energy Information Administration, "Nearly Half of U.S. Geothermal Power Capacity Came On Line in the 1980s," *Today in Energy*, November 20, 2019, https://www.eia.gov/todayinenergy/detail.php?id=42036.

45. US Department of Energy, GeoVision, Harnessing the Heat Beneath Our Feet, (US Department of Energy), https://www.energy.gov/eere/geothermal/downloads/geovision-harnessing-heat-beneath-our-feet.

46. *Geothermal Technologies Office, What IS an Enhanced Geothermal System (EGS)?* US Department of Energy, https://www.energy.gov/sites/prod/files/2016/05/f31/EGS%20Fact%20Sheet%20May%202016.pdf.

47. US Department of Energy, *GeoVision*, 38-39, https://www.energy.gov/eere/geothermal/downloads/geovision-harnessing-heat-beneath-our-feet.

48. For an optimistic take on the technology, see David Roberts, "Geothermal Energy Is Poised For a Big Breakout," *Vox*, October 21, 2020, https://www.vox.com/energy-and-environment/2020/10/21/21515461/renewable-energy-geothermal-egs-ags-supercritical.

49. For basic information on nuclear fission, see "How Does a Nuclear Reactor Work?," World Nuclear Association, https://www.world-nuclear.org/nuclear-essentials/how-does-a-nuclear-reactor-work.aspx. For information on types of nuclear reactors, see "Nuclear Power Reactors," World Nuclear Association, April 2020, https://www.world-nuclear.org/information-library/nuclear-fuel-cycle/nuclear-power-reactors/nuclear-power-reactors.aspx.

50. "Nuclear Explained: Where Our Uranium Comes From," US Energy Information Administration, https://www.eia.gov/energyexplained/nuclear/where-our-uranium-comes-from.php.

51. "Supply of Uranium," World Nuclear Association, May 2020, https://www.world-nuclear.org/information-library/nuclear-fuel-cycle/uranium-resources/supply-of-uranium.aspx.

52. "Nuclear Explained: The Nuclear Fuel Cycle," US Energy Information Administration, https://www.eia.gov/energyexplained/nuclear/the-nuclear-fuel-cycle.php.

53. Pushker Karecha and James Hansen, *Coal and Gas Are Far More Harmful Than Nuclear Power* (NASA Goddard Institute for Space Studies, 2013), https://www.giss.nasa.gov/research/briefs/kharecha_02/.

54. See Spiro and Stigliani, *Chemistry of the Environment*, 60-62 for descriptions of the Three Mile Island and Chernobyl accidents.

55. "Fukushima Daiichi Accident," World Nuclear Association, last modified May, 2020, https://www.world-nuclear.org/information-library/safety-and-security/safety-of-plants/fukushima-daiichi-accident.aspx.

56. Rob Nikolewski, "Regulators Vote to Shut Down Diablo Canyon, California's Last Nuclear Power Plant," *Los Angeles Times*, January 11, 2018, https://www.latimes.com/business/la-fi-diablo-canyon-nuclear-20180111-story.html.

57. "Nuclear Reprocessing: Dangerous, Dirty and Expensive," Union of Concerned Scientists, April 5, 2011, https://www.ucsusa.org/resources/nuclear-reprocessing-dangerous-dirty-and-expensive.

58. "Storage of Spent Nuclear Fuel," United States Nuclear Regulatory Commission, https://www.nrc.gov/waste/spent-fuel-storage.html.

59. For a unique perspective on how humanity can live in harmony with nature that includes a role for nuclear power, see *An Ecomodernist Manifesto*, http://www.ecomodernism.org.

60. Peter Behr, "Can a Hydrogen Makeover Save Nuclear Power?" *E&E News*, July 29, 2020, https://www.eenews.net/stories/1063640735.

61. See https://www.nuscalepower.com for a description of prospective reactors from the NuScale company, which presently leads US efforts in new nuclear technology.

62. Mike Price, "A Look at the NuScale Small Modular Nuclear Reactor Project," *East Idaho News*, August 22, 2019, https://www.eastidahonews.com/2019/08/a-comprehensive-look-at-the-nuscale-small-modular-reactor-project/. For the NRC approval, see "NRC Approves First U.S. Small Modular Reactor Design," US Department of Energy Office of Nuclear Energy, September 2, 2020, https://www.energy.gov/ne/articles/nrc-approves-first-us-small-modular-reactor-design. For more details about small modular reactors, see "Small Nuclear Power Reactors," World Nuclear Association, https://www.world-nuclear.org/information-library/nuclear-fuel-cycle/nuclear-power-reactors/small-nuclear-power-reactors.aspx.

63. Julian Spector, "What Comes Next After Batteries Replace Gas Peakers?," *Greentech Media*, July 1, 2019, https://www.greentechmedia.com/articles/read/california-clean-power-outlook-what-comes-after-shorter-duration-batteries.

64. Chapter 10 describes natural and technological methods for atmospheric carbon drawdown.

65. For analysis of tidal power, see Talal Husseini, "Riding the Renewable Wave: Tidal Energy Advantages and Disadvantages," *Power Technology*, October 26, 2018, https://www.power-technology.com/features/tidal-energy-advantages-and-disadvantages/. For a description of nuclear fusion, see Tom Clynes, "5 Big Ideas for Making Nuclear Fusion a Reality," IEEE Spectrum, January 28, 2020, https://spectrum.ieee.org/energy/nuclear/5-big-ideas-for-making-fusion-power-a-reality.

66. "Biomass and Waste Fuels Made Up 2% of US Electricity Generation in 2016," US Energy Information Administration, November 27, 2017, https://www.eia.gov/todayinenergy/detail.php?id=33872.

67. William H. Schlesinger, "Are Wood Pellets a Green Fuel?" *Science 359*, no. 6382 (2018): https://science.sciencemag.org/content/359/6382/1328.

68. See the discussions about biomass uses in Chapters 9 and 10.

69. David Roberts, "Getting to 100% Renewables Requires Cheap Energy Storage. But How Cheap?" *Vox*, September 20, 2019, https://www.vox.com/energy-and-environment/2019/8/9/20764886/renewable-energy-storage-cost-electricity.

70. Elliot Negin, "Clean Energy + Battery Storage = Game Changer," *Catalyst 19* (2019): 16, https://www.ucsusa.org/sites/default/files/2019-10/catalyst-fall-2019.pdf.

71. "Pumped-Storage Hydropower," Department of Energy, https://www.energy.gov/eere/water/pumped-storage-hydropower.

72. For a brief review of present options for storage, see Cabe Atwell, "6 Promising Energy Storage Options to Tie Into the Grid," *Power Electronics*, February 13, 2018, https://www.powerelectronics.com/technologies/alternative-energy/article/21864101/6-promising-energy-storage-options-to-tie-into-the-grid.

73. "Vehicle-to-Grid: Everything You Need to Know," Virta, https://www.virta.global/vehicle-to-grid-v2g.

74. "Energy Storage," Environmental and Energy Storage Institute, February 2019, https://www.eesi.org/papers/view/energy-storage-2019. See also National Renewable Energy Laboratory, Declining Renewables Costs Drive Focus on Energy Storage, January 2, 2020, https://www.nrel.gov/news/features/2020/declining-renewable-costs-drive-focus-on-energy-storage.html.

75. This process is described in Chapter 9.

76. Peter Fairley, "Can the Grid Work With 100% Renewables? There's a Scientific Fight Brewing," *IEEE Spectrum*, June 19, 2017, https://spectrum.ieee.org/energywise/energy/renewables/can-the-us-grid-work-with-100-renewables.

77. Nuscale, *Frequently Asked Questions*, https://www.nuscalepower.com/about-us/faq#EC9.

78. "Plummeting Solar, Wind and Battery Costs Can Accelerate Our Clean Future," Goldman School of Public Policy, University of California, Berkeley, June 8, 2020, https://gspp.berkeley.edu/news/news-center/the-us-can-reach-90-percent-clean-electricity-by-2035-dependably-and-without-increasing-consumer-bills.

79. T. Mai et al., *Renewable Electricity Futures Study: Executive Summary* (National Renewable Energy Laboratory, 2012), 14-17, https://www.nrel.gov/docs/fy13osti/52409-ES.pdf. See Peter Fox-Penner, Power After Carbon: Building a Clean, Resilient Grid (Cambridge, MA: Harvard University Press, 2020), for an accessible, book-length study of the future of the electricity grid.

80. Although limited in duration, these credits nonetheless help preserve the recent explosive growth of solar and wind. For a summary of renewable energy tax credits in the omnibus funding bill, see "Congress Extends Renewable Energy Tax Credits in 2021 Omnibus Spending Bill," SheppardMullin, December 23, 2020, https://www.jdsupra.com/legalnews/congress-extends-renewable-energy-tax-98223/.

81. For a fascinating tour of electricity generation by all 50 states, see Nadja Popovich, "How Does Your State Make Electricity?" *New York Times*, October 28, 2020, https://www.nytimes.com/interactive/2020/10/28/climate/how-electricity-generation-changed-in-your-state-election.html.

82. Harvey, *Designing Climate Solutions*, Chapters 4 and 5, offers detailed and lucid discussions of policy principles.

83. "State Renewable Portfolio Standards and Goals," National Conference of State Legislatures, April 17, 2020, https://www.ncsl.org/research/energy/renewable-portfolio-standards.aspx.

84. Jenny Heeter and Lori Bird, *Including Alternative Resources in State Renewable Portfolio Standards: Current Design and Implementation Experience* (National Renewable Energy Laboratory, 2012), https://www.nrel.gov/docs/fy13osti/55979.pdf.

85. "State Renewable Portfolio Standards and Goals," State Renewable Portfolio, National Conference of State Legislatures, April 17, 2020, https://www.ncsl.org/research/energy/renewable-portfolio-standards.aspx. For clean energy mandates in the states, see *Progress Toward 100% Clean Energy* (UCLA Luskin Center for Innovation, 2019), https://innovation.luskin.ucla.edu/wp-content/uploads/2019/11/100-Clean-Energy-Progress-Report-UCLA-2.pdf.

86. Harvey, *Designing Climate Solutions*, 80-81.

87. "Renewable Electricity: How Do You Know You Are Using It?," National Renewable Energy Laboratory, https://www.nrel.gov/docs/fy15osti/64558.pdf.

88. "Renewable Energy Certificates (RECs)," Environmental Protection Agency, https://www.epa.gov/greenpower/renewable-energy-certificates-recs.

89. "What is a REC and How do They Work?," Urban Grid, June 25, 2019, https://www.urbangridsolar.com/what-is-a-rec-how-do-they-work/.

90. Harvey, *Designing Climate Solutions*, 84-85.

91. See Committee on Energy & Commerce, *Summary of the Climate Leadership and Environmental Action for Our Nation's (CLEAN) Future Act*, (Committee on Energy and Commerce, 2020), https://energycommerce.house.gov/sites/democrats.energycommerce.house.gov/files/documents/Section-by-Section%20of%20CLEAN%20Future%20Act%20.pdf.

92. For an analysis of clean energy standards, see Kathryne Cleary, Karen Palmer, and Kevin Rennert, *Clean Energy Standards* (Resources for the Future, 2019), https://media.rff.org/documents/CleanEnergy-Issue20Brief_2.pdf.

93. Julian Brave NoiseCat, Leah C. Stokes, and Narayan Subramanian, "The Fingerprints of the Green New Deal Are All Over the Clean Future Act," Data for Progress, February 18, 2020, https://www.dataforprogress.org/blog/2020/2/18/the-fingerprints-of-the-green-new-deal-are-all-over-the-clean-future-act.

94. Noah Kaufman, "A Clean Electricity Standard's Weaknesses May be its Biggest Strengths," Columbia | SIPA Center on Global Energy Policy, May 8, 2019, https://energypolicy.columbia.edu/research/commentary/clean-electricity-standards-weaknesses-may-be-its-biggest-strengths. See Chapter 6 for discussion of the politics of carbon pricing.

95. "New U.S. Power Plants Expected to be Mostly Natural Gas Combined Cycle and Solar PV," US Energy Information Administration, March 8, 2019, https://www.eia.gov/todayinenergy/detail.php?id=38612.

96. Nadja Popovich, "How Does Your State Make Electricity?" *New York Times*, October 28, 2020, https://www.nytimes.com/interactive/2020/10/28/climate/how-electricity-generation-changed-in-your-state-election.html.

97. For a detailed study concluding that stranding of all US fossil fuel-fired electricity generation by 2035 would be less costly than in other countries, see Emily Grubert, "Fossil Electricity Retirement Deadlines for a Just Transition," *Science* 370, no. 6521 (2020): 1171-1173, https://science.sciencemag.org/content/370/6521/1171.

98. See Chapter 10 for a discussion of carbon capture and storage.

99. For a list of EEI member companies, see https://www.eei.org/about/members/uselectriccompanies/Documents/memberlist_print.pdf. For a list of the 19 EEI members with zero carbon goals, see Peter Behr et al., "Is Biden's 100% Clean Electricity Plan Doable?," *E&E News*, July 15, 2020, https://www.eenews.net/stories/1063565769.

100. "Investor-Owned Utilities Served 72% of U.S. Electricity Customers in 2017," US Energy Information Administration, August 15, 2019, https://www.eia.gov/todayinenergy/detail.php?id=40913. The term *public utility* is also often used to refer to electricity providers. Confusingly, this does not necessarily refer to a *publicly owned utility*, but is used generally, often with respect to IOUs.

101. PUCs always regulate IOUs, and sometimes also regulate municipally owned utilities and electric cooperatives, at least for some of their activities. See "Engagement Between Public Utility Commissions and State Legislatures," National Council on Electricity Policy, https://www.ncsl.org/Portals/1/Documents/energy/NCSL_NARUC_Engage_Leg_PUCs_34251.pdf.

102. "How Utilities Determine Generation and Distribution Rates (Ratemaking)," *Electric Choice*, https://www.electric-choice.com/blog/how-utilities-determine-generation-and-distribution-rates-ratemaking/.

103. Herman K Trabish, "As the Power Sector Transforms, Can Utilities and Customers Find Common Ground on Ratemaking?," *Utility Dive*, July 2, 2018, https://www.utilitydive.com/news/as-the-power-sector-transforms-can-utilities-and-customers-find-common-gro/526399/.

104. See "Deregulated Energy States & Markets," Electric Choice, last modified 2020, https://www.electricchoice.com/map-deregulated-energy-markets/.

105. For a clear description of how vertically integrated utilities inhibit the transition to clean electricity, see David Roberts, "The US South Could Save Money by Cleaning Up Its Power Grid," *Vox*, September 1, 2020, https://www.vox.com/energy-and-environment/2020/9/1/21407275/duke-energy-southern-company-renewable-power-solar-wind-market-competition.

106. Benjamin Storrow, "Many Utilities Have a 'Must Run' Policy. One Broke the Rule," *E&E News*, January 14, 2020, https://www.eenews.net/stories/1062072877.

107. Joel Makower, "The Growing Concern Over Stranded Assets," *GreenBiz*, September 10, 2019, https://www.greenbiz.com/article/growing-concern-over-stranded-assets. See Chapter 6 for discussion of fossil fuel company business models.

108. Ryna Yiyun Cui et al., "Quantifying Operational Lifetimes for Coal-Fired Power Plants Under the Paris Goals," *Nature Communications* 10 (2019): https://www.nature.com/articles/s41467-019-12618-3.

109. Jeffrey Tomich, "'Stranded Costs' Mount as Coal Vanishes from the Grid," *E&E News*, May 29, 2019.

110. Jeffrey Tomich, "The Next Coal War? Industry Versus Its Customers," *E&E News*, August 2, 2019, https://www.eenews.net/stories/1060826471.

111. Synapse Energy Economics Inc., *A Brief Survey of State Integrated Resource Planning Rules and Requirements*, (Synapse Energy Economics, Inc., 2011), http://www.synapse-energy.com/sites/default/files/SynapseReport.2011-04.ACSF_.IRP-Survey.11-013.pdf.

112. Nate Berg, "How the US Power Grid Is Evolving to Handle Solar and Wind," *GreenBiz*, October 16, 2019, https://www.greenbiz.com/article/how-us-power-grid-evolving-handle-solar-and-wind.

113. American Wind Energy Association, *Grid Vision: The Electric Highway to a 21st Century Economy*, (American Wind Energy Association, 2019), https://www.awea.org/Awea/media/Resources/Publications%20and%20Reports/White%20Papers/Grid-Vision-The-Electric-Highway-to-a-21st-Century-Economy.pdf.

114. David Roberts, "We've Been Talking About a National Grid for Years. It Might be Time to Do It," *Vox*, August 3, 2018, https://www.vox.com/energy-and-environment/2018/8/3/17638246/national-energy-grid-renewables-transmission.

115. Federal Regulatory Energy Commission, *Energy Primer: A Handbook of Energy Market Basics*, (Federal Regulatory Energy Commission, 2015), 43-44, https://www.ferc.gov/market-assessments/guide/energy-primer.pdf.

116. See Americans for a Clean Energy Grid, https://cleanenergygrid.org.

117. See Macro Grid Initiative, *Macro Grids in the Mainstream: An International Survey of Plans and Progress* (Macro Grid Initiative, 2020), https://acore.org/macro-grid-initiative/.

118. This is the same FERC that regulates the interstate transport of natural gas and oil; see Chapter 5. Federal Energy Regulatory Commission, "What FERC Does," https://www.ferc.gov/about/ferc-does.asp.

119. Steven Weissman and Rena Kakon, "Phasing Out the Use of Fossil Fuels for the Generation of Electricity," *Legal Pathways to Deep Decarbonization in the United States*, ed. Michael B. Gerrard and John C. Dernbach (Washington, D.C.: Environmental Law Institute, 2019): 629-630.

120. Avi Zevin et al., "Building a New Grid Without New Legislation: A Path to Revitalizing Federal Transmission Authorities," Columbia | SIPA Center on Global Energy Policy, December 2020, https://www.eenews.net/assets/2020/12/15/document_ew_02.pdf.

121. Federal Regulatory Energy Commission, *Energy Primer*, 39-40, https://www.ferc.gov/market-assessments/guide/energy-primer.pdf.

122. "Regional Transmission Organizations (RTO)/Independent Systems Operators (ISO)," Federal Energy Regulatory Commission, https://www.ferc.gov/industries/electric/indus-act/rto.asp.

123. ISO New England, *2017 Report of the Consumer Liaison Group* (ISO New England, 2018), https://www.iso-ne.com/static-assets/documents/2018/02/2017_report_of_the_consumer_liaison_group_final.pdf.

124. "Electricity Explained," US Energy Information Administration, https://www.eia.gov/energyexplained/electricity/electricity-in-the-us-generation-capacity-and-sales.php.

125. "Levelized Cost of Energy and Levelized Cost of Storage 2019," Lazard, November 7, 2019, https://www.lazard.com/perspective/lcoe2019.

126. "Net Metering," Solar Energy Industries Association, https://www.seia.org/initiatives/net-metering.

127. "Should You Install a Solar Battery for Home Use?" Energysage, last modified July 15, 2020, https://www.energysage.com/solar/solar-energy-storage/how-do-solar-batteries-work/.

128. Herman K Trabish, "An Emerging Push for Time-of-Use Rates Sparks New Debates About Customer and Grid Impacts," *Utility Dive*, January 28, 2019, https://www.utilitydive.com/news/an-emerging-push-for-time-of-use-rates-sparks-new-debates-about-customer-an/545009/.

129. "State Net Metering Policies," National Conference of State Legislatures, November 20, 2017, https://www.ncsl.org/research/energy/net-metering-policy-overview-and-state-legislative-updates.aspx.

130. "An Overview of the California Solar Mandate," Energysage, https://news.energysage.com/an-overview-of-the-california-solar-mandate/.

131. David Roberts, "California Now Requires Solar Panels on All Homes. That's Not Necessarily a Good Thing," *Vox*, January 2, 2020, https://www.vox.com/energy-and-environment/2018/5/15/17351236/california-rooftop-solar-pv-panels-mandate-energy-experts.

132. Kathryne Cleary and Karen Palmer, *Renewables 101: Integrating Renewable Energy Resources into the Grid* (Resources for the Future, 2020), https://www.rff.org/publications/explainers/renewables-101-integrating-renewables/.

133. "Community Solar: What Is It?" Energysage, March 6, 2020, https://www.energysage.com/solar/community-solar/community-solar-power-explained/.

134. "About Microgrids," Grid Integration Group, Berkley Lab, https://building-microgrid.lbl.gov/about-microgrids.

135. Several states require utilities to invest in storage technology, although these mandates have been carried out outside of the state RPS laws. See Warren Leon, "Becoming More Aggressive: States Implement Ambitious Goals and Standards," *Renewable Energy World*, March 18, 2020, https://www.renewableenergyworld.com/2020/03/18/becoming-more-aggressive-states-implement-ambitious-goals-and-standards/#gref.

136. "Using the Solar Investment Tax Credit for Energy Storage," Energysage , May 27, 2020, https://www.energysage.com/solar/solar-energy-storage/energy-storage-tax-credits-incentives/.

137. See *An Ecomodernist Manifesto*, https://thebreakthrough.org/articles/an-ecomodernist-manifesto.

138. "Dams: Impacts on Salmon and Steelhead," Northwest Power and Conservation Council, https://www.nwcouncil.org/reports/columbia-river-history/damsimpacts.

139. US Army Corps of Engineers, Bureau of Reclamation, and Bonneville Power Administration, *Executive Summary, Columbia River System Operations Environmental Impact Statement* (US Army Corps of Engineers, Bureau of Reclamation, and Bonneville Power Administration, 2020), https://usace.contentdm.oclc.org/utils/getfile/collection/p16021coll7/id/14957.

140. US Army Corps of Engineers, Bureau of Reclamation, and Bonneville Power Administration, "The Columbia River System Inside Story," (US Army Corps of Engineers, Bureau of Reclamation, and Bonneville Power Administration, 2001), https://www.bpa.gov/news/pubs/GeneralPublications/edu-The-Federal-Columbia-River-Power-System-Inside-Story.pdf.

141. See https://gorgefriends.org.

Chapter 9 Notes

1. Friedlingstein, P., O'Sullivan, M., Jones, M. W., Andrew, R. M., Hauck, J., Olsen, A., Peters, G. P., Peters, W., Pongratz, J., Sitch, S., Le Quéré, C., Canadell, J. G., Ciais, P., Jackson, R. B., Alin, S., Aragão, L. E. O. C., Arneth, A., Arora, V., Bates, N. R., Becker, M., Benoit-Cattin, A., Bittig, H. C., Bopp, L., Bultan, S., Chandra, N., Chevallier, F., Chini, L. P., Evans, W., Florentie, L., Forster, P. M., Gasser, T., Gehlen, M., Gilfillan, D., Gkritzalis, T., Gregor, L., Gruber, N., Harris, I., Hartung, K., Haverd, V., Houghton, R. A., Ilyina, T., Jain, A. K., Joetzjer, E., Kadono, K., Kato, E., Kitidis, V., Korsbakken, J. I., Landschützer, P., Lefèvre, N., Lenton, A., Lienert, S., Liu, Z., Lombardozzi, D., Marland, G., Metzl, N., Munro, D. R., Nabel, J. E. M. S., Nakaoka, S.-I., Niwa, Y., O'Brien, K., Ono, T., Palmer, P. I., Pierrot, D., Poulter, B., Resplandy, L., Robertson, E., Rödenbeck, C., Schwinger, J., Séférian, R., Skjelvan, I., Smith, A. J. P., Sutton, A. J., Tanhua, T., Tans, P. P., Tian, H., Tilbrook, B., van der Werf, G., Vuichard, N., Walker, A. P., Wanninkhof, R., Watson, A. J., Willis, D., Wiltshire, A. J., Yuan, W., Yue, X., and Zaehle, S.: Global Carbon Budget 2020, Earth Syst. Sci. Data, 12, 3269–3340, https://doi.org/10.5194/essd-12-3269-2020, 2020.

2. The Environmental Protection Agency publishes a cumulative, quantitative inventory of all US greenhouse gas emissions. See *Environmental Protection Agency, Inventory of US Greenhouse Gas Emissions and Sinks: 1990-2018* (Environmental Protection Agency, 2020), https://www.epa.gov/ghgemissions/inventory-us-greenhouse-gas-emissions-and-sinks-1990-2018.

3. Environmental Protection Agency, "Chapter 2: Trends in Greenhouse Gas Emissions," *Inventory of U.S. Greenhouse Gas Emissions and Sinks: 1990–2018* (Environmental Protection Agency, 2020), https://www.epa.gov/sites/production/files/2020-04/documents/us-ghg-inventory-2020-chapter-2-trends.pdf.

4. Jeff Deason et al., *Electrification of Buildings and Industry in the United States* (Lawrence Berkeley National Laboratory, 2018), http://ipu.msu.edu/wp-content/uploads/2018/04/LBNL-Electrification-of-Buildings-2018.pdf.

5. Stern, *Why Are We Waiting?*, 46-50. Also see *Energy Efficiency is the Cornerstone for Building a Secure and Sustainable Energy System,* International Energy Agency, 2018, https://www.iea.org/news/energy-efficiency-is-the-cornerstone-for-building-a-secure-and-sustainable-energy-system.

6. "Overview of CHP technologies," US Department of Energy, November 2017, https://www.energy.gov/sites/prod/files/2017/12/f46/CHP%20Overview-120817_compliant_0.pdf.

7. Jeffrey Rissman et al., "Technologies and Policies to Decarbonize Global Industry: Review and Assessment of Mitigation Drivers Through 2070," *Applied Energy* 266 (2020), https://www.sciencedirect.com/science/article/pii/S0306261920303603?via%3Dihub#s0260.

8. J.P. Morgan Asset Management, *Mountains and Molehills: Achievements and Distractions on the Road to Decarbonization*, J.P. Morgan Asset Management, 2019, https://am.jpmorgan.com/content/dam/jpm-am-aem/global/en/insights/market-insights/mountainsmolehillsfinal.pdf. For cost information, see "Average Energy Prices for the United States, Regions Census Divisions, and Selected Metropolitan Areas," US Bureau of Labor Statistics, https://www.bls.gov/regions/midwest/data/averageenergyprices_selectedareas_table.htm.

9. "Electrify Your Home," *Electrify Now*, https://electrifynow.net/electrify-your-home.

10. Our discussion here omits the emissions generated from the production of fossil fuels, such as fugitive methane released from the pipeline infrastructure. See Chapter 6 for more information on these emissions.

11. Environmental Protection Agency, "Chapter 2," in *Inventory of US Greenhouse*.

12. Environmental Protection Agency, "Chapter 2," in *US Greenhouse*.

13. This process is described in more detail in Chapter 6.

14. Jeffrey Rissman et al., "Technologies and Policies."

15. See Chapter 10 for discussion of carbon capture approaches.

16. Harvey, *Designing Climate Solutions*, 243-244.

17. Jeffrey Rissman et al., "Technologies and Policies."

18. In addition to being used in cement manufacturing, lime is also produced for a variety of applications in environmental remediation, metallurgy, and the chemicals industry. Of course, this adds to the process emissions of carbon dioxide. See Environmental Protection Agency, "Chapter 4: Industrial Processes and Product Use," in *Inventory of US Greenhouse*, https://www.epa.gov/sites/production/files/2020-04/documents/us-ghg-inventory-2020-chapter-4-industrial-processes-and-product-use.pdf.

19. Jocelyn Timperley, "Q&A: Why Cement Emissions Matter for Climate Change," *CarbonBrief*, September 13, 2018, https://www.carbonbrief.org/qa-why-cement-emissions-matter-for-climate-change.

20. Timperley, "Q&A: Why Cement Emissions."

21. Jeffrey Rissman et al., "Technologies and Policies."

22. "Biomass Feedstocks," US Department of Energy Office of Energy Efficiency and Renewable Energy, https://www.energy.gov/eere/bioenergy/biomass-feedstocks.

23. David Roberts, "This Climate Problem Is Bigger Than Cars and Much Harder to Solve," *Vox*, January 31, 2020, https://www.vox.com/energy-and-environment/2019/10/10/20904213/climate-change-steel-cement-industrial-heat-hydrogen-ccs.

24. Ben King et al., "Clean Products Standard: A New Approach to Industrial Decarbonization," Rhodium Group, December 9, 2020, https://rhg.com/research/clean-products-standard-industrial-decarbonization/.

25. Colin Cunliff and David Hart, *Global Energy Innovation Index* (Information Technology Innovation Foundation, 2019), http://www2.itif.org/2019-global-energy-innovation-index.pdf.

26. For comparisons of clean energy investments among nations, see Cunliff and Hart, *Global Energy Innovation Index*. For a roadmap to clean energy investment in the United States, see Varun Sivaram et al., *Energizing America, A Roadmap to Launch a National Energy Innovation Mission* (Columbia | SIPA Center on Global Energy Policy, 2020), https://www.energypolicy.columbia.edu/sites/default/files/file-uploads/EnergizingAmerica_FINAL_EXECUTIVE%20SUMMARY.pdf.

27. See the discussion of the California Low Carbon Fuel Standard later in this chapter.

28. See the discussion of electricity policy in Chapter 7. For a more detailed analysis of industry emissions and policy, see "ICEF Industrial Heat Decarbonization Roadmap," Innovation for Cool Earth Forum, December 2019, https://www.icef-forum.org/pdf2019/roadmap/ICEF_Roadmap_201912.pdf.

29. "Climate-Friendly Alternatives to HFCs," European Commission, https://ec.europa.eu/clima/policies/f-gas/alternatives_en.

30. Coral Davenport, "Climate Change Legislation Included in Coronavirus Relief Deal," *New York Times*, December 23, 2020, https://www.nytimes.com/2020/12/21/climate/climate-change-stimulus.html.

31. See Chapter 2 for discussion of the Montreal Protocol. For the Kigali Amendment, see Faye Leone, "Kigali Amendment Enters Into Force, Bringing Promise of Reduced Global Warming," International Institute for Sustainable Development, January 8, 2019, https://sdg.iisd.org/news/kigali-amendment-enters-into-force-bringing-promise-of-reduced-global-warming/.

32. Michael J. Altieri, "Ozone's Cure is Climate's Scourge—Northeast States to Ban Use of Hydrofluorocarbons," Hunton Andrews Kurth, March 30, 2020, https://www.huntonnickelreportblog.com/2020/03/ozones-cure-is-climates-scourge-northeast-states-to-ban-use-of-hydrofluorocarbons/.

33. Composting is beneficial if done aerobically, in the presence of oxygen. Methane emissions from composting reflect improper practices in which the compost pile becomes anaerobic. See Paul Hawken, *Drawdown: The Most Comprehensive Plan Ever Proposed to Reverse Global Warming* (New York, NY: Penguin Group, 2017), 62.

34. Environmental Protection Agency, "Chapter 2," in *Inventory of US Greenhouse*. See Chapter 2 for information on natural and anthropogenic sources of global methane; see Chapters 6 and 10 for discussion of methane emissions from fossil fuel operations and agriculture, respectively.

35. "Municipal Solid Waste Landfill New Source Performance Standards and Emissions Guidelines," Harvard Law School Environmental and Energy Law Program, last modified December 17, 2019, https://eelp.law.harvard.edu/2017/09/municipal-solid-waste-landfill-new-source-performance-standards-and-emissions-guidelines/.

36. Erica Geis, "Landfills Have a Huge Greenhouse Gas Problem. Here's What We Can Do About It," *Ensia*, October 25, 2016, https://ensia.com/features/methane-landfills/. For related information about anaerobic biodigesters, see Chapter 10.

37. "Biomass Explained: Landfill Gas and Biogas," US Energy Information Administration, https://www.eia.gov/energyexplained/biomass/landfill-gas-and-biogas.php. Biogas from landfills is counted among the "biomass" sources of electricity generation. See Chapter 8.

38. Nearly 20 trillion cubic feet of natural gas was used in the commercial, residential, and industrial sectors in 2018, so landfills can only provide a small fraction of this. For a guide to how communities can develop renewable natural gas resources, see the work of the nonprofit advocacy group Energy Vision, https://energy-vision.org/wp-content/uploads/2019/10/EV-RNG-Community-Guide.pdf.

39. International Energy Agency, *Outlook for Biogas and Biomethane: Prospects for Organic Growth*, (International Energy Agency, 2020), https://www.iea.org/reports/outlook-for-biogas-and-biomethane-prospects-for-organic-growth. See Chapter 10 for methane recovery from livestock.

40. For descriptions of the roles of renewable hydrogen in the coming carbon-free economy, see the website of the non-profit Hydrogen Energy Center, https://www.hydrogenenergycenter.org/benefits-of-the-hydrogen-economy. See "About Methanol," Methanex, https://www.methanex.com/about-methanol/how-methanol-used. See Chapter 10 for a description of how hydrogen is used to make fertilizers.

41. See "About Methanol," Methanex, https://www.methanex.com/about-methanol/how-methanol-used. See Chapter 10 for a description of how hydrogen is used to make fertilizers.

42. "Hydrogen Production: Natural Gas Reforming," US Department of Energy Office of Energy Efficiency and Renewable Energy, https://www.energy.gov/eere/fuelcells/hydrogen-production-natural-gas-reforming.

43. Loz Blain, "New Water-Splitting Process Could Kick Start 'Green' Hydrogen Economy," *New Atlas*, December 17, 2019, https://newatlas.com/energy/water-splitting-electrolysis-hydrogen-efficient-cheap/.

44. See Chapter 8 for a discussion of "dispatchability."

45. S. Julio Friedman et al., *Low-Carbon Heat Solutions for Heavy Industry: Sources, Options and Costs Today* (Columbia | SIPA Center on Global Energy Policy, 2019), https://energypolicy.columbia.edu/sites/default/files/file-uploads/LowCarbonHeat-CGEP_Report_100219-2_0.pdf.

46. See Chapter 10 for a discussion of CCS.

47. "Could Hydrogen Help Save Nuclear?" US Department of Energy Office of Nuclear Energy, June 24, 2020, https://www.energy.gov/ne/articles/could-hydrogen-help-save-nuclear.

48. "350 ppm Pathways for the United States," Evolved Energy Research, May 8, 2019, https://www.evolved.energy/post/2019/05/08/350-ppm-pathways-for-the-united-states.

49. See Chapter 10 for a description of this technology, which is known as direct air capture (DAC). The reaction to make methane from hydrogen and carbon dioxide has been known for over 100 years and is called the Sabatier reaction. See Charlotte Vogt et al., "The Renaissance of the Sabatier Reaction and Its Applications on Earth and in Space," *Nature Catalysis* 2 (2019): 188-197, https://www.nature.com/articles/s41929-019-0244-4.

50. Methane is converted to syngas first because its other reactions are hard to control, but syngas generation is not economical unless done at a large scale. New approaches to methane chemistry provide alternatives. See Ferdi Schuth, "Making More from Methane," *Science* 363 (2019): 1282-1283, https://science.sciencemag.org/content/363/6433/1282.

51. Michael Talmadge et al., *Syngas Upgrading to Hydrocarbon Fuels Technology Pathway* (National Renewable Energy Laboratory, 2013), https://www.nrel.gov/docs/fy13osti/58052.pdf.

52. David Roberts, "The Missing Puzzle Piece for Getting to 100% Clean Power," *Vox*, March 28, 2020, https://www.vox.com/energy-and-environment/2020/3/28/21195056/renewable-energy-100-percent-clean-electricity-power-to-gas-methane.

53. Wärtsilä Corporation, "Wärtsilä Details Faster, Cleaner and Cheaper Path to Reach California's Climate Goals Using Wind, Solar, Storage and Flexible Generation," Wärtsilä, March 19, 2020, https://www.wartsila.com/media/news/19-03-2020-wartsila-details-faster-cleaner-and-cheaper-path-to-reach-californias-climate-goals-using-wind-solar-storage-and-flexible-generation.

54. Industry, transportation, agriculture, and commercial/residential are all *end use* sectors, while the electricity sector enables the whole economy. For data on emissions, see "Sources of Greenhouse Gas Emissions," Environmental Protection Agency, https://www.epa.gov/ghgemissions/sources-greenhouse-gas-emissions. We will discuss agriculture separately in Chapter 10.

55. Reid Ewing et al., *Growing Cooler: The Evidence on Urban Development and Climate Change* (Urban Land Institute, 2007), 12, https://www.nrdc.org/sites/default/files/cit_07092401a.pdf.

56. "Sources of Greenhouse Emissions: Transportation Sector Emissions," Environmental Protection Agency, https://www.epa.gov/ghgemissions/sources-greenhouse-gas-emissions#transportation.

57. For biomass feedstocks, see "Biomass Feedstocks," Office of Energy Efficiency and Renewable Energy, https://www.energy.gov/eere/bioenergy/biomass-feedstocks. Also see *The Promise of Biomass: Clean Power and Fuel, if Handled Right* (Union of Concerned Scientists, 2012), https://www.ucsusa.org/sites/default/files/2019-09/Biomass-Resource-Assessment.pdf.

58. In 2018, the US consumed 143 billion gallons of gasoline and 64 billion gallons of diesel fuel. See https://www.eia.gov/tools/faqs and https://www.eia.gov/energyexplained/oil-and-petroleum-products/use-of-oil.php. Data on US gasoline and diesel fuel consumption over time can be found at https://www.statista.com/statistics/189410/us-gasoline-and-diesel-consumption-for-highway-vehicles-since-1992/.

59. Data on the use of the corn crop is available at https://afdc.energy.gov/data/10339. Data for US ethanol production is from https://www.eia.gov/totalenergy/data/monthly/pdf/sec10_7.pdf.

60. See https://www.epa.gov/renewable-fuel-standard-program/overview-renewable-fuel-standard.

61. "Gasoline Explained: Gasoline and the Environment," US Energy Information Administration, https://www.eia.gov/energyexplained/gasoline/gasoline-and-the-environment.php.

62. National Academy of Sciences, *Renewable Fuel Standard: Potential Economic and Environmental Effects of U.S. Biofuel Policy* (National Academy of Sciences, 2011), https://www.nap.edu/resource/13105/Renewable-Fuel-Standard-Final.pdf.

63. Mario Loyola, "Stop the Ethanol Madness," *The Atlantic*, November 23, 2019, https://cei.org/content/stop-ethanol-madness.

64. "Renewable Fuel Standard Program Unlikely to Meet Its Targets for Reducing Greenhouse Gas Emissions," US Government Accountability Office, December 1, 2016, https://www.gao.gov/products/GAO-17-264T.

65. Union of Concerned Scientists, *The Promise of Biomass* (Union of Concerned Scientists, 2012), https://www.ucsusa.org/sites/default/files/2019-09/Biomass-Resource-Assessment.pdf.

66. The yearly schedule for required production volumes of renewable fuels is listed in the Clean Air Act; see Clean Air Act 42 U.S.C. §7545(o) (2)(B)(i)(III). The revised volumes for 2019 through 2021, necessitated by the failure to produce enough cellulosic ethanol, are found at https://www.epa.gov/renewable-fuel-standard-program/final-renewable-fuel-standards-2020-and-biomass-based-diesel-volume.

67. "Biodiesel Blends," US Department of Energy Office of Energy Efficiency and Renewable Energy, http://www.afdc.energy.gov/fuels/biodiesel_blends.html.

68. See https://www.epa.gov/renewable-fuel-standard-program/overview-renewable-fuel-standard.

69. This estimate is made from data available at Cropwatch, University of Nebraska, https://cropwatch.unl.edu.

70. Suzanne Paulson, *Biodiesel Fuel* (UCLA Institute of the Environment, 2010), https://www.ioes.ucla.edu/wp-content/uploads/Biodiesel-Fuel.pdf. See Chapter 2 for a discussion of how black carbon and particulate matter impact climate.

71. Abrahm Lustgarten, "Palm Oil Was Supposed to Help Save the Planet. Instead, It Unleashed a Catastrophe," *New York Times*, November 20, 2018, https://www.nytimes.com/2018/11/20/magazine/palm-oil-borneo-climate-catastrophe.

72. See https://www.worldwildlife.org/industries/soy.

73. See https://www.eia.gov/todayinenergy/detail.php?id=39292.

74. James Mulligan et al., *Technological Carbon Removal in the United States* (Washington, D.C., World Resources Institute, 2018), 8-14, https://www.wri.org/publication/tech-carbon-removal-usa. Also see the discussion on sustainable forest management in Chapter 10.

75. For an overall description of the LCFS, see https://ww2.arb.ca.gov/our-work/programs/low-carbon-fuel-standard/about.

76. See Chapter 2 for discussions of energy units and carbon dioxide equivalents.

77. "Lifecycle Analysis of Greenhouse Gas Emissions Under the Renewable Fuel Standard," Environmental Protection Agency, https://www.epa.gov/renewable-fuel-standard-program/lifecycle-analysis-greenhouse-gas-emissions-under-renewable-fuel.

78. For an overview of California's many climate change policies, see https://ww2.arb.ca.gov/our-work/programs/climate-change-programs.

79. California Air Resources Board, *LCFS Basics* (California Air Resources Board), 14, https://ww2.arb.ca.gov/resources/documents/lcfs-basics.

80. Megan Boutwell, "LCFS 101—A Beginner's Guide," Stillwater Associates, February 28, 2017, https://stillwaterassociates.com/lcfs-101-a-beginners-guide/?cn-reloaded=1.

81. See https://ww3.arb.ca.gov/fuels/lcfs/dashboard/dashboard.htm.

82. For a full table of CI values for all LCFS-approved pathways, see https://ww3.arb.ca.gov/fuels/lcfs/fuelpathways/pathwaytable_test3.htm.

83. Union of Concerned Scientists, *The Promise of Biomass*.

84. See https://ww3.arb.ca.gov/fuels/lcfs/fuelpathways/pathwaytable_test3.htm.

85. Clifford Krauss, "Oil Refineries see Profit in Turning Kitchen Grease into Diesel," *New York Times*, December 3, 2020.

86. "Natural Gas Buses—A Cost, Operational and Environmental Alternative," *The Road Ahead Blog*, Clean Energy, June 11, 2018, https://www.cleanenergyfuels.com/blog/clean-energy-natural-gas-fuels-canadian-transportation-fleets.

87. "How Do Natural Gas Vehicles Work?" US Department of Energy Office of Energy Efficiency and Renewable Energy, https://afdc.energy.gov/vehicles/how-do-natural-gas-cars-work.

88. "Natural Gas Vehicles," US Department of Energy Office of Energy Efficiency and Renewable Energy, https://afdc.energy.gov/vehicles/natural_gas.html.

89. California Air Resources Board, *LCFS Basics*, 14.

90. See https://ww3.arb.ca.gov/fuels/lcfs/fuelpathways/pathwaytable_test3.htm.

91. "Hydrogen Production: Biomass Gasification," US Department of Energy Office of Energy Efficiency and Renewable Energy, https://www.energy.gov/eere/fuelcells/hydrogen-production-biomass-gasification.

92. See the Interlude and Williams et al., *Pathways to Deep Decarbonization in the United States*, The US report of the Deep Decarbonization Pathways Project of the Sustainable Development Solutions Network and the Institute for Sustainable Development and International Relations, Energy and Environmental Economics, Inc., 2014, revision with technical supplement, November 16, 2015, https://usddpp.org/downloads/2014-technical-report.pdf.

93. US Department of Energy, *2016 Billion-Ton Report* 9 US Department of Energy, 2016), https://www.energy.gov/eere/bioenergy/2016-billion-ton-report. A pilot study in Sweden demonstrating methane production from biomass-derived syngas was reasonably promising, suggesting that perhaps half of US natural gas needs could be met by a billion tons of biomass (author's calculation). See Anton Larsson et al., *The GoBiGas Project* (Göteborg Energi), https://research.chalmers.se/publication/509030/file/509030_Fulltext.pdf.

94. Union of Concerned Scientists, *The Promise of Biomass*.

95. See Chapter 8 for discussion of renewable hydrogen generation and Chapter 10 for discussion of carbon capture and sequestration.

96. "Oregon Clean Fuels Program," Oregon.gov, https://www.oregon.gov/deq/aq/programs/Pages/Clean-Fuels.aspx.

97. California, Oregon, Washington, and British Columbia have created the Pacific Coast Collaborative to foster a low-carbon regional economy, with shared goals for greenhouse gas reduction and development of green technology. See http://pacificcoastcollaborative.org/about/.

98. California Air Resources Board, *LCFS Basics*, 13.

99. John Perona, "Biodiesel for the 21st Century Renewable Energy Economy," *Energy Law Journal* 38 (2017): 165-212, http://www.eba-net.org/assets/1/6/23-165-212-Perona-[FINAL].pdf.

100. For a look at the company that is furthest ahead in developing algal biodiesel, see "Algal Cell Factories," Synthetic Genomics, https://syntheticgenomics.com/algal-cell-factories/. For a review of the field, see Perona, "Biodiesel for the 21st Century."

101. The Renewable Fuels Association is a prominent ethanol industry trade group. For its 2019 report, which unabashedly promotes corn ethanol without acknowledging its many harms, see https://ethanolrfa.org/wp-content/uploads/2019/02/RFA2019Outlook.pdf.

102. See https://www.reuters.com/article/us-usa-ethanol-lawsuit/u-s-refiner-group-sues-trump-epa-over-high-ethanol-gasoline-idUSKCN1TB2HP.

103. "Tesla Overtakes Toyota to Become World's Most Valuable Carmaker," *BBC*, July 1, 2020, https://www.bbc.com/news/business-53257933.

104. "FOTW#1136, June 1, 2020: Plug-in Vehicle Sales Accounted for About 2% of Light-Duty Vehicle Sales in the United States in 2019," US Department of Energy, Office of Energy Efficiency & Renewable Energy, https://www.energy.gov/eere/vehicles/articles/fotw-1136-june-1-2020-plug-vehicle-sales-accounted-about-2-all-light-duty.

105. Alexandre Milovanoff, Daniel Posen, and Heather MacLean, "Electrification of Light-Duty Vehicle Fleet Alone Will Not Meet Mitigation Targets," *Nature: Climate Change* 10 (2020): doi: 10.1038/s41558-020-00921-7.

106. Kathryne Cleary, "Electrification 101," Resources for the Future, December 5, 2019, https://media.rff.org/documents/Electrification_Explainer_101_odobEoP.pdf.

107. "Types of Electric Vehicles," EVgo, https://www.evgo.com/why-evs/types-of-electric-vehicles/.

108. "Developing Infrastructure to Charge Plug-In Electric Vehicles," US Department of Energy Office of Energy Efficiency and Renewable Energy, https://afdc.energy.gov/fuels/electricity_infrastructure.html.

109. Ann Steffora Mutschler, "Why EV Battery Design Is So Difficult," Semiconductor Engineering, October 3, 2019, https://semiengineering.com/why-ev-battery-design-is-so-difficult/.

110. Mutschler, "Why EV Battery Design."

111. Amit Katwala, "The Spiralling Environmental Cost of Our Lithium Battery Addiction," *Wired*, August 5, 2018, https://www.wired.co.uk/article/lithium-batteries-environment-impact.

112. George Crabtree, "The Coming Electric Vehicle Transformation," *Science* 366, no. 6464 (2019): 422-424, doi: 10.1126/science.aax0704.

113. Nicole Kobie, "A Cobalt Crisis Could Put the Brakes on Electric Car Sales," *Wired*, February 22, 2020, https://www.wired.com/story/a-cobalt-crisis-could-put-the-brakes-on-electric-car-sales/.

114. Maddie Stone, "'Million Mile' Batteries Are Coming. Are They a Revolution?," *Grist*, July 6, 2020, https://grist.org/energy/million-mile-batteries-are-coming-are-they-really-a-revolution/.

115. "Why Electric Buses Haven't Taken Over the World—Yet," *Wired*, June 7, 2019, https://www.wired.com/story/electric-buses-havent-taken-over-world/.

116. Julian Spector, "3 Trends Making the Case for Bus Electrification," *Greentech Media*, October 4, 2019, https://www.greentechmedia.com/articles/read/3-trends-making-the-case-for-bus-electrification. For other information on the adoption of EVs by US cities, see Amanda Levin, *Seventh Annual Energy Report: Clean Energy Opportunities and Dirty Energy Challenges* (National Resources Defense Council), 9, https://www.nrdc.org/media/2019/191106-0.

117. See https://www.solutionaryrail.org.

118. Umair Irfan, "Forget Cars. We Need Electric Airplanes," *Vox*, April 9, 2019, https://www.vox.com/2019/3/1/18241489/electric-batteries-aircraft-climate-change.

119. Daniel Oberhaus, "Want Electric Ships? Build a Better Battery," March 19, 2020, https://www.wired.com/story/want-electric-ships-build-a-better-battery/.

120. Union of Concerned Scientists, *Ready for Work: Now is the Time for Heavy-Duty Electric Vehicles* (Union of Concerned Scientists, 2019), https://www.ucsusa.org/sites/default/files/2019-12/ReadyforWorkFullReport.pdf.

121. "Advanced Clean Trucks," California Air Resources Board, https://ww2.arb.ca.gov/our-work/programs/advanced-clean-trucks.

122. "How Do Fuel Cell Electric Vehicles Work Using Hydrogen?," US Department of Energy Office of Energy Efficiency and Renewable Energy, https://afdc.energy.gov/vehicles/how-do-fuel-cell-electric-cars-work.

123. PG&E Gas R&D and Innovation, *Pipeline Hydrogen* (PG&E, 2018), https://www.pge.com/pge_global/common/pdfs/for-our-business-partners/interconnection-renewables/interconnections-renewables/Whitepaper_PipelineHydrogen.pdf.

124. "LCFS Electricity and Hydrogen Provisions," California Air Resources Board, https://ww2.arb.ca.gov/resources/documents/lcfs-electricity-and-hydrogen-provisions.

125. Bloomberg New Energy Finance, *Electric Vehicle Outlook 2020 Executive Summary* (Bloomberg New Energy Finance, 2020), https://bnef.turtl.co/story/evo-2020/. The International Energy Agency, however, projects that just 8 percent of US new car sales will be electric in 2030. See International Energy Agency, *Global EV Outlook 2019* (International Energy Agency, 2019), https://www.iea.org/reports/global-ev-outlook-2019.

126. Xavier Mosquet et al., "Who Will Drive Electric Cars to the Tipping Point?" Boston Consulting Group, January 2, 2020, https://www.bcg.com/en-us/publications/2020/drive-electric-cars-to-the-tipping-point.aspx.

127. "Costs and Benefits of Electric Cars vs. Conventional Vehicles," Energysage, April 22, 2020, https://www.energysage.com/electric-vehicles/costs-and-benefits-evs/evs-vs-fossil-fuel-vehicles/.

128. Robert Walton, "Biden Plan to Electrify Federal Fleet Will Boost EV Market, But Many Questions Remain, Experts Say," *UtilityDive*, January 27, 2021, https://www.utilitydive.com/news/biden-plan-to-electrify-federal-fleet-will-boost-ev-market-but-many-questi/594029/.

129. See Chapter 8 for discussion of electric utilities.

130. Lia Cattaneo, "Investing in Charging Infrastructure for Plug-In Electric Vehicles," Center for American Progress, July 30, 2018, https://www.americanprogress.org/issues/green/reports/2018/07/30/454084/investing-charging-infrastructure-plug-electric-vehicles/. For a comprehensive consumer advocacy guide to EVs, see Citizens Utility Board, *The ABCs of Evs* (Citizen Utility Board, 2017), https://citizensutilityboard.org/wp-content/uploads/2017/04/2017_The-ABCs-of-EVs-Report.pdf.

131. "Electric Car Tax Credits & Incentives," EnergySage, June 30, 2020, https://www.energysage.com/electric-vehicles/costs-and-benefits-evs/ev-tax-credits/.

132. Harvey, *Designing Climate Solutions*, 158-163.

133. "What is ZEV?" Union of Concerned Scientists, last modified September 12, 2019, https://www.ucsusa.org/resources/what-zev.

134. Virginia McConnell et al., "California's Evolving Zero Emission Vehicle Program: Pulling New Technology into the Market," Resources for the Future, November 20, 2019, https://www.rff.org/publications/working-papers/californias-evolving-zero-emission-vehicle-program/.

135. David Ferris, "'Achilles Heel': How Charging Hobbles the Electric Truck," *E&E News*, October 16, 2020, https://www.eenews.net/stories/1063716351.

136. "A Brief History of US Fuel Efficiency Standards," Union of Concerned Scientists, last modified December 6, 2017, https://www.ucsusa.org/resources/brief-history-us-fuel-efficiency.

137. "Corporate Average Fuel Economy," US Department of Transportation, https://www.nhtsa.gov/laws-regulations/corporate-average-fuel-economy.

138. For a detailed look at which Trump administration rules will likely become targets for change under the Biden administration, see Hogan Lovells, "Environmental Law Outlook Under a Biden Administration," November 9, 2020, https://www.lexology.com/library/detail.aspx?g=d709f96f-8d10-41db-b2fa-8ebaae16e632.

139. Debra Kahn, "How California Will Shape U.S. Environmental Policy Under Biden," *Politico*, November 10, 2020, https://www.politico.com/states/california/story/2020/11/10/how-california-will-shape-us-environmental-policy-under-biden-1335423.

140. See Chapter 7 for description of this carbon pricing program.

141. Transportation & Climate Initiative of the Northeast and Mid-Atlantic States, https://www.transportationandclimate.org/content/about-us.

142. By the end of 2020, formal pledges to enter into the program had been made by Massachusetts, Rhode Island, Connecticut, and the District of Columbia. "Memorandum of Understanding," Transportation and Climate Initiative Program, https://www.transportationandclimate.org/sites/default/files/TCI%20MOU%2012.2020.pdf.

143. Kim Parker et al., "What Unites and Divides Urban, Suburban and Rural Communities," Pew Research Center, May 22, 2018, https://www.pewsocialtrends.org/2018/05/22/what-unites-and-divides-urban-suburban-and-rural-communities/.

144. See https://www.wearestillin.com/signatories.

145. See "Electrify Your Home," Electrify Now, https://electrifynow.net/electrify-your-home.

146. Deason et al., "Electrification of Buildings," http://ipu.msu.edu/wp-content/uploads/2018/04/LBNL-Electrification-of-Buildings-2018.pdf. For a global perspective, see UN Environment Programme, "2020 Global Status Report for Buildings and Construction," https://globalabc.org/news/launched-2020-global-status-report-buildings-and-construction.

147. David Iaconangelo, "As Natural Gas Bans Go National, Can Cities Fill the Gap?" *E&E News*, August 3, 2020, https://www.eenews.net/stories/1063674561.

148. See Harvey, *Designing Climate Solutions*, 201-212, for good policy principles on enhancing building and appliance energy efficiencies.

149. Thomas A. Deetjen et al., "Review of Climate Action Plans in 29 Major U.S. Cities: Comparing Current Policies to Research Recommendations," *Sustainable Cities and Society* 41 (2018): 711-727.

150. "Building Efficiency Accelerator," World Resources Institute, https://wrirosscities.org/our-work/project-city/building-efficiency-accelerator.

151. See https://globalabc.org.

152. Hawken, *Drawdown*, 210-211.

153. Hallie Busta, "Mass Timber 101: Understanding the Emerging Building Type," *Construction Dive*, https://www.constructiondive.com/news/mass-timber-101-understanding-the-emerging-building-type/443476/.

154. David Roberts, "The Hottest New Thing in Sustainable Building Is, Uh, Wood," *Vox*, January 15, 2020, https://www.vox.com/energy-and-environment/2020/1/15/21058051/climate-change-building-materials-mass-timber-cross-laminated-clt. This article contains a very useful reference list for further reading.

155. Meaghan O'Neill, "The World's Tallest Timber-Framed Building Finally Opens Its Doors," *Architectural Digest*, March 22, 2019, https://www.architecturaldigest.com/story/worlds-tallest-timber-framed-building-finally-opens-doors.

156. "What Is the Current Status of Tall Mass Timber Buildings in the Building Code?" WoodWorks, https://www.woodworks.org/experttip/current-status-tall-mass-timber-buildings-building-code/.

157. See Chapter 10 for discussion of sustainable forestry practice.

158. "Appliance Standards," American Council for an Energy-Efficient Economy, https://www.aceee.org/sites/default/files/pdf/fact-sheet/appliance-standards-031119.pdf.

159. "ENERGY STAR," American Council for an Energy-Efficient Economy, https://www.aceee.org/sites/default/files/pdf/fact-sheet/energy-star-031119.pdf.

160. For a review of progress in urban climate planning, see Samuel A. Markolf et al., "Pledges and Progress: Steps Toward Greenhouse Gas Emissions Reductions in the 100 Largest Cities Across the US," Brookings Institute, October 2020, https://www.brookings.edu/research/pledges-and-progress-steps-toward-greenhouse-gas-emissions-reductions-in-the-100-largest-cities-across-the-united-states/.

161. Harvey, *Designing Climate Solutions*, 173-187. This discussion also includes distinct recommendations for cities of varying sizes, since not all policies are appropriate for small and large cities alike.

162. Nadja Popovich and Denise Lu, "The Most Detailed Map of Auto Emissions in America," *New York Times*, October 10, 2019, https://www.nytimes.com/interactive/2019/10/10/climate/driving-emissions-map.html.

163. See "What Is Metro?" Metro, https://www.oregonmetro.gov/regional-leadership/what-metro.

164. Thomas A. Deetjen et al., "Review of Climate Action."

165. See the discussion on equity in Chapter 5.

166. Alexa Waud, "The Urban Heart of a Just Transition: How Cities Plan for Social Justice in Climate Action," *Metropolitics*, April 3, 2018, https://www.metropolitiques.eu/The-Urban-Heart-of-a-Just-Transition-How-Cities-Plan-for-Social-Justice-in.html.

167. Brad Plumer and Nadja Popovich, "How Decades of Racist Housing Policy Left Neighborhoods Sweltering," *New York Times*, August 24, 2020, https://www.nytimes.com/interactive/2020/08/24/climate/racism-redlining-cities-global-warming.html.

168. Maria Hart et al., "How to Prevent City Climate Action from Becoming 'Green Gentrification,'" World Resources Institute, December 12, 2019, https://www.wri.org/blog/2019/12/how-prevent-city-climate-action-becoming-green-gentrification.

169. United States Conference of Mayors, *Mayors and Climate Protection Best Practices* (United States Conference of Mayors, 2018), 5-6, http://www.usmayors.org/wp-content/uploads/2018/06/climateawards2018.pdf.

170. "Inclusive Climate Action," C40 Cities, https://resourcecentre.c40.org/resources/inclusive-climate-action.
171. For a detailed roadmap on energy innovation that includes strategies for bridging the valley of death, see *Energizing America*, Center on Global Energy Policy, https://spark.adobe.com/page/Azf8uWSlPJOo9/.
172. Daniel T. Plunkett and Erin M. Minor, "Briefing Papers 1, 3–4," *Public-Private Partnerships: Primer, Pointers and Potential Pitfalls* (2013), 13–7.
173. Chris Galford, "Public-Private Partnership Yields New 2.5 MW Battery Project in Iowa," *Daily Energy Insider*, July 23, 2020, https://dailyenergyinsider.com/news/26440-public-private-partnership-yields-new-2-5-mw-battery-project-in-iowa/.
174. Perona, "Biodiesel for the 21st Century."
175. Richard Nunno, "Electrification of US Railways: Pie in the Sky or Realistic Goal?" Environmental and Energy Study Institute, May 30, 2018, https://www.eesi.org/articles/view/electrification-of-u.s.-railways-pie-in-the-sky-or-realistic-goal.
176. See https://www.solutionaryrail.org.
177. Bill Moyer, Patrick Mazza, and the Solutionary Rail Team, *Solutionary Rail: A People-Powered Campaign to Electrify America's Railroads and Open Corridors to a Clean Energy Future,* Create Space (2016), 55-59. This book is available for download at https://www.solutionaryrail.org/buybook.

Chapter 10 Notes

1.   See Chapter 3 and the Interlude for a full discussion.
2.   G.F. Nemet et al., "Negative Emissions—Part 3: Innovation and Upscaling," *Environmental Research: Letters* 13 (2018), https://iopscience.iop.org/article/10.1088/1748-9326/aabff4/meta.
3.   For a summary of the National Resources Management Act, see https://www.energy.senate.gov/public/index.cfm?a--files.serve&File_id=6AE823EB-7FE1-42BF-987E-A2234BAEA46F. The bill passed with a 92-8 majority in the Senate and a 363-62 margin in the House.
4.   Dan Harsha, "The Biggest Land Conservation Legislation in a Generation," *Harvard Gazette*, July 27, 2020, https://news.harvard.edu/gazette/story/2020/07/the-likely-impact-of-great-american-outdoors-act/.
5.   See the discussion in Chapters 1 and 2.
6.   See Figure 3.2.
7.   Nicolas Gruber et al., "The Oceanic Sink for Anthropogenic CO2 From 1994 to 2017," *Science* 363, no. 6432 (2019): 1193-1199, https://science.sciencemag.org/content/363/6432/1193?rss=1. Of course, this CO2 uptake comes with the high price of acidification (see Chapter 4).
8.   See Chapter 4 for a discussion of tipping points.
9.   For a comprehensive analysis of how long it may take to establish negative emissions technologies at a large scale, see G.F. Nemet et al., "Negative Emissions—Part 3."
10. Rogelj, "Mitigation Pathways Compatible," 101.
11. See Chapter 1 for a discussion of the gigaton unit. For comparison, in 2019 worldwide carbon dioxide emissions from fossil fuels and the industrial sector were 37 GtCO2, while the estimated contribution from human land use change was an additional 6 GtCO2. Sabine Fuss et al., "Negative Emissions—Part 2: Costs, Potentials and Side Effects," *Environmental Research Letters* 13, no. 6 (2018): 4, https://iopscience.iop.org/article/10.1088/1748-9326/aabf9f/meta.
12. Global CCS Institute, *Global Status of CCS 2020*, Global CSS Institute, 13, https://www.globalccsinstitute.com/resources/global-status-report/.
13. James Mulligan et al., *CarbonShot: Federal Policy Options for Carbon Removal in the United States* (World Resources Institute, 2020), 3, https://www.wri.org/publication/carbonshot-federal-policy-options-for-carbon-removal-in-the-united-states.
14. Howard Herzog, *Carbon Capture* (Cambridge, MA: MIT Press, 2018), 84.
15. Mulligan et al., *CarbonShot: Federal Policy*, 10, 12,
16. For a discussion of compatibility of carbon sequestration with healthy forests, see Polly C. Buotte et al., "Carbon Sequestration and Biodiversity Co-Benefits of Preserving Forests in the Western United States," *Ecological Applications* (2019), https://esajournals.onlinelibrary.wiley.com/doi/full/10.1002/eap.2039. For a description of soil management practices that retain carbon consistent with healthy farmland, see Mulligan et al., *CarbonShot: Federal Policy*, 40-41.
17. See the discussion in IPCC, 2019: Summary for Policymakers. In: *Climate Change and Land: an IPCC special report on climate change, desertification, land degradation, sustainable land management, food security, and greenhouse gas fluxes in terrestrial ecosystems* [P.R. Shukla, J. Skea, E. Calvo Buendia, V. Masson-Delmotte, H. – O. Pörtner, D. C. Roberts, P. Zhai, R. Slade, S. Connors, R. van Diemen, M. Ferrat, E. Haughey, S. Luz, S. Neogi, M. Pathak, J. Petzold, J. Portugal Pereira, P. Vyas, E. Huntley, K. Kissick, M. Belkacemi, J. Malley, (eds.)]. In press. https://www.ipcc.ch/site/assets/uploads/sites/4/2020/02/SPM_Updated-Jan20.pdf, sections B1 and B2.
18. A quantitative assessment of natural land management approaches in the US is J.E. Fargione et al., "Natural Climate Solutions for the United States," *Science Advances* 4, no. 11 (2018): https://advances.sciencemag.org/content/4/11/eaat1869. The World Resources Institute (WRI) has released a set of four comprehensive (and quite readable) reports that concentrate on US options. See https://www.wri.org/publication-series/carbonshot-creating-options-carbon-removal-scale-united-states. For global analysis, see B.W. Griscom et al., "Natural Climate Solutions," *Proceedings of the National Academy of Sciences of the United States of America* 114, no. 44 (2017): 11645-11650, https://www.pnas.org/content/pnas/114/44/11645.full.pdf. Another good synthesis is Jan C. Minx et al., "Negative Emissions—Part 1: Research Landscape and Synthesis," *Environmental Research Letters* 13 (2018). Finally, for a useful high-level summary of adaptation and mitigation response options, see IPCC, *Climate Change and Land*, 19-30.
19. J.E. Fargione et al., "Natural Climate Solutions for the United States," *Science Advances* 4 (2018), https://advances.sciencemag.org/content/4/11/eaat1869.
20. This is the estimate given in Fargione et al., "Natural Climate Solutions." For a review of global forest sequestration potential, see Sabine Fuss et al., "Negative Emissions—Part 2: Costs, Potentials and Side Effects", *Environmental Research Letters* 13 (2018): 14-16, https://iopscience.iop.org/article/10.1088/1748-9326/aabf9f/meta.

21. Fargione et al., "Natural Climate Solutions."

22. M.D. Nelson et al., *National Report on Sustainable Forests—2015: Conservation of Biological Diversity* (US Forest Service, 2015), https://www.nrs.fs.fed.us/pubs/50436.

23. K. Zhu et al., "Limits to Growth of Forest Biomass Carbon Sink Under Climate Change," Nature Communications (2018): https://doi.org/10.1038/s41467-018-05132-5.

24. Fargione et al., "Natural Climate Solutions."

25. For more detail, see the discussion in Chapter 4.

26. W.R.L. Anderegg et al., "Climate-Driven Risks to the Climate Mitigation Potential of Forests," *Science* 368, no.6497 (2020): https://science.sciencemag.org/content/368/6497/eaaz7005.

27. Annie Sneed, "Ask the Experts: Does Rising CO2 Benefit Plants?" *Scientific American*, January 23, 2018, https://www.scientificamerican.com/article/ask-the-experts-does-rising-co2-benefit-plants1/.

28. Net primary productivity (NPP) also accounts for CO2 loss from plants when they metabolize sugars and starch for energy. For a brief description, see "Net Primary Productivity," NASA Earth Observatory, https://earthobservatory.nasa.gov/global-maps/MOD17A2_M_PSN.

29. C.M. Gough et al., "High Rates of Primary Production in Structurally Complex Forests," *Ecology* (2019): https://doi.org/10.1002/ecy.2864. This study focused specifically on US forests.

30. D.C. McKinley et al., "A Synthesis of Current Knowledge on Forests and Carbon Storage in the United States," *Ecological Applications* (2011): 1902-1924, https://www.fs.fed.us/rm/pubs_other/rmrs_2011_mckinley_d001.pdf. See also N. Seddon et al., "Grounding Nature-Based Climate Solutions in Sound Biodiversity Science," *Nature Climate Change 9* (2019): 82-87, https://www.nature.com/articles/s41558-019-0405-0.

31. See Chapter 1 for a discussion of how Earth's albedo influences the greenhouse effect.

32. Gabriel Popkin, "How Much Can Forests Fight Climate Change?," *Nature*, January 15, 2019, https://www.nature.com/articles/d41586-019-00122-z.

33. National Association of State Foresters, *State Foresters by the Numbers: Data and Analysis from the 2014 NASF State Forestry Statistics Survey* (2015), www.stateforesters.org/sites/default/files/publication-documents/2014%20State%20Foresters%20by%20the%20Numbers%20FINAL.pdf.

34. Brook J. Detterman and Kirsten K. Gruver, "EPA Announces Plan to Classify Wood-Based Power as Carbon Neutral," *National Law Review*, April 9, 2019, https://www.natlawreview.com/article/epa-announces-plan-to-classify-wood-based-power-carbon-neutral.

35. William H. Schlesinger, "Are Wood Pellets a Green Fuel?" *Science* 359 (2018): https://science.sciencemag.org/content/359/6382/1328. For a description of life-cycle analysis, see Chapter 9.

36. See https://www.dogwoodalliance.org/2020/05/enviva-continues-to-destroy-natural-forests/. Enviva's practices are also described in *Climate Consequences of Current Carbon Accounting Practices for Bioenergy: A Case Study on Wood-Pellet Manufacturer Enviva*, Security & Exchange Commission, https://www.sec.gov/rules/petitions/2019/ptn4-741-exa.pdf.

37. A brief homage to the importance of these laws is Martin Nie, "Two Lesser-Known Reasons to Celebrate Our Public Lands," *High Country News*, November 22, 2016, https://www.hcn.org/articles/flpma-deserves-a-party-too.

38. The Biden administration's initiative to protect 30 percent of US land and coastal ocean by 2030 could produce major climate benefits. See Sarah Gibbens, "The U.S. Commits to Tripling Its Protected Lands. Here's How It Could Be Done," *National Geographic*, January 27, 2021, https://www.nationalgeographic.com/environment/2021/01/biden-commits-to-30-by-2030-conservation-executive-orders/.

39. See Chapter 6 for a discussion of agency regulations.

40. For information about regulations under NFMA, see https://www.fs.fed.us/emc/nfma/index.shtml. The Bureau of Land Management writes and enforces the rules under FLPMA; see https://www.blm.gov/about/laws-and-regulations.

41. See https://www.fire.ca.gov/programs/resource-management/forest-practice/.

42. James Mulligan et al., "Carbon Removal in Forests and Farms in the United States," 9.

43. Mulligan et al., "CarbonShot: Federal Policy," 22-24.

44. See https://www.conservationeasement.us/what-is-a-conservation-easement/.

45. US Forest Service, "Forest Products," United States Department of Agriculture. https://www.fs.usda.gov/science-technology/energy-forest-products/forest-products.

46. "12 Uses of Wood Product in Everyday Items," Canadian Institute of Forestry, https://www.cif-ifc.org/2018/07/12-uses-of-wood-product-in-everyday-items/.

47. See Chapter 9 for a description of life-cycle analysis.

48. Forest Stewardship Council, "Forest Management Certification," https://fsc.org/en/forest-management-certification. For sustainable management principles, see https://fsc.org/en/document-centre/documents/resource/392.

49. "Detailed Comparison of FSC and LCA as Sustainability Assessment Tools for Forest Products," World Wildlife Federation, https://c402277.ssl.cf1.rackcdn.com/publications/878/files/original/Detailed_Comparison_of_FSC_and_LCA_as_Sustainability_Assessment_Tools.pdf?1463666601.

50. Jennifer Ortman and David Raglin, *More Than 30% of Homes Heated by Wood in some Counties*, US Census Bureau, February 26, 2018, https://www.census.gov/library/stories/2018/02/who-knew-wood-burning-fuel.html.

51. John Gulland, "An Environmentalists Guide to Responsible Wood Heating," Woodheat.org, 2004, https://woodheat.org/responsible-heating.html.

52. See https://www.nature.org/en-us/.

53. Fargione et al., "Natural Climate Solutions." See Table S1 in the supplementary material for quantitative estimates.

54. Jennifer Oldham, "Expanding Efforts to Keep 'Cows Over Condos' Are Protecting Land Across the West," *Washington Post*, April 11, 2020, https://www.washingtonpost.com/national/expanding-efforts-to-keep-cows-over-condos-are-protecting-land-across-the-west/2020/04/10/96ec2f80-79c6-11ea-9bee-c5bf9d2e3288_story.html.

55. Sustainable Agricultural Research and Education (SARE) offers grants and education for the farming community. See https://www.sare.org/Learning-Center/Books/Building-Soils-for-Better-Crops-3rd-Edition/Text-Version. See Fred Magdoff and Harold van Es, *Building Soils for Better Crops*,3rd ed. (Sustainable Agriculture Research and Education, 2010), https://www.sare.org/Learning-Center/Books/Building-Soils-for-Better-Crops-3rd-Edition/Text-Version/Cover-Crops/Types-of-Cover-Crops.

56. Magdoff and van Es, *Building Soils*, https://www.sare.org/Learning-Center/Books/Building-Soils-for-Better-Crops-3rd-Edition/Text-Version/Cover-Crops/Cover-Crop-Management.

57. See "Mission 2014: Feeding the World," *Organic Industrial Agriculture*, http://12.000.scripts.mit.edu/mission2014/solutions/organic-industrial-agriculture.

58. Magdoff and van Es, *Building Soils*, https://www.sare.org/Learning-Center/Books/Building-Soils-for-Better-Crops-3rd-Edition/Text-Version/Nutrient-Management-An-Introduction/Using-Fertilizers-and-Amendments.

59. Mulligan et al., "CarbonShot: Federal Policy," 40.

60. For a description of agroforestry, see "Agroforestry," US Department of Agriculture, https://www.usda.gov/topics/forestry/agroforestry.

61. See Chapter 9 for a discussion of life-cycle analysis. The International Biochar Initiative provides a great deal of information about the technology; see "Biochar is a Valuable Soil Amendment," International Biochar Initiative, https://biochar-international.org/biochar/. For limitations, see Mulligan et al/, "CarbonShot: Federal Policy," 41.

62. Mulligan et al., "CarbonShot: Federal Policy," 39-49.

63. See https://www.nrcs.usda.gov/wps/portal/nrcs/main/national/programs/financial/eqip/.

64. See https://www.nrcs.usda.gov/wps/portal/nrcs/main/national/programs/financial/csp/.

65. See "What is the Farm Bill?" Congressional Research Service, last modified September 26, 2019, https://fas.org/sgp/crs/misc/RS22131.pdf.

66. See https://www.soil4climate.org/news/healthy-soils-legislation-update-may-2019.

67. *Overview of Greenhouse Gas Emissions: Nitrous Oxide Emissions*, Environmental Protection Agency, https://www.epa.gov/ghgemissions/overview-greenhouse-gases#nitrous-oxide.

68. Three US agencies, including the Food and Drug Administration, are collaborating on a strategy to reduce food waste. See "Food Loss and Waste," US Food and Drug Administration, https://www.fda.gov/food/consumers/food-loss-and-waste.

69. Hannah Ritchie, "Food Production Is Responsible for One-Quarter of the World's Greenhouse Gas Emissions," Our World in Data, November 6, 2019, https://ourworldindata.org/food-ghg-emissions. For a recent study emphasizing the importance of minimizing food system emissions, see Michael A. Clark et al., "Global Food System Emissions Could Preclude Achieving the 1.5° and 2°C Climate Change Targets," *Science* 370 (2020): 705-708.

70. See https://sustainableagriculture.net.

71. See Chapter 2 for a discussion of methane sources and sinks.

72. See "Lowering Methane Emissions," *FutureLearn*, https://www.futurelearn.com/courses/climate-smart-agriculture/0/steps/26578.

73. Environmental Protection Agency, "How Does Anaerobic Digestion Work?," *AgStar*, https://www.epa.gov/agstar/how-does-anaerobic-digestion-work.

74. See "Why Are CAFOs Bad?," Sierra Club, Michigan Chapter, https://www.sierraclub.org/michigan/why-are-cafos-bad.

75. Kelsey Piper, "Want to Help Animals? Focus on Corporate Decisions, Not People's Plates," Vox, January 29, 2019, https://www.vox.com/future-perfect/2018/10/31/18026418/vegan-vegetarian-animal-welfare-corporate-advocacy.

76. Global CCS Institute, *Introduction to Industrial Carbon Capture and Storage* (Global CCS Institute, 2016), https://www.globalccsinstitute.com/wp-content/uploads/2019/08/Introduction-to-Industrial-CCS.pdf. The Global CCS Institute is an international think tank devoted to accelerating CCS deployment. It offers many resources for learning about CCS. See https://www.globalccsinstitute.com.

77. See Chapters 7 and 9 for discussions of how carbon pricing and low carbon fuel standards can promote CCS expansion. See Chapter 9 for a discussion of industry emissions.

78. International Energy Agency, *Energy Technology Perspectives 2020: Special Report on Carbon Capture Utilization and Storage* (International Energy Agency), 13-15, https://www.iea.org/reports/ccus-in-clean-energy-transitions. See Chapter 9 for a discussion of renewable hydrogen.

79. See Herzog, *Carbon Capture*, 39-66, for a good description of the CCS process.

80. See "Carbon Storage FAQS," National Energy Technology Laboratory, https://www.netl.doe.gov/coal/carbon-storage/faqs/carbon-storage-faqs. Also, see Herzog, *Carbon Capture,* 67-78.

81. For a brief description of federal regulations and links to sites describing state policies, see "Carbon Capture," Center for Climate and Energy Solutions, https://www.c2es.org/content/carbon-capture/.

82. See "Underground Injection Control. Class VI—Wells Used for Geologic Sequestration of CO_2," Environmental Protection Agency, https://www.epa.gov/uic/class-vi-wells-used-geologic-sequestration-co2.

83. As of 2020, the total number of commercial CCS facilities worldwide was 65, of which 26 were operating, three were under construction, and 34 were in advanced or early-stage development. The US leads the world in new projects, initiating 12 of the 17 such projects that began in 2020. Global CCS Institute, *Global Status of CCS 2020* (Global CCS Institute,) 18-19, 70-71, https://www.globalccsinstitute.com/resources/global-status-report/.

84. For a description of EOR, see "Enhanced Oil Recovery," US Department of Energy Office of Fossil Energy, https://www.energy.gov/fe/science-innovation/oil-gas-research/enhanced-oil-recovery. For analysis, see Mulligan et al., "CarbonShot: Federal Policy," 34-35.

85. See Chapter 9 for discussions of biofuels and electric vehicles.

86. Perona, "Biodiesel for the 21st Century."

87. For examples of carbon dioxide utilization, see *Case Studies: Real-World Companies That Are Pioneering Direct Air Capture Technology and Market Applications of Carbon Dioxide* (Bipartisan Policy Center), https://bipartisanpolicy.org/wp-content/uploads/2018/09/Bipartisan_Energy-DAC-Fact-Sheet-Part-3_R01_5.3.2020edits.pdf.

88. See Chapter 1 for a discussion of carbon chemistry.

89. See Herzog, *Carbon Capture*, 91-94. For a more comprehensive (and optimistic) overview of CCU's potential, see Jeffrey Bobeck et al., *Carbon Utilization: A Vital and Effective Pathway for Decarbonization* (Center for Climate and Energy Solutions, 2019), https://www.c2es.org/document/carbon-utilization-a-vital-and-effective-pathway-for-decarbonization/.

90. Herzog, *Carbon Capture*, 128-129.

91. James Mulligan et al., "Technological Carbon Removal in the United States" (World Resources Institute, 2018), 7, https://www.wri.org/publication/tech-carbon-removal-usa.

92. Herzog, *Carbon Capture*, 51-66.

93. See https://www.climeworks.com/news/renewable-jet-fuel-from-air.

94. See Chapter 9 for discussions of renewable hydrogen and carbon.

95. Mulligan et al., "Carbonshot: Federal Policy," 27.

96. See Chapter 1 for a discussion of weathering and the timescales of the carbon cycle.

97. David Beerling, "Guest Post: How 'Enhanced Weathering' Could Slow Climate Change and Boost Crop Yields," *CarbonBrief*, February 19, 2018, https://www.carbonbrief.org/guest-post-how-enhanced-weathering-could-slow-climate-change-and-boost-crop-yields.

98. Mulligan et al., "CarbonShot: Federal Policy," 49-53.

99. See Christopher Consoli, *2019 Perspective: Bioenergy and Carbon Capture and Storage* (Global CCS Institute, 2019), 3-4, https://www.globalccsinstitute.com/wp-content/uploads/2019/03/BECCS-Perspective_FINAL_18-March.pdf.

100. See "Biomass Resources," US Department of Energy Office of Energy Efficiency & Renewable Energy, https://www.energy.gov/eere/bioenergy/biomass-resources. See Chapter 8 for a discussion of energy crops and ethanol production from corn and cellulosic biomass.

101. See the discussion on using biomass for electricity generation in Chapter 8.

102. Simon Evans, "World Can Limit Global Warming to 1.5 °C 'without BECCS,'" *CarbonBrief*, April 13, 2018, https://www.carbonbrief.org/world-can-limit-global-warming-to-onepointfive-without-beccs. Also see National Academies of Sciences, Engineering, and Medicine, *Negative Emissions Technologies and Reliable Sequestration: A Research Agenda* (The National Academies Press, 2019), 137, https://doi.org/10.17226/25259.

103. See "Regional Carbon Sequestration Partnerships (RCSP) Initiative," National Energy Technology Laboratory, https://www.netl.doe.gov/coal/carbon-storage/storage-infrastructure/regional-carbon-sequestration-partnerships-initiative.

104. Christopher Consoli, *2019 Perspective*, 4-6, 39-41.

105. Jennifer Christensen, "Before & After: How the FUTURE Act Reformed the 45Q Carbon Capture and Storage Tax Credit," Great Plains Institute, March 5, 2018, https://www.betterenergy.org/blog/future-act-reformed-45q-carbon-capture-storage-tax-credit/. For some details on the implementation of the law via regulations, see Carlos Anchondo, "Trump's CCS Rule: Details, Doubts and EPA Disputes," *E&E News*, June 1, 2020, https://www.eenews.net/energywire/2020/06/01/stories/1063286747. The 45Q tax credit was extended by two years as part of the omnibus budget and Coronavirus relief bill passed by Congress in the last days of 2020.

106. See National Defense Authorization Act For Fiscal Year 2020, S. 1790, 116th Congress. (2019-2020), https://www.congress.gov/bill/116th-congress/senate-bill/1790/text?q=%7B%22search%22%3A%5B%22%5C%22blue+carbon%5C%22%22%22%5D%7D&r=7&s=1.

107. The Fifth Amendment to the US Constitution allows government to take private property for a public use, as long as just compensation is provided to the landowner. See the discussion in Chapter 6 on siting for oil pipelines and terminals, which is not regulated by the federal government. The lack of federal involvement favors climate advocates' efforts to block the pipelines.

108. Mulligan et al., "Carbonshot: Federal Policy," 70-72.

109. Noah Kaufman et al., *An Assessment of the Energy Innovation*, 13-17, .https://energypolicy.columbia.edu/sites/default/files/file-uploads/EICDA_CGEP-Report.pdf. See Chapter 7 for a discussion of the federal 2019 EICDA carbon tax bill, the Energy Innovation and Carbon Dividend Act. See Color Plate 19 for an illustration of how the steep federal carbon tax incentivizes CCS at natural gas-fired electricity plants.

110. See Chapters 8 and 9 for discussions of the renewable portfolio standard and the low carbon fuel standard. For a detailed policy plan to accelerate the expansion of CCS in California, see Energy Futures Initiative, *An Action Plan for Carbon Capture and Storage in California: Summary for Policymakers*, (Energy Future Initiative, 2020), https://sccs.stanford.edu/sites/g/files/sbiybj7741/f/efi-stanford-ca-ccs-sfpm-rev1-10.25.20_0.pdf.

111. See Chapter 6 for a discussion of the Clean Power Plan and general federal regulation.

112. Mulligan et al, *Foundational Questions on Carbon Removal in the United States* (World Resources Institute, 2018), https://www.wri.org/publication/foundational-questions-on-carbon-removal-usa.

113. See the discussion on albedo in Chapter 1.

114. Minx et al., "Negative Emissions—Part 1," 3-5.

115. For a good overview of solar engineering approaches, see Daisy Dunne, "Explainer: Six Ideas to Limit Global Warming With Solar Engineering," *CarbonBrief*, September 5, 2018, https://www.carbonbrief.org/explainer-six-ideas-to-limit-global-warming-with-solar-geoengineering.

116. Douglas G. MacMartin and Ben Kravitz, "Mission-Driven Research for Stratospheric Aerosol Geoengineering," *Proceedings of the National Academy of Sciences of the United States of America* 116 (2019): 1089, https://www.pnas.org/content/116/4/1089.

117. DeAngelo, B., J. Edmonds, D.W. Fahey, and B.M. Sanderson, 2017: Perspectives on Climate Change Mitigation. In: *Climate Science Special Report: Fourth National Climate Assessment, Volume I* [Wuebbles, D.J., D.W. Fahey, K.A. Hibbard, D.J. Dokken, B.C. Stewart, and T.K. Maycock (eds.)]. U.S. Global Change Research Program, Washington, DC, USA, pp. 393-410, doi: 10.7930/J0M32SZG, section 14.3.

118. Daisy Dunne, "Explainer: Six Ideas"

119. See the discussion of the Paris Agreement in Chapter 3 and in the Interlude.

120. Willam S. Eubanks II, "A Brief History of U.S. Agricultural Policy and the Farm Bill", in *Food, Agriculture and Environmental Law* by Mary Jane Angelo, Jason Czarnezki, and W.S. Eubanks II (Washington D.C.: Environmental Law Institute, 2013), 4-6.

121. Michael Pollan, *The Omnivore's Dilemma: A Natural History of Four Meals* (New York, NY: Penguin Random House, 2006), 123-133.

122. Spiro and Stigliani, *Chemistry of the Environment*, 463-467.

123. See the discussion of hydrogen production in the second section of Chapter 9.

124. "Demand for Nature-Based Solutions for Climate Drives Voluntary Carbon Markets to a Seven-Year High," Forest Trends, December 9, 2019, https://www.forest-trends.org/pressroom/demand-for-nature-based-solutions-for-climate-drives-voluntary-carbon-markets-to-a-seven-year-high/.

125. The Growing Climate Solutions Act of 2020, H.R. 7393, 116th Cong. (2019-2020), https://www.braun.senate.gov/sites/default/files/2020-06/Growing%20Climate%20Solutions%20Act%20One%20Pager_0.pdf.

126. Chris Clayton, "Groups Support Ag Carbon Credit Bill," Progressive Farmer, April 21, 2021, https://www.dtnpf.com/agriculture/web/ag/news/world-policy/article/2021/04/21/growing-climate-solutions-act-bi.

127. See https://www.agriculture.senate.gov/imo/media/doc/GCSA%20Act%20Supporters%20FINAL.pdf

bibliography

Angelo, Mary Jane, Jason J. Czarnezki, and William S. Eubanks II. *Food, Agriculture, and Environmental Law*. Washington, D.C.: Environmental Law Institute, 2013.

Archer, David. *The Global Carbon Cycle*. Princeton Primers in Climate. Princeton, NJ: Princeton University Press, 2010.

Archer, David. *Global Warming: Understanding the Forecast*. Hoboken, NJ: Wiley, 2012.

Bowen, Mark. *Censoring Science: Inside the Political Attack on Dr. James Hansen and the Truth of Global Warming*. New York: Dutton, 2008.

Bowen, Mark. *Thin Ice: Unlocking the Secrets of Climate in the World's Highest Mountains*. New York: Henry Holt, 2005.

Chomsky, Noam, and Robert Pollin. *Climate Crisis and the Global Green New Deal: The Political Economy of Saving the Planet*. New York: Verso, 2020.

Cole, David. *Engines of Liberty: How Citizen Movements Succeed*. New York: Basic Books, 2016.

Edwards, Paul N. *A Vast Machine: Computer Models, Climate Data, and the Politics of Global Warming*. Infrastructures. Cambridge, MA: MIT Press, 2010.

Emanuel, Kerry. *What We Know about Climate Change*. Updated ed. Cambridge, MA: MIT Press, 2018.

Esty, Daniel C. ed., *A Better Planet: Big Ideas for a Sustainable Future*. New Haven: Yale University Press, 2019.

Fox-Penner, Peter. *Power After Carbon: Building a Clean, Resilient Grid*. Cambridge, MA: Harvard University Press, 2020.

Goodell, Jeff. *Big Coal: The Dirty Secret Behind America's Energy Future*. Boston: Mariner Books, 2007.

Goodell, Jeff. *The Water Will Come: Rising Seas, Sinking Cities, and the Remaking of the Civilized World*. New York: Little, Brown, 2017.

Goodstein, David. *Out of Gas: The End of the Age of Oil*. New York: W. W. Norton, 2004.

Gore, Al. *Earth in the Balance: Ecology and the Human Spirit*. New York: Earthscan, 1992.

Gore, Al. *An Inconvenient Truth: The Planetary Emergency of Global Warming and What We Can Do About It*. Emmaus, PA: Rodale, 2006.

Haidt, Jonathan. *The Righteous Mind: Why Good People Are Divided by Politics and Religion*. New York: Vintage Books, 2012.

Harvey, Hal, Robbie Orvis, and Jeffrey Rissman. *Designing Climate Solutions: A Policy Guide for Low-Carbon Energy*. Washington, D.C.: Island Press, 2018.

Hawken, Paul. *Drawdown: The Most Comprehensive Plan Ever Proposed to Reverse Global Warming*. New York: Penguin Books, 2017.

Henson, Robert. *The Thinking Person's Guide to Climate Change*. 2nd ed. Boston: American Meteorological Society Press, 2019.

Herzog, Howard J. *Carbon Capture*. Cambridge, MA: MIT Press, 2018.

Hochschild, Arlie Russell. *Strangers in Their Own Land: Anger and Mourning on the American Right*. New York: New Press, 2016.

Holthouse, Eric. *The Future Earth: A Radical Vision for What's Possible in the Age of Warming*. San Francisco: HarperOne, 2020.

Houghton, John. *Global Warming: The Complete Briefing*. 5th ed. Cambridge: Cambridge University Press, 2015.

Hsu, Shi-Ling. *The Case for a Carbon Tax: Getting Past Our Hang-Ups to Effective Climate Policy*. Washington, D.C.: Island Press, 2011.

Jackson, Tim. *Prosperity without Growth: Economics for a Finite Planet*. New York: Earthscan, 2009.

Klein, Naomi. *This Changes Everything: Capitalism vs. the Climate*. New York: Simon & Schuster, 2014.

Kolbert, Elizabeth. *The Sixth Extinction: An Unnatural History*. New York: Henry Holt, 2014.

Lomborg, Bjorn. *False Alarm: How Climate Change Panic Costs Us Trillions, Hurts the Poor, and Fails to Fix the Planet*. New York: Basic Books, 2020.

Lovelock, James. *Gaia: A New Look at Life on Earth*. Oxford: Oxford University Press, 1979.

Mann, Michael E. *The Hockey Stick and the Climate Wars: Dispatches from the Front Lines*. New York: Columbia University Press, 2012.

Mann, Michael E. *The New Climate War: The Fight to Take Back Our Planet*. New York: PublicAffairs, 2021.

Mayer, Jane. *Dark Money: The Hidden History of the Billionaires behind the Rise of the Radical Right*. New York: Doubleday, 2016.

McKibben, Bill. *American Earth: Environmental Writing Since Thoreau*. New York: Library of America, 2008.

McKibben, Bill. *Deep Economy: The Wealth of Communities and the Durable Future*. New York: Times Books, 2007.

McKibben, Bill. *Falter: Has the Human Game Begun to Play Itself Out?* New York: Henry Holt, 2019.

Nordhaus, William. *The Climate Casino: Risk, Uncertainty, and Economics for a Warming World*. New Haven, CT: Yale University Press, 2013.

Oreskes, Naomi, and Erik M. Conway. *Merchants of Doubt: How a Handful of Scientists Obscured the Truth on Issues from Tobacco Smoke to Global Warming*. London: Bloomsbury Press, 2010.

Pettifor, Ann. *The Case for the Green New Deal*. New York: Verso, 2019.

Pollan, Michael. *The Omnivore's Dilemma: A Natural History of Four Meals*. New York: Penguin Press, 2006.

Rabe, Barry G. *Can We Price Carbon?* American and Comparative Environmental Policy. Cambridge, MA: MIT Press, 2018.

Rich, Nathaniel. *Losing Earth: A Recent History*. New York: MCD Books, 2019.

Romm, Joseph. *Climate Change: What Everyone Needs to Know*. Oxford: Oxford University Press, 2016.

Rosling, Hans. *Factfulness: Ten Reasons We're Wrong about the World—and Why Things Are Better Than You Think*. New York: Flatiron Books, 2018.

Ruddiman, William F. *Earth's Climate: Past and Future*. 3rd ed. New York: W. H. Freeman, 2014.

Salzman, James, and Barton H. Thompson Jr. *Environmental Law and Policy*. 4th ed. Concepts and Insights. St. Paul, MN: Foundation Press, 2014.

Schumacher, E. F. *Small Is Beautiful: Economics as if People Mattered*. Vancouver, BC: Hartley & Marks Publishers, 1999.

Simmons, Matthew R. *Twilight in the Desert: The Coming Saudi Oil Shock and the World Economy*. Hoboken, NJ: Wiley, 2005.

Smil, Vaclav. *Energy Transitions: Global and National Perspectives*. 2nd ed. Westport, CT: Praeger, 2017.

Smil, Vaclav. *Harvesting the Biosphere: What We Have Taken from Nature*. Cambridge, MA: MIT Press, 2012.

Spiro, Thomas G., and William M. Stigliani. *Chemistry of the Environment*. Subsequent ed. New York: Pearson College Division, 2002.

Stern, Nicholas. *Why Are We Waiting? The Logic, Urgency, and Promise of Tackling Climate Change*. Lionel Robbins Lectures. Cambridge, MA: MIT Press, 2015.

Stokes, Leah C. *Short Circuiting Policy: Interest Groups and the Battle Over Clean Energy and Climate Policy in the American States*. Studies in Postwar American Political Development. Oxford: Oxford University Press, 2020.

Thunberg, Greta. *No One Is Too Small to Make a Difference*. New York: Penguin Books, 2018.

Wallace-Wells, David. *The Uninhabitable Earth: Life After Warming*. New York: Tim Duggan Books, 2019.

Weart, Spencer R. *The Discovery of Global Warming*. New Histories of Science, Technology, and Medicine. Cambridge, MA: Harvard University Press, 2008.

Wold, Chris, David Hunter, and Melissa Powers. *Climate Change and the Law*. 2nd ed. New York: LexisNexis, 2013.

Wood, Mary Christina. *Nature's Trust: Environmental Law for a New Ecological Age*. Cambridge: Cambridge University Press, 2013.

Yergin, Daniel. *The Prize: The Epic Quest for Oil, Money & Power*. New York: Free Press, 2008.

index

Page numbers in italics refer to boxes, figures, illustrations, and tables.

ooligan press

Ooligan Press is a student-run publishing house rooted in the rich literary culture of the Pacific Northwest. Founded in 2001 as part of the Portland State University's Department of English, Ooligan is dedicated to the art and craft of publishing. Students pursuing master's degrees in book publishing staff the press in an apprenticeship program under the guidance of a core faculty of publishing professionals.

Project Managers
Callie Brown
Julie Collins
Devyn Yan Radke

Acquisitions
Michael Shymanski
Kim Scofield
Des Hewson
Jennifer Ladwig

Digital
Chris Leal
Megan Crayne-Bell
Amanda Hines

Design
Morgan Ramsey
Denise Morales-Soto
Katherine Flitsch

Marketing
Sydnee Chesley
Hannah Boettcher
Sarah Moffatt

Social Media
Alix Martinez
Riley Robert
Faith Muñoz

Research Assistants
Bailey Potter
Kelly Zatlin

Editorial
Emma Wolf
Erica Wright
Melinda Crouchley
Olivia Rollins
Rachel Lantz
Rachel Howe

Publicity
Alex Gonzales
Emma St. John

Book Production

Alex Burns
Alexa Schmidt
Alexandra Magel
August Amoroso
Bayley McComb
Claire Plaster
Cole Bowman
Courtney La Verne
Courtney Young
Eily McIlvain
Elle Klock
Emily Plate
Frances Frangela

Giacomo Ranieri
Grace Hansen
Jenna Admunson
Jill Bowen
John Huston
Kaitlyn Sheehee
Kendra Ferguson
Laura Mills
Luis Ramos
Matthew McDonald
Nif Lindsay
Phoebe Whittington
Rachel Howe

Rebecca Gordon
Rosina Miranda
Rylee Warner
Scott Fortman
Shalyn Shipper
Shannon Gilb
Siri Vegulla
Stephanie Johnson Lawson
William York
Wren Haines
Xian Wang